PALEOECOLOGY OF AFRICA
VOLUME 12

ERTS satellite photograph taken on 11 February 1974 (01-A-2 0025 1911) showing Saharan dust which forms an anticyclonic vortex near the beginning of its Atlantic transit (reproduced with permission of Eros Data Center, Sioux Falls, USA). Original image from the U.S. National Aeronautics Space Administration.

PALAEOECOLOGY OF AFRICA
AND THE SURROUNDING ISLANDS

E.M.VAN ZINDEREN BAKKER SR
& J.A.COETZEE(editors)

Volume 12

SAHARA AND SURROUNDING SEAS
Sediments and climatic changes

*Proceedings of an international symposium,
Akademie der Wissenschaften und der Literatur Mainz,
1-4 April 1979*

edited by
M.SARNTHEIN/E.SEIBOLD
Geol.Pal.Inst., Univ. Kiel

P.ROGNON
Dépt. Géographie Physique, Univ. Paris VI

A.A.BALKEMA/ROTTERDAM/1980

ISBN 90 6191 050 1

© 1980, A.A.Balkema, P.O.Box 1675, Rotterdam, Netherlands

In USA & Canada: 99 Main Street, Salem, NH 03079

Printed in the Netherlands

CONTENTS

V

METEOROLOGY AND CLIMATE

CLIMATIC FLUCTUATIONS IN THE ARID BELT OF THE 'OLD WORLD' SINCE THE LAST GLACIAL MAXIMUM; POSSIBLE CAUSES AND FUTURE IMPLICATIONS

HERMANN FLOHN / SHARON NICHOLSON
University of Bonn

The main content has already been presented by one of the authors (H.F.) at a symposium on the Meteorology of Semi-Arid areas organized by the American and the Israel Meteorological Societies at Tel Aviv, 31 October 1977. This lecture was devoted to the memory of my dear friend, N.Rosenan, eminent climatologist from Israel.

1 INTRODUCTION

The occurrence of two separate wet periods in the Sahara during the Holocene was one of the great surprises emerging from intense paleoclimatological and archeological research during two decades. Their interpretation as based on present knowledge of the general atmospheric circulation has been controversial; investigations of the recent Sahel drought period based partly on insufficient data added to the controversy. The first of these moist periods coincides with the warmest Holocene period in southern latitudes (around 9 500 B.P.); the second, with the Holocene thermal maximum in Europe (Atlanticum, around 6 000 B.P.).

If, as expected, CO_2 and other infrared-absorbing gases continue to increase and their greenhouse effect of global warming reaches certain levels – this problem is now under intense discussion – thermal conditions similar to those of the past could be reached in the first decades of the next century. Can we then expect a similar beneficial evolution of the hydrological cycle? Are the climatic boundary conditions now the same as 6-10 000 years ago? With such questions this seemingly academic problem of climatic history becomes quite current, attaining high socio-economic dimensions.

The answer needs a look at the mechanism of the general atmospheric circulation, especially the interaction between tropical and extratropical weather systems (Chapter 2). Then a short survey of the climatic history of that area during the last 20 000 years (Chapter 3) is given, without plunging deeply into incompletely solved local and regional problems. Finally a new interpretation of the observed facts, especially of the simultaneous intensification of

3

the rainfall on both flanks of the Sahara, is proposed, using statistical evidence (see Appendix A) and taking into account changes of the boundary conditions during the last 10 000 years.

2 CLIMATIC BELTS AND ATMOSPHERIC CIRCULATION MODES

The general atmospheric circulation is characterized, near the surface, by a simple pattern of wind systems and cellular pressure belts:

1. Subpolar low pressure belt with cyclonic centers.
2. Zone of surface westerlies with travelling mid-latitude cyclones.
3. Subtropical high pressure belt with quasistationary anticyclones, most pronounced over the oceans.
4. Zone of surface easterlies (trade-winds) most regular over the oceans.
5. Intertropical convergence zone (ITCZ) with variable winds and frequent rains.

These belts are in continuous motion. A remarkable feature is the large-scale tropical monsoon system extending in the tropics between West Africa and the Philippines. In this area zone (5) is displaced, during northern summer, to Latitude 20-30°N and a belt of moisture-laden westerlies develops on its equatorward flank, superimposed by a permanent Tropical easterly jet (Flohn 1964, 1975). Normally, however, zone (2) broadens in the upper troposphere (9-14 km) and extends, above zone (3) and (4), towards the equator; here large meanders with a sequence of troughs and ridges travel from west to east and interact with low-level disturbances frequently observed in zone (5).

Two interacting branches or modes of circulation can be distinguished. In the Tropics, the dominant circulation is the Hadley circulation described as a screw-like (helical) cell in a meridional-vertical plane. Ascending motion prevails in the ITCZ, while subsidence occurs in the anticyclonic cells (zone (3)) which are separated, in the upper troposphere, by troughs in the westerlies.

The second circulation mode is represented by the extratropical westerlies (zone (2)) which extend, in the upper troposphere near 200 mb (~12 km) as a large irregular vortex around one (or two) centers in polar regions, with a continuous chain of wave-like meandering distortions. The meridional transport of heat and momentum occurs in travelling or quasi-stationary waves, i.e. in a nearly horizontal plane and not in an idealized helical cell. The subtropical anticyclonic belt (zone (3)) coinciding with the most intense westerly flow aloft − the subtropical jet-stream − is common to both modes; a complete theoretical interpretation (Palmén and Newton) − which is not intended here − contains processes from both sides. In a latitudinal average, the center of zone (3) varies seasonally between Latitude 32°N (January) and 42°N (July), and parallel to this between Latitudes 35°S and 27°S; annual averages

4

are Latitudes 37°N and 31°S. Subsidence leads to aridity, even if the large-scale vertical components of motion are only on the order of cm/sec or less. Rainfall is correlated with ascending motions concentrated, in the tropics, in meso-scale disturbances (with a diameter of 100-300 km) or in the middle and higher latitudes with travelling cyclones with a diameter between 500 and 3 000 km. These climatic belts are displaced seasonally towards the actual summer hemisphere; this shift is much greater over the monsoon section Africa-Southern Asia/Indian Ocean-Australia than above the Pacific.

A noteworthy atmospheric feature relates to the thermal contrast between the Antarctic continent with its huge ice-sheet and the Arctic Ocean, where only a few meters of broken drifting ice separate the unfrozen ocean from the atmosphere: the markedly different heat budgets produce Arctic-Antarctic temperature differences which reach, in the annual average, 20-30°C near the surface and 11-12°C in the troposphere (Flohn 1967, 1978). Consequently the southern hemisphere circulation is distinctly stronger than that of the northern hemisphere; the above-mentioned circulation belts are displaced towards the equator; and the average annual position of the ITCZ (or the meteorological equator) is near Latitude 6°N, reaching in northern summer/southern winter as far as 15°N. The seasonal thermal character of the two hemispheres is also different because of the distribution of land and sea: the northern hemisphere is 40 per cent land-covered, the southern hemisphere only to 19 per cent, and in subpolar latitudes the contrast rises to 80 per cent versus 0 per cent. The main consequences are strong zonal flow patterns at the southern hemisphere, with prevailing stormy westerlies (zone (2)), and

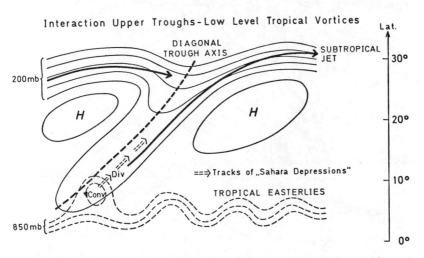

Figure 1. Interaction between high-tropospheric troughs (200 mb ~ 12 km) and low-level tropical easterlies (850 mb ~ 1.5 km). Origin and track of Saharan depressions (vortices).

5

more meridional flow patterns of the northern hemisphere caused by the seasonal varying thermal contrast between ocean and continents. Due to this seasonal displacement of all pressure/wind belts – with an amplitude of 10-15° Latitude – the southern flank of the Sahara is affected by the offshoots of tropical summer-rains (zone (5)), the northern flank by those of the extra-tropical rains (zone (2)) during the cool season.

In addition to this seasonal displacement of climatic belts, a more permanent interaction between the two modes of the atmospheric circulation exists. In subtropical regions, this interaction (Flohn 1975) is characterized by a large-scale meandering motion of the upper westerlies at the level 150-300 mb extending above the tropical easterlies deeply into low latitudes; during southern winter the southern westerlies extend even beyond the equator. These meanders, more exactly their tilted troughs, frequently trigger, at their equatorial spurs (Figure 1) the development of the tropical easterly waves (zone (4)) into cyclonic vortices, which travel ahead of the upper troughs northeastward across the Sahara. They are accompanied in low lati-

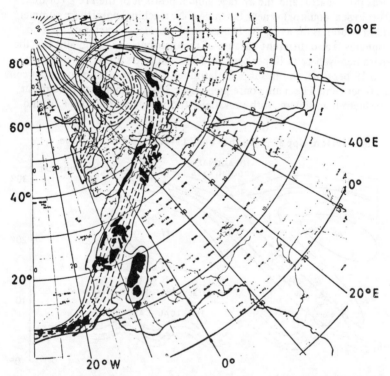

Figure 2. Cloud band (dashed: Cirrostratus, Altostratus; black: Cumulonimbus or Nimbostratus) in the water-vapour absorption band 3.4-4.2 μm (Nimbus III, 21 September 1969; Flohn 1975).

6

tudes by heavy showers and clusters of thunderstorms; after raining out, their convective activity may be reduced to heavy duststorms, but can again produce heavy rains after entraining moist air from the Mediterranean. The frequency of the penetrating 'diagonal' troughs diminishes across North Africa from west to east. In the oases of the interior Maghreb (northwestern Sahara) the rare rainfall events associated with such a mechanism occur most frequently in autumn and spring, with a relative minimum during winter.

Remarkable recent examples are the catastrophic rains in September/ October 1969 in the southern Maghreb (Figure 2), caused by such an interaction (Flohn 1975). These 'Saharan depressions' were well-known to French and British meteorologists in the 1930's; they are by no means rare. In 10-15 per cent of all days between September and May, upper winds above southern Algeria indicate the occurrence of such tilted troughs, and the average annual frequency of Saharan depressions (Meteor. Office 1962) was evaluated to be 14 (see also Pedgley 1970 for Northeast Africa). While the great frequency of these cases in the transitional seasons is difficult to interpret, their absence during summer (mid-June to end of August) is directly related to the dominance of extended anticyclonic cells at 100-300 mb, coinciding with the regular occurrence of the Tropical Easterly Jet with its ageostrophic northward component in its exit region (Flohn 1964, 1975) (see Appendix B).

3 SURVEY OF CLIMATIC HISTORY OF THE LAST 20 000 YEARS

The aridity of the equatorial tropics and their semiarid margins during and especially after the peak of the last glaciation (see Table 1: 20-14 000 years B.P.) is well documented. While dune systems spanned the entire east-west subsaharan belt, indicating a progression of the desert southward into the humid tropics near 10°N, semiarid vegetation and sandy soils covered wide areas of the present tropical rainforests, which were reduced to a few residual areas (Livingstone 1975, Shackleton 1977) in the wettest spots. Since at the same time the equatorial ocean surface temperatures were 4-7°C lower than today (except the Indian Ocean, where they dropped only 0-2°C), the phenomenon of equatorial upwelling must have been stronger and more frequent than now, at least in the Pacific and Atlantic. Recent investigations of the fluxes of latent and sensible heat – based on the application of the bulk aerodynamic formula to individual measurements (Trempel 1978) – in these upwelling regions indicate that the sensible heat flux is directed downwards, while evaporation reduces to 25-30 per cent of the normal value over tropical oceans. A cautious extrapolation of these results to sea surface temperatures around 14-16°C, as during the last glacial, led to the hypothesis that the evaporation should have been reduced to zero or even negative values (mist and condensation at the cold water surface). Recent theoretical studies (e.g. Wyrtki 1975), indicate an intensification of upwelling with increasing intensity of the

7

Sahara : Climatic and Cultural History (tentative)

1000 yrs BP (approx.)	Climate North	South	Cultures	Large Scale Events
0				
2			Metal Age, Camels	Warmer and colder periods (Neoglacials)
			Irrigation Farming, Horses	
4	Desiccation	Hyperarid	Retreat → Valleys, Old+Middle Empire, Egypt	⌠Hol. Opt. Arctic ⌡Disappearance Laur. Ice
6	wet, warm	wet, warm (exc. humid Tropics)	Saharan Neolithic, Cattle-raising nomads	⌡Hol. Opt. Europe ⌠(Atlanticum)
?		Dry interval	Early Neolithic	⌠Disintegration Hudson B. Ice ⌡Disappearance Scand. Ice
8	NW dry, NE wet (?) Highlands wet, cool	Lacustrine Period all-season rains	Epi-Paleolithic	← Hol. Opt. Subantarctic
10				Holocene ↑
	Semiarid with Fluctuations	Wetter with Fluctuations	Game hunters	Ice Retreat Rapid Fluctuations
12				
14		Hyperarid (exc. mountains) cool		
	Semiarid cold		?	
16				⌠Max. Contin. Ice-Sheets ⌡(NHemisphere)
18	Transition → more humid	Transition → hyperarid		⌠Max. Extension Drift-Ice ⌡(both hemispheres)
20				

Table 1. Simplified climatic (and cultural) history of the Sahara, in comparison with large-scale events since 20 000 B.P.

tropospheric easterlies; evidence for this during the last glacial maximum has been found at the West African coast (Diester-Haass 1976, Giresse 1977), simultaneous with a disappearance of runoff from Senegal river. Apparently, during this time equatorial downwelling (= El Niño along the west coast of South America) either completely disappeared or at least was drastically reduced.

This reduction of ocean evaporation over areas in the order of 120 x 20 meridian degrees in the Pacific and 55 x 20 degrees in the Atlantic (more than 40×10^6 km^2) affected not only the tropical belt, but also the global water budget. A recent estimate (Flohn 1979) suggests an 18 per cent reduction of the oceanic evaporation during the last glacial maxima with a regional reduction near 70 per cent in the equatorial belt between ± 10°Latitude: this should be the most effective cause of the aridity. In addition to this, the intensification and equatorward concentration of the Hadley cells of both hemisphere may have suppressed, especially during northern summer, the development of the shallow Southwest Monsoon above Africa. However, no evidence for this has yet been found.

In some areas (also on other continents) the available evidence seems to indicate a marked climatic change around the *glacial* peak ~18 000 B.P. This change must have been part of a geophysical mechanism: a transition from prevailing ice accretion (with a warm Atlantic) towards ice ablation, i.e. an annihilation of the well-known positive albedo-ice-temperature feedback

8

lasting 5-10 millennia. Recent evidence (McIntyre et al. 1978) is available for a 5 000 year lag between the increase of the global ice volume and the cooling of the Atlantic Ocean, which provided the moisture source for both the Laurentide and Scandinavian ice. Toward 18 000 B.P., the Arctic drift-ice spread with polar water southward and covered, at least seasonally, the Atlantic Ocean up to about Latitude 43°N, reducing sharply its evaporation. The aridity of the ice margins may then have contributed − by adding dust to the ice-surface (Hoinkes 1961) and by increasing solar absorption and heating the air (as in the case of Martian dust-storms, Gierasch & Goody 1972) − to the retreat of the ice which is otherwise difficult to understand.

The peak of aridity of the southern Sahara was reached after 18 000 B.P. probably correlated with the minimum evaporation of equatorial oceans. At the same time, the northern flank of the desert tended to become more humid: this can be interpreted as a result of the southward displacement of the Atlantic baroclinic zone and of the accompanying rain-producing eddies, supported by surface cooling and lower local evaporation. After 16 000 B.P., these mediterranean rains reached even the Tibesti Mountains (Jäkel 1978).

The transition period between 14 000 and 10 000 years B.P. is characterized by a sequence of quite abrupt climatic changes, of warm episodes with accelerated ice retreat and of cool episodes with marked ice advances and vegetation changes. Some of these changes extended into the Andes of Colombia near Latitude 5°N (Schreve-Brinkman 1978), into the Danakil Desert (representing the water budget of the eastern Ethiopian highlands) (Grasse & Delibrias 1976) and even into the Antarctic (Lorius et al. 1978).

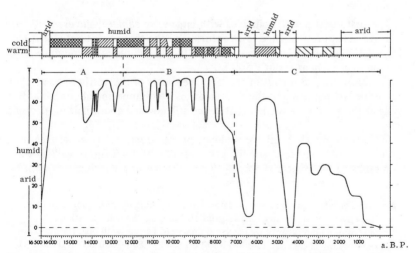

Figure 3. Chronodiagram of humid and arid phases in Tibesti: length of river sedimentation at the northern margin of Tibesti Mountains; shaded area at the top: cold and warm phases (Jäkel 1978).

9

Geomorphological studies in the Tibesti Mts (Jäkel 1978) indicate also short-period changes between cool/moist and warm/dry climates (Figure 3). In other mountains of the subtropical belt (e.g. Mexico) the peak of the last glaciation was delayed until this time (Heine 1974), probably because of changes of the wind-regime together with increasing moisture supply. A detailed climatological interpretation of this period seems to be premature as yet.

Of special interest are the *two moist periods* in the history of the Sahara, especially the first one between about 10 000 and 8 000 B.P., with series of extended lakes along the southern Saharan margin. One example is Lake Mega-Chad (about 320 000 km^2), with a sea-level more than 40 m higher than today, controlled by a temporary overflow* into the Benuë-Niger catchment (Maley 1977). Similarly, Lake Turkana (= Rudolf) stood 70 m higher than now, with an overflow to the Nile (Butzer 1972). Even in the hyperarid desert center between Kufra Oasis and Tibesti Mts, − now with less than 5 mm rain annually − permanent rivers were flowing, indicating 250-400 mm of rainfall, and the grasslands were occupied by a relatively dense population of game hunters (Gabriel 1977, Museen Köln 1978). Similar evidence is available from more than 40 sites between Mauretania (Longitude 17°W) and Rajasthan (Longitude 77°E) including interior Arabia. The climatic history of Rajasthan, now with an average rainfall near 250 mm, has been investigated by Singh (1974, Bryson-Murray 1977). Here a hyperarid desert period was followed, after 10 300 B.P., by a moist period with 400-800 mm rainfall, and a significant contribution of winter rains; this lasted with many fluctuations (Figure 4) until about 3 600 B.P. This result appears to be consistent with an early Holocene maximum of monsoonal flow along the Arabian coast (Prell 1978).

The first moist period coincides with large-scale *climatologic events:*

1. An abrupt retreat (in a few centuries) of the thin subantarctic drift-ice (Hays 1978) after the last glaciation, resulting in a climatic optimum (warmer than now) as early as about 9 500 B.P.

2. A slow gradual retreat of the large ice-sheets of the northern hemisphere, where the last remnants of the Scandinavian ice disappeared about 8 000 B.P., while at the same time the Laurentide ice still covered about half of its original area.

Unfortunately, the available evidence of the *equatorial ocean* surface temperatures is insufficient to check the assumption that dominant *upwelling* was then replaced by equatorial *downwelling*, with warm water, high evaporation

---

* A remarkable verification of this overflow has been found in the oceanic sedimentation off the Niger delta: between 11 500 and 10 900 B.P., the sedimentation rate increased from 0.3 mm/year to more than 6 mm/year. Periods with high freshwater discharge from the Niger were 13 000-11 800 B.P. and 11 500-4 500 B.P. (L.Pastouret, H.Chamley, G.Delibrias, J.C.Duplessy, J.Thiede: Late Quaternary climatic changes in Western Tropical Africa deduced from deep-sea sedimentation off the Niger delta. Oceanologica Acta 1(1978), 217-232).

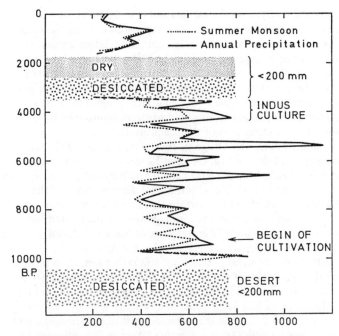

Figure 4. Rainfall estimates for Central Rajasthan (courtesy R.Bryson): pollen data (Singh 1974) evaluated with transfer functions (full line: annual rainfall; dashed: summer monsoon rainfall).

and high rainfall (more than today). The warming of the southern hemisphere should have lead to a weakening of the southern Hadley cell (southeast-trades), most effective just south of the equator, where today the maximum of up-welling occurs. Lacking definitive evidence, the rapid reforestation of Africa and South America after 13 000 B.P. (Shackleton 1977) supports the assumption of predominant oceanic downwelling, probably including a re-establishment of the African southwest monsoon in an area even larger than now. Since at that time the northern Hadley cell migrated poleward, in relation to the gradual warming of the North Atlantic (Ruddiman & McIntyre 1977) and the slow retreat of continental ice-sheets, the frequency and intensity of the upper troughs in the northern westerlies should have diminished only slightly. Together these events produce frequent interactions between tropical and subtropical systems, as described above, with Saharan depressions crossing the desert more frequently. The most essential consequence would be an extension of the rainy season over the whole year (Rognon 1976), with some of the characteristics of the extratropical rainfall during the cool seasons. The hypothesis of a northward slope of the ITCZ during this time (Rognon 1976) is based on the (qualitative) air-mass concept, while the above-mentioned

11

interpretation is based on dynamical concepts and is also consistent with recent anomalies (see Appendix A). Until the disappearance of the Scandinavian Ice the temperatures were comparatively cool. The relative dryness of Northwest Africa may be interpreted as a continental extension of the Azores anticyclone, which should have moved poleward parallel to the northward motion of the oceanic polarfront (Ruddiman & McIntyre 1977). In this position, it was quasi-stationary, associated with an upper trough above the central and eastern Mediterranean (as now during cold winters) maintained by the cold air flux from the remaining part of the Scandinavian ice.

Convincing evidence is available (Rognon 1976, Livingstone 1975, Fairbridge 1976) for a marked *dry interval* (Museen Köln 1978), in many areas lasting less than 1 000 years, between 8 000 and 6 500 B.P. and peaked around 7 000 B.P. Its onset seems to coincide with the final disappearance of the Scandinavian ice-sheet — which lead to surface characteristics of Europe similar to the present one — and with the disintegration of the Laurentide ice-sheet at Hudson Bay. A possible interpretation of this arid interval — which has no counterpart in the Rajasthan record — see Appendix C.

After 6 500 B.P., under boundary conditions still rather similar to those of 8 000 B.P. (with slow shrinking of the remaining three North American ice-sheets at Keewatin, Labrador and Baffin Island), a *second wet period* set in; it affected the whole Sahara and Sahel zone, only slightly less humid than the first, until about 5 000 B.P. Then the Saharan grasslands nourished a relatively dense Neolithic population of cattle-raising (!) nomads — evidenced by rock-paintings, fossil bones and numerous stone-places (Gabriel 1977), even in the now most arid parts. Rainfall could not have much differed from that of the first period, since evidence for higher temperatures (Jäkel 1978) indicates also higher evapotranspiration. Apparently, the vegetation had the character of a dry savanna with trees along the rivers. The simultaneous occurrence of increased rainfall on both flanks of the Sahara requires an interpretation rather similar to that of the first period. Probably the existence of vast snow and ice fields during the warm season in eastern Canada maintained (as now during spring and fall) the occurrence of deep tilted secondary troughs above Europe and North Africa, triggering cyclonic activity at the belt of tropical easterlies and Saharan depressions, without preference of a selected longitude.

During this neolithic moist period, the surface albedo of the savanna vegetation should have been distinctly lower than now, soil moisture higher than now: the man-made desertification process — investigated by circulation models (e.g. Charney 1975) — had obviously not yet become effective. Under such conditions, the climate might have been similar to the present one in Australia, where in the arid center 100-150 mm of rain falls annually, without definite seasonal preference. Then, simultaneous with the final disappearance of the Labrador ice and the maximum warming in the Arctic (Barry et al. 1977), a dry or even *hyperarid* period started toward 5 000 B.P. (Jäkel 1978, Dumont 1978).

12

The desiccation forced the nomadic herdsmen to settle in the remaining permanent river valleys, particularly in the Nile valley. This apparently coincides with the start of the first high civilization in Egypt: the Old Empire with the step pyramid of Sakkara. A stepwise lowering of the Nile level has been described during the first dynasties (see H.Lamb 1977, Figure 13.37) near Memphis, followed some centuries later by famines. Somewhat later (Bryson & Murray) the decline of the Indus Culture coincides with the desiccation of Rajasthan: such coincidences deserve more interest than hitherto, both from historians and climatologists.

During the last 4 500 years, in temperate latitudes colder phases (Neoglacials) and warmer phases (as the present one) have alternated. Not strictly periodical, these phases generally lasted several centuries, but shorter (decadal) fluctuations are intercalated. One such Neoglacial – the Little Ice Age (Nicholson 1976, 1979) – probably coincided with a humid period along both flanks (at least at the northern) of the Sahara. However, there is no evidence to suggest such a general correlation and it would be premature to give here a comprehensive and consistent survey of these highly effective climatic changes at a 'human' time-scale.

4 CHANGE OF BOUNDARY CONDITIONS AS RELATED TO A FUTURE CLIMATE SCENARIO

The continuous increase of CO_2 due to fossil fuel and deforestation is expected to lead, in the first decades of the next century, to a global warming – see J.Williams (ed.) 1978; World Climate Conference, WMO, Geneva, February 1979. One of the nearest examples of a slightly warmer climate was the Holocene 'optimum' (Hypsithermal, Atlanticum) centered about 6 000 B.P. This optimum has been used in a scenario of the future evolution of our climate (Flohn 1977, Kellogg 1977); it is characterized by an average global temperature increase of 1-1.5°C and a sharp reduction of arid areas (Sarnthein 1978).

Looking into the future: can such a climate repeat itself, with a conversion of the Sahara into pastoral grassland with permanent rivers and lakes? Unfortunately due to the drastic change of boundary conditions since the end of the (Neolithic) moist period, a return to a similar climate in the arid belt of the 'Old World' is very unlikely. The following changes are most essential:

1. Disappearance of the continental ice-sheets in North America and Europe since 5 000 B.P. Any new formation of such ice-sheets could occur only after a prolonged neoglacial period of the type 'Little Ice Age' (ca. 1550-1850). With a global warming the probability of such a transition (triggered by *natural* cooling processes) decreases more and more.

2. Degradation of the natural grassland, during and after the neolithic epoch, by extermination of the original fauna, overgrazing by domestic animals and all other anthropogenic desertification processes (Hare 1978). These processes could be reversed only through a concerted effort of all nations to

13

protect any possible recovering of vegetation, with serious socio-economic consequences.

In this context the spatial and temporal coherency of climatic variability and change in the Sahara and its teleconnections deserve high research priority.

APPENDIX A: SIGNIFICANT RAINFALL AND RUNOFF TELECONNECTIONS

At the Meteorological Institute of the University of Bonn, a research program of large-scale climatic teleconnections has been sporadically carried out since 1968, with most investigations concentrated on tropical rainfall anomalies. Such studies should not be based only on the records of individual stations, or of a sample of partly incoherent station records – as e.g. used by Winstanley (1973), disregarding the lack of longitudinal coherency in the belt of Mediterranean winter rains – but rather on quasi-homogenous series averaged over spatially coherent areas. The selection of areas is based on correlations of simultaneous monthly rainfall data (Helbig 1976) or of annual data (Nicholson). Representative area averages are selected around central Reference Stations (RS); the following monthly series are considered here:

a) Maghreb Coast (Longitude 1°W-13°E): 5 stations, available 1854-1973, RS Oran and Malta.
b) Maghreb Interior (Longitude 3°W-11°E, Latitude 30-33°N): 8 stations, available 1889-1973, RS El Oued.
c) 'Sahel' = Sahel and Sudan Belt of West and Central Africa (Longitude 15°W-16°E, Latitude 9-18°N): 20 stations, available 1905-1973, RS Niamey.
d) 'Sudan' = Central part of Republic of Sudan (Longitude 23-37°E, Latitude 10-16°N): 8 stations, available 1902-1973, RS El Obeid.

In addition to these, Jerusalem (1846-1976) and Tripoli (1892-1973) are used as individual stations. Results from Eastern and Southern Africa are disregarded for the sake of brevity. Sea temperatures in the Gulf of Guinea and in the Angola upwelling area show marked differences between wet and dry years in (c) (P.J.Lamb 1978). A more complete investigation is now in preparation (Sh.Nicholson); a world-wide study based on nearly 300 long station records within the Tropics was completed by H.Fleer (1978).

In nearly all above-mentioned series, November-April and May-October respectively yield more than 70 per cent and less than 30 per cent of the annual rainfall and can therefore be taken as representative for the rainy respective dry season. In some areas, the use of annual (instead of seasonal) means cannot be recommended: winter rains of the northern hemisphere as well as tropical summer rains of the southern hemisphere are split into two adjacent years which then contain parts of two different rainy seasons separated by a half year. Due to frequent intermonthly persistence, seasonal or

14

half year values are more meaningful. Auto- and cross-correlation analysis confirms the validity of this concept, which necessitates, however, the use of seasonal lag correlations.

Here mainly correlations between series (a) to (d) are considered. In spite of the longitudinal spread of more than 5 000 km (as confirmed by more detailed investigations by Sh.Nicholson) the zonal correlations between simultaneous values of (c) and (d) are highly significant. All monthly anomalies together yield a correlation coefficient of + 0.340; the rainy seasons (May-October), + 0.468 (both significant at the 0.1 per cent level). There exists a tendency toward persistence: correlation between (c) and (d) with one year lag yields still 0.204; respectively (d) with (c), 0.263 (5 per cent level).

Due to interannual storage as groundwater, runoff data usually yield higher correlations than rainfall. These have been compiled from Sircoulon (1976) who published long records for numerous large catchments, using the hydrological year (varying with climatological regime) and from World Weather Records which contain monthly values of Nile discharge. Mr Taba (Cairo), then President of the World Meteorological Organization, provided data to extend the Nile series to 1972. This series, for which only the second half year July-December (7-12) is used, represents Ethiopian summer rains. Some typical correlation coefficients are selected, arranged in a zonal direction:

Senegal	– Chari	(43 pairs)	+ 0.571
Ubangi	– Chari	(40 pairs)	+ 0,518
Nile (7-12)	– Chari	(41 pairs)	+ 0.739
Nile (7-12)	– Senegal	(70 pairs)	+ 0.438
Sudan (d)	– Chari	(42 pairs)	+ 0.654
Sudan (d)	– Senegal	(72 pairs)	+ 0.496

All these coefficients are significant to the 0.1 per cent level; however, due to persistence the number of independent pairs and thus the number of degrees of freedom is somewhat lower than indicated by the number of pairs.

A comparison of the Mediterranean winter rains with the tropical summer rains south of the Sahara, must take into account a seasonal difference of six months. Semiannual totals yield the following correlation coefficients (significant at the 5 per cent level):

c) with a) one half year later:	+ 0.203
d) with a) one half year later:	+ 0.290
c) with Jerusalem one half year later:	– 0.345
d) with Jerusalem one half year later:	– 0.279

These results are only partly consistent with those reported by Winstanley based on incoherent series; all correlations are weak (r^2 mostly ≤ 0.10). The correlation with (c) and (d) is positive for the Maghreb coast, but negative for Jerusalem. The hypothesis of a parallel shift of climatic belts during climatic history – similar to their regular seasonal displacement – would result

15

in a negative interseasonal correlation. This is the case for Jerusalem as detected already by Rosenan (1953) for the relation Jerusalem-Khartoum. In contrast to this, positive correlations (as with (a)) indicate a tendency for expansion and shrinking of the desert, as suggested by archeologists (before a reliable chronology was available); such a tendency can be interpreted as outlined in Chapter B (Figures 1-2). Other correlations, including the Maghreb interior (b) with Tripoli, remain below the 5 per cent significance level.

These positive correlations seem to indicate that in the *western* half of Northern Africa near-simultaneous shrinking and expanding of the desert prevails. This is also demonstrated by a (simultaneous) correlation between rainfall January-May in the interior Maghreb (b) and Sahel (c) with + 0.338 (significant above 0.01 level). Since in (c) rains before May are quite negligible, the onset of the tropical rains is related to the intensity of the winter-spring rains south of Atlas mountains, caused by Sahara depressions (Figure 1).

This important result, which should be checked more thoroughly, indicates a circulation pattern responsible for simultaneous shrinking and expansion of the desert in the historical past.

Such small but significant correlations discourage any direct use for long-range forecasting — here they are used in a purely climatological sense. They have been selected from many similar coefficients incorporating time-lags of ± five years. A sample of 660 coefficients yields 77 (~ 12 per cent) significant at the 5 per cent level, i.e. more than twice the number expected from randomly selected pairs.

The essential results derived from representative recent data can be summarized as follows:

1. In western North Africa a weak statistical tendency for near-simultaneous shrinking and expanding of the arid belt exists, working probably mainly in the transitional seasons with a mechanism as described in Figure 1.

2. While the longitudinal coherency of rainfall in the Sudan-Sahel belt is surprisingly large, it is restricted, in the Mediterranean belt of winter rains, to areas with a diameter of 500-1 000 km, which can be simultaneously controlled by a quasi-stationary trough-ridge pattern.

The authors appreciate gratefully the co-operation of Dipl. Met. B.Schweitzer, who has revised and checked the rainfall series and their correlations.

APPENDIX B: THE ROLE OF THE TROPICAL EASTERLY JET

In a comprehensive review of the Tropical Easterly Jet (Flohn 1964), it has been shown that during northern summer a cross-circulation in the exit region of this jet (TEJ) above Northern Africa forces large-scale subsidence (~ 600 m/day) in the belt 14°N-25°N, accompanied by large-scale lifting (~ 1 000 m/day) in the belt 7°N-14°N. The latter belt coincides with the tropical summer rain belt, while in the first belt convective activity is suppressed above the northern area of the humid southwest Monsoon, including the ITCZ (Flohn

1966). This mechanism has been interpreted as responsible for the surprisingly small poleward extension of the tropical summer rains (in comparison to South Africa (24°S), South America (21°S), Indo-Pakistan (35°N) and Australia (ca. 25°S)). The physical properties of the TEJ depend on the unique heating of the Tibetan Plateau (Flohn 1968), with a quite slow deceleration of the TEJ above the Arabian Sea and North Africa over a distance of about 100° Longitude. Above the Atlantic a similar asymmetric distribution of the divergence of the wind field and the vertical motion on both flanks of the ITCZ had also been calculated (Flohn 1957).

It was hoped that the course of climatic history would verify that the TEJ is responsible for sub-Saharan rainfall patterns. However, the expansion of tropical rains into the central Sahara during the two moist periods of the Holocene apparently contradicts this conclusion. Some interpretations may be offered, from which only (c) seems to be convincing.

a) Tectonic lifting of the Tibetan Plateau since the late Pleistocene could have intensified the TEJ. However, it is very unlikely that this process operating since the late Tertiary should have been larger than, say, 5 cm/year equivalent to 500 m during 10 000 years, i.e. about 12 per cent of the total lifting. If lifting remains below 1 cm/year (as is more likely), the resulting intensification of the TEJ should have been negligible.

b) During the cold periods of the Pleistocene the Tibetan plateau may have been covered with an extensive snow-cover, which should have reduced the intensity of summer-time heating and thus the intensity of the TEJ. However, it is very unlikely that such a situation lasted until the end of the moist period around 5 000 B.P. Furthermore the intensity of summer monsoon rains in Rajasthan during this moist period (Singh 1974) suggests higher (instead of lower) convective activity and thus stronger heating of the plateau.

c) The role of the TEJ is now restricted to the season of its occurrence (June to August, above Africa also September: Strüning and Flohn 1969). If this seasonal distribution has not changed during the Holocene, the extension of tropical rains into the (now arid) zone north of 16°N during the *transitional seasons* is not impeded by the TEJ. This interpretation apparently does not contradict any observed facts. It should be added that now during March/ April and October/November neither the TEJ nor its weaker counterpart during southern summer (see Strüning-Flohn Figure 24) are developed. During these seasons upper troughs from both hemispheres most frequently reach the equatorial zone, with greatest efficiency in triggering low-level rain-producing eddies.

APPENDIX C: A TENTATIVE INTERPRETATION OF THE DRY EPISODE AFTER 8 000 B.P.

During the decay of the Laurentide ice between 14 000 and about 4 500 B.P., the catastrophic incursion of the sea into the present Hudson Bay (Ives et al.

17

Naturwissenschaften *62*, 1975, 118-125) shortly after 8 000 B.P. forms a climax related to several hitherto hardly noticed consequences. One is the rapid world-wide sea-level rise of 8 m in about 200 years (cf. Mörner, Palaeogeogr. etc. *19*, 1976, 63-85). An event of this type could hardly happen linearly with time — which would yield about 4 cm/a instead of 0.12 cm/a today — but should probably rise in some years to 10-20 cm/a, or even more.

This allows an estimate of the climatic changes (especially of the changes of oceanic evaporation) during the melting of this enormous ice volume: 8 m eustatic sea-level rise is equvalent to $2.88 \cdot 10^6$ km^3 water (or $3.16 \cdot 10^6$ km^3 ice). Assuming melting heat to be 80 cal/g and adding 20 cal/g for warming from $-10°C$ to $+10°C$, a total heat loss of $2.88 \cdot 10^{23}$ gcal can be estimated. The whole area of the Atlantic Ocean involved into the melting process, including the central and eastern part of the Gulf Stream and the Canary Stream, can be estimated to be $12 \cdot 10^6$ km^2. If we allow also 200 years for melting, the heat loss per cm^2 and year (a) is then

$$\frac{2.88 \cdot 10^{23}}{12 \cdot 10^{16} \cdot 200} \frac{gcal}{cm^2\, a} = 1.2 \cdot 10^4 \frac{gcal}{cm^2\, a} = 12 \text{ Kilolangleys/a}$$

The daily heat loss is 33 Ly/d ~ 16 Watt/m^2. Assuming the Bowen ratio to be 0.10, about 90 per cent of this net energy could be used for evaporation: taking condensation heat to be 590 gcal/g, this yields an average loss of evaporation of about 0.5 mm/d or 18 cm/a.

Comparing this value with the average oceanic evaporation in these areas (60-120 cm/a), this represents a regional loss of evaporation of 15-30 per cent, which may rise, during the climax of this event by a factor 3-4 lasting some consecutive years (say, a decade). Since these estimates are rather conservative — e.g. the area involved could be substantially smaller, the loss more intensive — it is possible to interpret, as a working hypothesis, the enigmatic dry period at the Sahara after 8 000 B.P. (which apparently did not exist in Rajasthan) as caused by the catastrophic Hudson Bay 'surge' of about 3 million km^3 of continental ice into the Atlantic. A plausible mechanism in the form of a 'calving bay' has been described by Hughes (Rev. Geophys. Space Physics *15*, 1977, 1-46).

SELECTED REFERENCES

Barry, R.G., et al. 1977. Environmental change and cultural change in the Eastern Canadian Arctic during the last 5 000 years. *Arctic and Alpine Res. 9:*193-210.
Bryson, R.A. & Th.J.Murray 1977. *The climate of hunger.* Univ. Wisconsin Press.
Butzer, K.W., et al. 1972. Radiocarbon dating of East African lake levels. *Science 175:* 1069-1076.
Charney, J.G. 1975. Dynamics of deserts and droughts in the Sahel. *Quart. J. R. Meteor. Soc. 101:*193-202.
Diester-Haass, L. 1976. Late Quaternary climatic variations in NW Africa deduced from East Atlantic sediment cores. *Quatern. Res. 6:*299-314.

Fairbridge, R.W. 1976. Effects of Holocene climatic change on some tropical geomorphic processes. *Quatern. Res. 6:*529-556.

Fleer, Heribert 1978. Statistische Analyse und Telekonnektionen der Niederschlagschwankungen in der Tropenzone. Diss. Univ. Bonn (in print).

Flohn, H. 1957. Studien zur Dynamik der äquatorialen Atmosphäre. *Beitr. Phys. Atmos. 30:*18-46.

Flohn, H. 1964. Investigations on the Tropical Easterly Jet. *Bonner Meteor. Abh. 4.*

Flohn, H. 1966. Warum ist die Sahara trocken? *Zeitschr. Meteor. 17:*316-320.

Flohn, H. 1967. Bemerkungen zur Asymmetrie der atmosphärischen Zirkulation. *Ann. Meteor. N.F. 3:*76-80.

Flohn, H. 1968. Contributions to a meteorology of the Tibetan Highlands. *Atmos. Sci. Paper 130,* Colorado State Univ., Ft. Collins, Colo.

Flohn, H. 1975. Tropische Zirkulationsformen im Lichte der Satellitenaufnahmen. *Bonner Meteor. Abh. 21.*

Flohn, H. 1977. Climate and energy: A scenario to a 21st century problem. *Climatic Change 1:*5-20.

Flohn, H. 1978. Comparison of Antarctic and Arctic climate and its relevance to climatic evolution. In: E.M.van Zinderen Bakker (ed.), *Antarctic glacial history and world palaeoenvironments.* A.A.Balkema, Rotterdam, pp.3-13.

Flohn, H. 1979. Possible climatic consequences of a man-made global warming. *Rep. Intern. Inst. Appl. Systems Analysis,* WP-79-86, Laxenburg, Austria (XI + 103 pp.)

Gabriel, B. 1977. Zum ökologischen Wandel in der östlichen Zentralsahara. *Berl. Geogr. Abh. 27.*

Gasse, F. & G.Delibrias 1976. Les lacs d'Afar Central (Ethiopie et T.F.A.I.) au Pleistocène superieur. In: S.Horie (ed.), *Paleoclimatology of Lake Biwa.* Vol.4, pp.529-575.

Gierasch, P.J. & R.M.Goody 1972. The effect of dust on the temperature of the Martian atmosphere. *J.Atmos. Sci. 29:*400-402.

Giresse, P. 1978. Le contrôle climatique de la sédimentation marine et continentale en Afrique Centrale Atlantique à la fin du Quaternaire-problèmes de correlation. *Palaeogeogr. Palaeoclim. Palaeoecol. 23:*57-77.

Hare, F.K. 1977. Climate and desertification. In: *Desertification: Its causes and consequences.* UN Conf. on Desertification, Nairobi, pp.63-167.

Hays, J.D. 1978. A review of the Late Quaternary climatic history of the Antarctic Seas. In: E.M.van Zinderen Bakker (ed.), *Antarctic glacial history and world palaeoenvironments.* A.A.Balkema, Rotterdam, pp.57-71.

Heine, K. 1974. Bemerkungen zu neueren chronostratigraphischen Daten zum Verhältnis glazialer und pluvialer Klimabedingungen. *Erdkunde 28:*303-312 (see also *Erdkunde 31 (1977):*161-178).

Helbig, M. 1976. Korrelations- und Spektrumsanalyse von Niederschlags- und Abflussreihen aus Afrika und Reduktion der Niederschlagsdaten durch Gebietsmittelung. Dipl. Thesis, Univ. Bonn.

Hoinkes, H. 1961. Die Antarktis und die geophysikalische Erforschung der Erde. *Naturwissenschaften 48:*354-374.

Jäkel, D. 1978. Eine Klimakurve für die Zentralsahara. In: Museen der Stadt Köln, *Sahara, 10 000 Jahre zwischen Weide und Wüste.* Köln, pp.382-396 (cf. also *Palaeoecology of Africa 11,* 1979).

Kellogg, W.W. 1977. *Effects of human activities on global climate.* WMO, Techn. Note 165 (WMO Nr. 486).

Lamb, H.H. 1977. *Climate: Present, past and future,* Vol. II. Methuen, London.

Lamb, P.F. 1978. Large-scale Tropical Atlantic surface circulation pattern associated with Subsaharan weather anomalies. *Tellus 30:*240-251.

Livingstone, D.A. 1975. Late Quaternary climatic change in Africa. *Ann. Rev. Ecology Systematics 6:*249-280.

Lorius, C., et al. 1978. Climatic changes in Antarctica during the last 30 000 years. In: *CNES: Evolution des atmosphères planétaires et climatologie de la terre.* Coll. Intern. Nice (Oct. 1978), pp.71-82.

Maley, J. 1977. Palaeoclimates of central Sahara during the early Holocene. *Nature 269:* 573-577.

McIntyre, A., et al. 1978. The role of the North Atlantic Ocean in rapid glacial growth. In: *CNES: Evolution des atmosphères planétaires et climatologie de la terre.* Coll. Intern. Nice (Oct. 1978), pp.113-120.

Meteorological Office 1962. *Weather in the Mediterranean,* Vol. I, 2nd ed.

Museen der Stadt Köln 1978. *Sahara. 10 000 Jahre zwischen Weide und Wüste.* Ausstellungskatalog. Köln.

Nicholson, Sh.E. 1976. A climatic chronology for Africa: Synthesis of geological, historical and meteorological information and data. Ph.D. Thesis, Dept. of Meteor. Univ. of Wisconsin, Madison, Wis.

Nicholson, Sh.E. 1979. Saharan climates in historic times. In: H.Faure & M.E.J.Williams (eds.), *The Sahara and the Nile.* A.A.Balkema, Rotterdam.

Palmén, E. & Ch.W.Newton 1969. *Atmospheric Circulation Systems.* Intern. Geophys. Series 13.

Pedgley, D.E. 1972. Desert depressions over Northeast Africa. *Meteor. Mag. 101:*228-244.

Prell, W.L. 1978. Glacial/interglacial variability of monsoonal upwelling: Western Arabian Sea. In; *CNES: Evolution des atmosphères planétaires et climatologie de la terre.* Coll. Intern. Nice (Oct. 1978), pp.149-156.

Rognon, P. 1976. Essai d'interpretation des variations climatiques au Sahara depuis 40 000 ans. *Rev. Géogr. phys. Géol. dyn. 18:*251-282.

Rosenan, N. 1951. Long-range forecasts of rainfall. *Bull. Res. Council Israel,* pp.33-35.

Ruddiman, W.F. & A.McIntyre 1977. Late Quaternary surface ocean kinematics and climatic change in the high-latitude North Atlantic. *J. Geophys. Res. 82:*3877-3887.

Sarnthein, M. 1978. Sand deserts during glacial maximum and climatic optimum. *Nature 272:*43-46.

Schreve-Brinkman, E.J. 1978. A palynological study of the upper Quaternary sequence in the El Abra corridor and rock shelters (Colombia). *Palaeogeogr., Palaeoclim., Palaeoecol. 25:*1-109.

Shackleton, N.J. 1977. Carbon-13 in Uvigerina: Tropical rainforest history and the Equatorial Pacific carbonate dissolution cycles. In: N.R.Andersen & A.Malakoff (eds.), *The fate of fossil fuel CO_2 in the oceans.* Marine science Vol. 6, pp.401-427.

Singh, G., et al. 1974. Late Quaternary history of vegetation and climate of the Rajasthan Desert, India. *Phil. Trans. R. Soc. London B 267:*467-501.

Sircoulon, J. 1976. Les données hydropluviométriques de la sécheresse récente en Afrique intertropicale. Comparison avec les sécheresses '1913' at '1940'. *Cahiers ORSTOM, Sér. Hydrol. 13:*75-174.

Strüning, J.O. & H.Flohn 1969. Investigations on the atmospheric circulation above Africa. *Bonner Meteor. Abh. 10.*

Trempel, U. 1978. Eine klimatologische Auswertung der meteorologischen Beobachtungen deutscher Handelsschiffe vor der Westküste Südamerikas im Zeitraum 1869-1970. Dipl. Thesis, Univ. Bonn.

Williams, J. (ed.) 1978. *Carbon dioxide, climate and society*. IIASA Proceedings Series. Pergamon Press, Oxford.

Winstanley, D. 1973. Rainfall patterns and general atmospheric circulation. *Nature 245:* 190-194.

WMO 1979. *Proceedings of the World Climate Conference*, WMO-Nr. 537, XII + 791 pp.

Wyrtki, K. 1975. El Niño-The dynamic response of the equatorial Pacific Ocean to atmospheric forcing. *J. Phys. Oceanogr. 5:*572-584 (cf. loc. cit. *7*, 1977:779-787).

Wilson, J. T. (ed.) 1976. Continents adrift and continents aground. II, 1.2 Freeman, 2.3 Scale.
Paterson Press, Oxford ......
Wunsch, C. 1975. Internal tides... and equilibrium position of deposition. Rev ...
180-196 ...

WMO 1970. Proceedings of the World Climate Conference. WMO No. 537, XII + 791 p.
Wyrtki, K. 1975. Fluctuation of the dynamic ... of the Equatorial Pacific Ocean ...
during onset of ... geophysics ? 572-584 ...

SOME IMPLICATIONS OF GATE RESULTS FOR SAHARAN DUST TRANSPORTS ACROSS THE ATLANTIC

G. SIEDLER
Universität Kiel

INTRODUCTION

Evidence of Saharan dust being transported from the African continent into the North Atlantic region results from different observations: Turbidity measurements, aerosol observations, satellite photography and determinations of mineral particle components in shelf and deep-sea sediments.

Discussions of low visibility off the Northwest African coast and the related deposition of dust particles on the sea surface date back to the first part of this century (e.g. Jentzsch 1909, Semmelhack 1934). Most of the quantitative data on Saharan dust particles, however, have only been obtained during the last decade after the development of improved aerosol sampling systems that could be used on the ground or on aircraft (Prospero 1968, Jaenicke, Junge & Kanter 1971, Junge & Jaenicke 1971, Prospero & Carlson 1972, Jaenicke & Schütz 1978). Information about the horizontal extension and migration of dust-laden air parcels across the Atlantic has been obtained by satellite photography (Carlson & Prospero 1972, Szekielda 1978). The Saharan dust outfall at the sea surface will sink down to the bottom, possibly in the form of pellets, and can be detected in the shelf and deep-sea sediments (e.g. Johnson 1979, Kolla, Biscayne & Hanley 1979, Sarnthein & Koopmann 1979).

Temporal and spatial fluctuations of dust deposition will either be caused by the variability of mobilisation at the African source region or by changes in the atmospheric and oceanic transports. After having obtained an unprecedentedly dense meteorological and oceanographic data set for a large part of the tropical Atlantic and the surrounding land masses during GATE, the GARP Atlantic Tropical Experiment (GARP = Global Atmospheric Research Programme), in 1974, it seems appropriate to check whether the dominant scales of atmospheric flow patterns are consistent with the dominant scales found in aerosol observations at the eastern and western edges of the tropical Atlantic. This is the aim of the following discussion.

23

ATMOSPHERIC TRANSPORTS

The transport of Saharan dust can be expected to occur mainly in the layer of 'African air' above the Trades inversion between approximately 800 mb and less than 600 mb (Carlson & Prospero 1972), or roughly between a height of 1 and 5 km. Strong northeasterly or easterly winds, sometimes with a jet-like vertical structure, have frequently been observed in this nearly isentropic layer. Although a fairly dense network of radiosonde stations yielding information about this layer had been established during GATE (see Figures 2 and 3), a much better coverage in space and time of atmospheric motion was obtained by cloud wind data from the two geostationary satellites ATS III and SMS I. These data provide detailed information about the atmospheric motions at 850 mb, i.e. at the lower boundary of the layer supposedly carrying the Saharan dust. Although the transports will certainly differ from those at higher altitudes, it is nevertheless likely that similar scales will be present in the variation of the African air layer above the 850 mb level. Before discussing results from this experiment, it may be appropriate to give a short introduction to the objectives and the design of GATE.

The two basic objectives of GARP are (i) to improve the medium and long

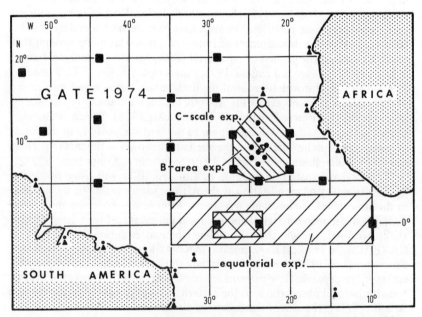

Figure 1. Area of GATE observational networks. Squares and small circles give positions of stationary ships, the large circle corresponds to Cape Verde Islands, and hatched surfaces indicate areas of intensive sub-experiments.

24

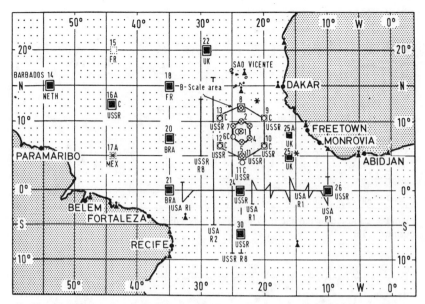

Figure 2. GATE research vessel distribution during phase I (26 June-16 July 1974). Squares and circles give stationary ships' positions, GATE numbers and nationality, hexagons relate ships in B-scale, other lines indicate observational sections performed with mobile ships.

range weather prediction and (ii) to facilitate the development of climate models. Since the convection over the tropical ocean is an important feature of the large-scale atmospheric circulation and since the tropical oceans had only been sparsely covered by measurements, the first major GARP experiment for the first objective was performed in the tropical Atlantic and the surrounding land masses. The primary aims of GATE were:

1. to develop methods that would allow parameterisation of small scale tropical weather systems — from the 'cloud clusters' with 100-1 000 km scale down to the 'hot towers' with 1-10 km scale — in order to incorporate their effects in the description of the synoptic scale circulation;

2. to advance the development of numerical models of the atmospheric circulation.

Although being mainly a meteorological programme, it included a fairly large oceanographic sub-programme aimed at describing the surface temperature field and at understanding the energy fluxes in the upper ocean related to atmospheric forcing. The experiment was performed in three phases from June to September 1974, with an international participation of 39 research vessels, 13 aircraft, several satellites and numerous ground stations and ships of opportunity.

25

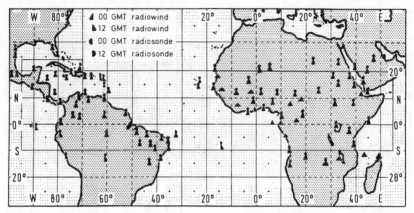

Figure 3. Land radiowind and radiosonde stations which are known to have functioned during GATE.

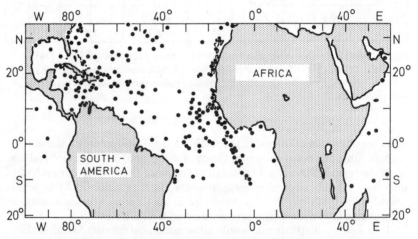

Figure 4. Distribution of merchant ships on a typical day of GATE.

The oceanic station network extended from the equatorial to the southern subtropical North Atlantic (Figure 1), with a particularly dense observation pattern in the Intertropical Convergence Zone (ITCZ) of the eastern Atlantic (B- and C-scale experiments) and at the equator (equatorial experiment). As examples, the positions of stationary research vessels during phase I are presented in Figure 2 and the positions of land-based stations for all three phases of GATE are given in Figure 3. The distribution of merchant vessels involved in GATE on a typical day is found in Figure 4.

At this season, the ITCZ was positioned between 5°N and 10°N, with the

26

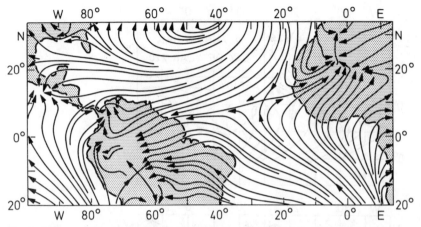

Figure 5. Mean surface streamlines for the 100 days of GATE (after Krishnamurti &
Krishnamurti 1978).

Northeast and Southeast Trades at either side and a monsoon system in the
Gulf of Guinea (Figure 5). It has to be kept in mind, however, that the actual
flow conditions on individual days will deviate considerably from the mean
field. Apart from a slow meridional displacement of the ITCZ, large wave-like
disturbances with westward propagation occur in this tropical convection
zone, the 'easterly waves'. An example of the perturbations that are part of
these easterly waves is presented in the upper part of Figure 6.

A detailed analysis of the wind fields at the surface and at 850 mb was
performed by a group at Florida State University (Krishnamurti, Levy & Hua-
Lu Pan 1975, T.N.Krishnamurti & R.Krishnamurti 1978). The meteorological
data used included two data sets per day from 25 research vessels, roughly
1 400 observations per day from merchant ships and cloud vector data from
two geo-stationary satellites with approximately 1 400 low cloud motion vectors per day. By using a man-machine mix of data handling, 2 per cent of
artificial data were inserted in gaps of the observations. Cloud winds were
related to surface winds by an objective analysis scheme not depending on
any specific planetary boundary layer model. As a result coherent flow patterns could be generated for the surface and the 850 mb level. Films were
produced to display the temporal and spatial variability of the flow field.

The analysis by Krishnamurti and co-workers leads to the following main
results. Near simultaneous surges are found in the Trades of the two hemispheres, with a phase shift occurring at the time of a particularly active phase
of African disturbances in August 1974. The zonal wind has considerable
variability at time scales of three to six days and 10 to 15 days. The three to
six day disturbances are apparently related to the easterly waves in the ITCZ.
The 10 to 15 day disturbances also appear to be westward propagating.

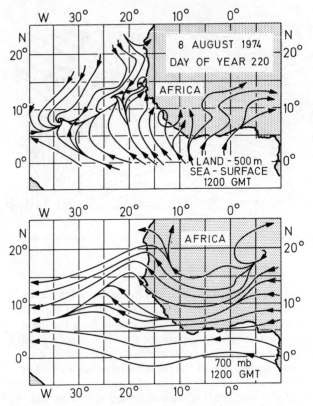

Figure 6. Stream lines at the surface and at 700 mb on 8 August 1974 (after GATE Report No. 17, p.94, WMO/ICSU, Geneva, 1975).

CONCLUSIONS

Studies of Saharan dust aerosol concentrations at selected positions showed large temporal variations in the concentration values. Data from the Cape Verde Islands in the eastern Atlantic in July 1973 (Jaenicke & Schütz 1978) and during GATE from July to August 1974 and western Atlantic samples collected at Barbados and Miami during GATE (Savoie & Prospero 1977) show high concentration events with a typical duration of two to four days. An example is given in Figure 7. These time scales coincide sufficiently well with the time scales of disturbances related to the easterly waves (Krishna-murti et al. 1978) to suggest that fluctuations of Saharan dust occurrence are to a large extent due to these variations in the atmospheric flow field over the Atlantic, at least in the northern hemisphere summer. It cannot be

28

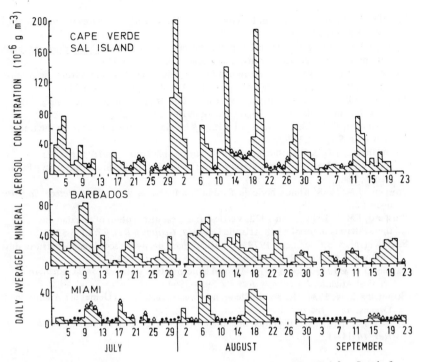

Figure 7. Daily averaged mineral aerosol concentrations during GATE (after Savoie & Prospero 1977).

concluded from such a qualitative discussion whether observed dust concentration changes are mainly due to speed variations or rather to meridional displacements of the dust transporting layer. Case studies using satellite photography as suggested by Carlson & Prospero (1977) will be required for this purpose.

REFERENCES

Carlson, T.N. & J.M.Prospero 1972. The large-scale movement of Saharan air outbreaks over the northern equatorial Atlantic. *J. Appl. Meteor. 11:*283-297.

Carlson, T.N. & J.M.Prospero 1977. *Saharan air outbreaks: Meteorology, aerosols and radiation.* Report US GATE Central Program Workshop NCAR, Boulder, Colorado, pp.57-78.

Jaenicke, R., C.Junge & H.J.Kanter 1971. Messungen der Aerosolgrösenverteilungen über dem Atlantik. *'Meteor' Forsch.-Ergebn. B 7:*1-54.

Jaenicke, R. & L.Schütz 1978. Comprehensive study of physical and chemical properties of the surface aerosols in the Cape Verde Islands region. *J.Geophys. Res. 83:*3585-3599.

29

Jentzsch, M. 1909. Staubfälle im Passatgebiet des Nordatlantischen Ozeans. *Ann. Hydrogr. Marit. Meteor. 37:*373-376.

Johnson, L.R. 1979. Mineralogical dispersal patterns of North Atlantic deep-sea sediments with particular reference to eolian dusts. *Mar. Geol. 29:*335-346.

Junge, C. & R.Jaenicke 1971. New results in background aerosol studies from the Atlantic expedition of the R.V.Meteor, spring 1969. *J. Aerosol Sci. 2:*305-314.

Kolla, V., P.Biscayne & A.Hanley 1979. Distribution of quartz in late Quaternary Atlantic sediments in relation to climate. *Quatern. Res. 11:*261-277.

Krishnamurti, T.N., C.E.Levy & Hua-Lu Pan 1975. On simultaneous surges in the trades. *J. Atmos. Sci. 32:*2367-2370.

Krishnamurti, T.N. & R.Krishnamurti 1978. *Surface meteorology over the GATE A-scale.* Report No. 78-6, Dept. Meteor. and Oceanogr., Florida State University, Tallahassee, Fla.

Prospero, J.M. 1968. Atmospheric dust studies on Barbados. *Bull. Amer. Meteor. Soc. 49:* 645-652.

Prospero, J.M. & T.N.Carlson 1972. Vertical and areal distribution of Saharan dust over the western equatorial north Atlantic Ocean. *J. Geophys. Res. 77:*5255-5265.

Sarnthein, M. & B.Koopmann 1979. Late Quaternary deep-sea record on NW African dust supply and wind circulation. *Palaeoecology of Africa 12:*239-253.

Savoie, D.L. & J.M.Prospero 1977. Aerosol concentration statistics for the northern tropical Atlantic. *J. Geophys. Res. 82:*5954-5964

Semmelhack, W. 1934. Die Staubfälle im nordwest-afrikanischen Gebiet des Atlantischen Ozeans. *Ann. Hydrogr. Marit. Meteor. 62:*273-277.

Szekielda, K.-H. 1978. Eolian dust into the northeast Atlantic. *Oceonogr. Marit. Biol. Ann. Rev. 16:*11-41.

METEOROLOGICAL PATTERNS AND THE TRANSPORT OF MINERAL DUST FROM THE NORTH AFRICAN CONTINENT

G. TETZLAFF / K. WOLTER
Universität Hannover

1. INTRODUCTION

The analysis and interpretation of indirect parameters is the only way to get evidence on climatic variations in the past. One of these possible parameters is the distribution of mineral dust deposits in the deep-sea.

Both the lower tropospheric Northeast Trades originating from the African continent and the well-known effect of the deflation of the Sahara in recent periods, starting several thousand years before present should result in dust deposits in the Atlantic Ocean (Schütz & Jaenicke 1979; Pachur 1974). Recently, samples of deep-sea sediment cores were investigated quantitatively on their dust fraction (Johnson 1979; Sarnthein & Koopman 1979). Results show distinct patterns in the amounts deposited and in the grain size distribution for different climatic periods. However, even the most modern sediments from the cores are composed of a mixture of materials deposited over an interval of several hundred years (Sarnthein & Koopman 1979). In this paper, we attempt to inspect more closely a horizontal pattern of dust deposit in the Atlantic – namely, a ridge of maximum dust deposits stretching hook-like from Nouadhibou to the Atlantic and then back to the African coast via the Canary Islands (Figure 1) – and to correlate that structure with meteorological patterns having occurred during the past 15 years.

2. METEOROLOGICAL DATA

Most of the climatic data are presented in the form of long term averages, assuming implicitly a very simple statistical distribution. Furthermore, it is generally agreed that an observed long term effect such as the deposition of dust, is approximately correlated with the long term mean pattern of the meteorological parameters. The smallest elements to be considered here are disturbances of the synoptic scale, high and low pressure systems of the tropics and mid-latitudes with a typical horizontal scale of 1 000 km and a time scale of some days. Information about them can be obtained in the

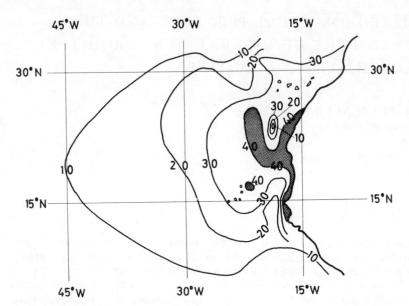

Figure 1. Relative data of recent mineral dust deposits in the Atlantic (after Sarnthein & Koopmann 1979).

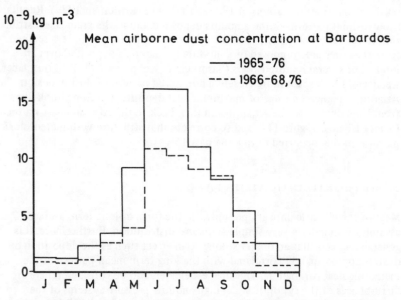

Figure 2. Mean airborne dust concentration at Barbados for non-drought years (dashed) and the overall period (solid) from 1965-1976.

easiest way from routine data sources, at least on a day by day basis. However, it must be inferred that during the integration interval for the dust deposits the general circulation with its characteristic sequence of individual meteorological events did not alter. In particular, the position of the ITCZ, the position of the heat low at the surface on the African continent south of the Atlas Mountains, the position of the heat high at levels between 700 mb and 500 mb are assumed to have remained unchanged. Very probable this is so, as there is no evidence that wind and precipitation experienced any major alteration during the last thousand years, though slight oscillations did occur (Jäkel 1977, Servant-Servant 1970, Nicholson 1978).

Therefore, the meteorological data of the last 15 years are examined here in order to investigate the possibility of the occurrence of a meteorological structure based on repeated events on a time scale of some days, resulting in the formation of a ridge of wind deposits (Figure 1). The data used are daily weather charts with the observational data derived especially from the London Centre for Overseas Pest Research and also from the 'Täglicher Wetterbericht' of the 'Deutscher Wetterdienst' and the 'Berliner Wetterkarte'.

3. THE GENERAL FEATURES OF THE METEOROLOGICAL AND DUST TRANSPORT PATTERNS

The transport of dust from North Africa is a long known and investigated phenomenon. Perhaps the most striking fact was that African mineral dust was observed at the West Indies in considerable concentration. Measurements taken there exhibited an annual variation of concentration (Figure 2) and are presented as monthly averages (Carlson & Prospero 1977), taken with some interruptions since about 1965. In this period, the years of Sahel drought are included and show considerably higher concentrations than the other years. In Figure 2, the non-drought years (1965-1968, 1976) are presented by a dashed line and the overall average by a solid line. In both cases, the structures are very similar, with a low winter-minimum and a summer-maximum one order of magnitude larger than the minimum. The non-drought years exhibit concentration data on two levels with short transitional periods of about one month in spring and autumn. The drought years do not extend the summer-maximum period but the transitional ones. Even during winters the dust concentration has been higher than in non-drought years.

The source of this dust is not really known. Prospero & Nees (1977) used satellite pictures to find the origin. They claim a more continuous source in the Sahel-zone just north of $15°N$ and a more incidental one in the northern flat desert plains. However, it is understood that clouds prevent a continuous record of satellite information. Therefore, a surface bound confirmation is desirable. The synoptic routine data of the local weather bureaux contain information about visibility. The horizontal visibility at the surface can only be taken into account, if the process of removing the dust from the surface

33

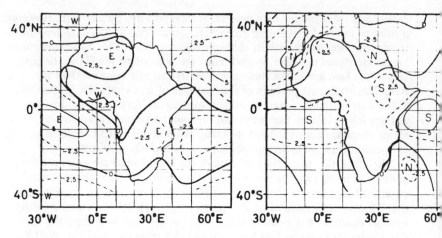

Figure 3a, b. Long-term summer season average values of the surface zonal (a) and meridional (b) wind component (after Newell et al. 1972).

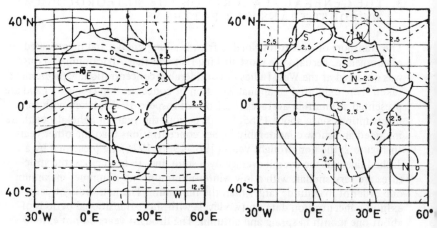

Figure 3c, d. As in Figure 3a, b for the 700 mb-level zonal (c) and meridional (d) wind component (after Newell et al. 1972).

34

is well-known. Since the dust propagation mainly occurs at a level of about 700 mb (Carlson & Prospero 1977) — that is about 3 000 m above sea level — this requires in addition the existence of a mechanism that moves the dust from the surface to the 700 mb-level.

Sometimes the diurnal variation of the mixing depth is quite large in the area south of the continental ITCZ (about 12-20°N in Summer). In the early afternoon, the static instability allows convective elements to penetrate even the inversion above the monsoonal air up to a height of about 5 000 m rather regularly. There is a supply of dust originating from the areas covered by downdrafts of thunderstorms correlated with deep convection further in the South. This mechanism would allow to transport dust more continuously into the easterlies at 700 mb. Other levels below like for example 850 mb, display a too perturbed wind field pattern in summer and generally do not allow a trajectory across the Atlantic towards Barbados (Krishnamurti 1979). On the other hand, the correlation between the 500 mb-level winds and the 700 mb-level winds is very close (Thompson 1965), thereby enabling us to analyse the 500 mb-charts.

The ITCZ propagation is coupled with the formation of a surface-heat generated low-pressure system and an upper air pressure system related to a heat high on the continent south of the Atlas Mountains. The average wind field for June-August reflects the northernmost position of the ITCZ (Figures 3a-d after Newell et al. 1972).

Figures 3a and 3b give the mean surface wind speeds (m s^{-1}) in zonal and meridional direction respectively and Figures 3c and 3d show the same parameters for the 700 mb-level. The strong easterly winds at 15°N and 700 mb are quite distinct, as well as the southerly winds at the surface over the continent, coinciding with the monsoonal winds in summer.

The pattern of marine mineral dust deposits (Figure 1 after Sarnthein & Koopmann 1979) exhibits a tongue stretching from the African coast westward at a latitude of about 15-20°N. Off Dakar, a minimum occurs probably due to ocean currents. A small maximum northeast of the Capeverde Islands is probably caused by the blocking effect of the islands on the air flow. The zone of maximum dust deposition displays a hook-like structure from the coast at 18°N to the west then gradually bends northwards, from about 24°N even eastwards, finally pointing back to the African coast via the Canary Islands (cf. Sarnthein & Koopmann, this volume). In the centre of that hook a distinct minimum is found.

The general wind pattern (Figures 3a-d) is held responsible for the outlined structure of the westward directed tongue with gradually decreasing rates of deposition. However, the smaller scaled structure of the hook containing the dust minimum, requires an additional explanation not met in applying average atmospheric flow patterns.

35

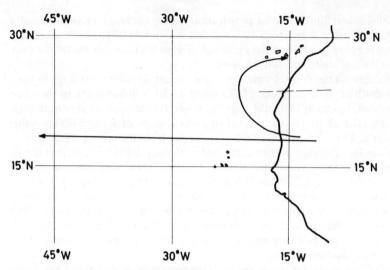

Figure 4. Prevailing directions of 700 mb-trajectories transporting dust from the Sahara. The hook-like trajectory is almost fixed in its position, the westward directed one may shift north- and squthwards.

4. METEOROLOGICAL FLOW PATTERNS

This hook-like ridge presents a pattern with a horizontal scale of at least 1 000 km. That is the order of magnitude of typical atmospheric vortices, cyclones and anticyclones. The oceanic sediment data confirm this structure by analysis of different variables and thus its existence may be regarded as proven. The 700 mb-level, responsible for the propagation of the dust, has to form a characteristic trajectory repeatedly during the summer season to enable the hook-like mineral dust transport events — the trans-oceanic conveyance occurring more continuously (Figure 4).

Considering Figures 3a-d, the hook-like trajectory is shown not to be an average structure. On the contrary, the 700 mb mean meridional winds are from the North where they are required to come from the south. On the other hand, the zonal components in the 700 mb-level are as demanded for the hook formation with easterly winds south of 26°N and westerly winds north of that latitude suggesting a variable flow pattern between both of them.

As no actual data on trajectories are available, the isohypses were taken as only approximations for streamlines, which seems to be sufficient as a first attempt. This method only fails in very unstable cases.

The histogram (Figure 5) is a summary of the analysis of ten years of weather charts. All the cases with an anticyclonic isohypse circling around

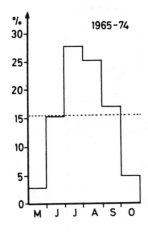

Figure 5. Mean monthly percentage of days with a hook-like trajectory (streamline) form coinciding with the ridge of the mineral dust deposits.

the minimum of dust deposits are included, as well as those with a similar flow pattern but without a closed isohypse in that region. Bearing in mind that this approach is rather subjective, we can expect that an objective analysis may yield less cases of exactly coinciding streamlines. Nevertheless, Figure 5 exhibits a distinct summer peak with a frequency of about one case per four days.

The differences in the data originating from the different sources are smaller than might be expected, also there seems to be little variation due to individual interpretation of the charts. The error is estimated to range about two days per month corresponding to about six per cent in the scale of Figure 5. That means that the realization of a hook-like streamline seems to occur in a minority of the days, but rather regularly during summer. It has to be kept in mind that the mere existence of that streamline only fulfills one necessary condition for the formation of the hook. There is no direct information available allowing to conclude whether dust is contained in the air following the streamline. Only by indirect methods more information can be gathered on that point.

The daily synoptic 500 mb-charts show that the upper air heat high is quasi-stationary and by reason of its existence limited to the African continent. The average climatic wind pattern (Newell et al. 1972) and the daily 500 mb-charts suggest that the high pressure systems of the Azores Islands and of the African continent pulsate, thus allowing sometimes the formation of a trough-like structure above the Canary Islands. The pulsation of the continental heat high just finds its average limit at the streamline of the ridge of deposits (Figure 6 is an illustrative example). Further investigations showed that the extension of the continental high-pressure cell was rather well correlated with the occurrence of a type of tropical disturbance — an eastern wave — further south.

37

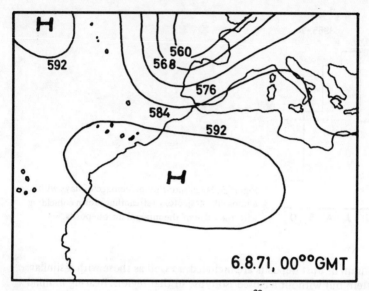

Figure 6. 500 mb-weather chart of 6 August 1971, 00°°GMT ('Deutscher Wetterdienst-Täglicher Wetterbericht') showing the synoptic situation with the characteristic upper air streamline.

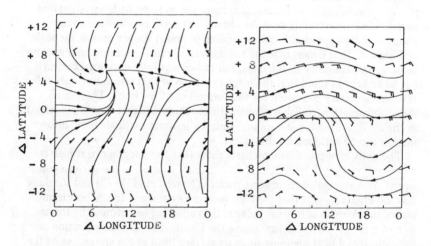

Figure 7a, b. Streamlines of the total surface (a) and 700 mb (b) wind field of a composed African Easterly Wave from Summer 1974. The path follows roughly a latitude of 12°N, both axes divided in degrees (latitude and longitude) (after Reed et al. 1977).

38

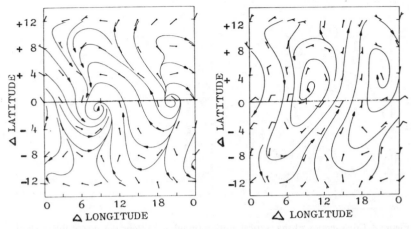

Figure 7c, d. As in Figure 7a, b for the perturbed surface (c) and 700 mb (d) wind field (after Reed et al. 1977).

5. DUST TRANSPORT AND EASTERLY WAVES

Easterly waves are common in the tropics and extensively investigated (Burpee 1974, Reed & Recker 1971, Carlson 1969, Reed et al. 1977). In midsummer seven to ten waves per month were observed passing over tropical Africa westward towards the Atlantic (Burpee 1972). The structure of African easterly waves during GATE was described by means of a composite technique by Reed et al. (1977). These waves show several properties in favour of an occurrence of the hook-like dust-laden streamlines: they provide a disturbance of the pressure field as far north as is required supplying a mechanism for the transport of dust from the Sahel area located south of the ITCZ across the inversion into the mid-tropospheric hot desert air with its easterly winds. Figures 7a-d (after Reed et al. 1977) show the flow patterns at the surface and at the 700 mb-level as composed of eight waves in 1974. In Figures 7c and 7d, the mean flow is removed and the pure disturbance flow remains. These figures show a full wave with at least 2 600 km extension from North to South, centred around the track of the waves (located at about 12°N in the average), and about the same extension from east to west. A distinct southerly air flow appears very clearly in the wake of the wave. Figure 8 shows the vertical motion – a light but steady upward one along the wave's track (solid line marked in Figures 7a-d). This upward motion supports the transport of dust in the hot upper air. In addition, this region of the waves is the area of minimum precipitation (Figure 9) favouring dust generation and transport.

The analysed frequency of the easterly waves coincides with the westward

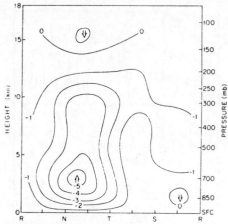

Figure 8. Cross-section of the vertical motion (in mb h⁻¹) along the axis of the track of the wave disturbance. Key: R = ridge; T = trough; S = southerly winds; N = northerly winds (after Reed et al. 1977).

expansion of the continental heat high with its hook-like streamline during the years for which both data sets were available. Synchronized with the passage of an easterly wave (with a phase shift of about one to two days), the Saharan upper air heat high extended to the Atlantic in about 75 per cent of all cases. Thereby, the formation of the hook-like streamline in the position of the ridge of silty deep-sea deposits can be understood.

6. CONCLUSIONS

The results at present point out that both the passage of a dust generating easterly wave and the expansion of the continental heat high to its western-most position, thus forming a dust-depositing hook-like streamline, are coupled. Although no direct evidence exists for this connection because hardly any direct upper air dust concentration measurements are available. However, the evaluation of further data from individual meteorological stations including visibility and the computation of trajectories may allow further confirmation of the existence of the outlined mechanism.

7. ACKNOWLEDGEMENTS

We wish to show our appreciation to Mr D.E.Pedgley for making available the archives of the Centre for Overseas Pest Research. The geological data came from Prof Sarnthein and Mr B.Koopmann. Mr W.Grusdat carefully drew the graphs. Their contributions are gratefully acknowledged.

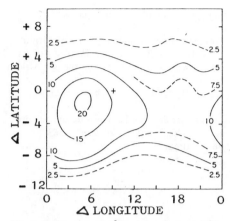

Figure 9. Average precipitation rate (mm d^{-1}), axes denominated as in Figure 7a-d (after Reed et al. 1977).

REFERENCES

Burpee, R.W. 1972. The origin and structure of easterly waves in the lower troposphere of North Africa. *J. Atmos. Sci. 29:*77-90.

Burpee, R.W. 1974. Characteristics of North African easterly waves during the Summers of 1968 and 1969. *J. Atmos. Sci. 31:*1556-1570.

Carlson, T.N. 1969. Some remarks on African disturbances and their progress over the tropical Atlantic. *Mon. Weath. Rev. 97:*716-726.

Carlson, T.N. & J.M.Prospero 1977. *Saharan air outbreaks: Meteorology, aerosols, and radiation.* Report US GATE Central Program Workshop (NCAR), Boulder, Colorado, pp.57-78.

Jäkel, D. 1977. Abfluss und fluviatile Formungsvorgänge im Tibestigebirge als Indikatoren zur Rekonstruktion einer Klimageschichte der Zentralsahara im Spätpleistozän und Holozän. Report at the X. INQUA-Congress-Birmingham, Present Environments, 1. *Climatology,* 20.8.77.

Johnson, L.R. 1979. Mineralogical dispersal patterns of North Atlantic deep-sea sediments with particular reference to eolian dusts. *Mar. Geol. 29:*335-346.

Krishnamurti, T.N. 1979. Atmospheric transports across the Atlantic. Films presented by G.Siedler at the Symposium Sahara und angrenzende Meere-Klimawechsel und Sedimente, Mainz, 1.-4.4.1979.

Newell, R.E., J.W.Kidson, D.G.Vincent & G.J.Boer 1972. *The general circulation of the tropical atmosphere,* Vol. 1. MIT Press, Cambridge, Mass. and London.

Nicholson, S.E. 1978. Climatic variations in the Sahel and other African regions during the past five centuries. *J. Arid Environ. 1:*3-24.

Pachur, H.-J. 1974. Geomorphologische Untersuchungen im Raum der Serir Tibesti (Zentralsahara). *Berliner Geogr. Abh. 17.*

Prospero, J.M. & R.T.Nees 1977. Dust concentration in the atmosphere of the equatorial North Atlantic: Possible relationship to the Sahelian drought. *Science 196:*1196-1198.

Reed, R.J. & E.E.Recker 1971. Structure and properties of synoptic-scale wave disturbances in the equatorial western Pacific. *J. Atmos. Sci. 28:*1117-1133.

Reed, R.J., E.E.Recker & D.C.Norquist 1977. The structure and properties of African wave disturbances as observed during phase III of GATE. *Mon. Weath. Rev. 105:*317-333.

Sarnthein, M. & B.Koopmann 1979. Late Quaternary deep-sea record on the NW African wind circulation. *Palaeoecology of Africa 12.*

Schütz, L. & R.Jaenicke 1979. Eolian dust from the Sahara desert. *Palaeoecology of Africa 12.*

Servant, M. & S.Servant 1970. Les formations lacustres et les diatomées du quaternaire récent du fond de la cuvette tchadienne. *Rev. Géogr. phys. Géol. dyn. 12:*63-76.

Thompson, B.W. 1965. *The climate of Africa.* Nairobi.

INDICATORS OF CLIMATE ON LAND

PLUVIAL AND ARID PHASES IN THE SAHARA: THE ROLE OF NON-CLIMATIC FACTORS

P. ROGNON

Université P. et M. Curie (Paris)

The Quaternary climatic history of the Sahara is characterized mainly by the distinction between pluvial and arid periods. A period is usually defined as pluvial if the hydrological and biogeographical conditions were wetter than at present, and as arid when the reconstruction of the palaeoenvironment indicates drier conditions. Pluvial and arid phases should thus be defined not only by the amount of rainfall, but also by the available water at a given site, either on the surface or in the soil.

This available water depends on many factors, which may be either climatic or non-climatic. Precipitation is the first climatic factor, its effect depending both upon its total quantity and its distribution throughout the year. Some authors consider that temperatures also are very important in defining a pluvial phase, since lower temperatures reduce the average evapotranspiration and consequently increase the amount of available water. These climatic factors are closely interconnected, making the precise definition of a pluvial phase even more difficult.

In reconstructing palaeoenvironments of the Sahara, however, non-climatic factors must also be taken into account before drawing palaeoclimatic inferences. It stands to reason that water is rare in arid countries. All geomorphological or hydrological processes, as well as the possibility of life, depend less upon the local rainfall itself than upon other factors which control the amount of available water, such as soil permeability, runoff from hillslopes, water bodies concentrated in the lower part of hydrological basins and level of the water table. Because of the different processes by which water may be stored, some environments in the Sahara are now supplied with a volume of water which is out of proportion to the average rainfall. This is the 'Oasis effect', which man has exploited and artificially increased by the creation of palm groves. Conversely, any factor contributing to wastage of water or increase of evaporation can bring about non-climatically controlled aridity.

The aim of this paper is to show that a given amount of rainfall may have very different effects upon the environment, according to whether the water is stored or wasted (Figure 1). Each non-climatic factor will be analysed, both for the present and for Quaternary times.

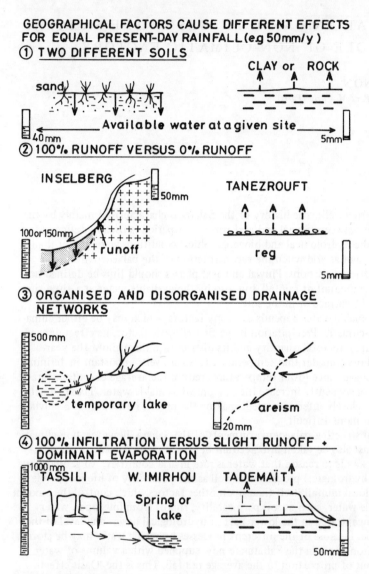

Figure 1. Geographical factors cause different effects for equal present-day rainfall (e.g. 50 mm/y):
1. Two different soils;
2. 100 per cent runoff versus 0 per cent runoff;
3. Organised and disorganised drainage networks;
4. 100 per cent infiltration versus slight runoff + dominant evaporation.

1. THE AVAILABILITY OF SOIL WATER

When travelling from the Saharan borders to the more arid regions, it is observed that vegetation gradually concentrates on the more permeable sandy soils and leaves the clayey soils, which are often more fertile but poor in water and toxic due to salts. For instance, Smith (1949) pointed out that in Sudan, Acacias mellifera or A.seyal grow with only 200-300 mm/y of rainfall on very sandy soils, but can live on soils containing 60 per cent of clay, only if rainfall reaches 600 to 800 mm/y. In dune sands, water infiltrates very rapidly, thus reducing the loss by evaporation. If rainfall is frequent enough to maintain a sufficient reserve of water, sandy soils will always bear a considerably denser plant cover than do stony or clayey soils.

This observation, when applied to the past, can explain the particular evolution of those large spaces of eolian sands called ergs (sandseas). These spread over vast areas, about 70 000 to 100 000 km² for each sandsea in the Northern Sahara. In the Southern Sahara, eolian sands driven by the continental tradewinds accumulated during the Quaternary. They extend almost continuously over several hundreds of kilometers from North to South, and over more than 4 000 km from West to East. The eolian sands may be 60 to 80 m thick, as for example to the North-East of Tumbuctu, or in the Chad Basin.

The sandseas are typical landscapes of most periods. As the plant cover becomes sparser the sand carried by the wind accumulates as dunes or eolian mantles, which are the typical landforms of arid periods. Vegetation rapidly disappears as the soil becomes more and more unstable and the water reserve deeper. During arid periods, sandseas seem to be very unfavourable to life, which explains why the South-Eastern part of the Erg oriental (Algeria) or Majâbat al Koubrâ (Mauritania) were among the last parts of the Sahara to be explored.

However, as soon as rainfall becomes more frequent, the quantity of available water increases because the rainwater infiltrates rapidly and escapes immediate evaporation. The sandseas are then very quickly colonized by vegetation and particularly by deep-rooted trees (Figure 2). The development of soils gradually makes the flanks of the dunes impermeable, and rills appear such as those that are observed in the sandseas of Aouker (Daveau 1965) or of Manga. Part of the rainwater flows down these rills towards the interdune hollows, where available water is also provided by the rising water-table. So, as a result of seepage or runoff, amounts of water disproportionately greater than actual precipitation concentrate in the interdune hollows. Above a certain threshold hydromorphic soils may appear, then seasonal ponds and finally lakes, as rainfall increases. The water-table rises almost simultaneously to the level of all the interdune hollows of the same sandsea. Thus we can understand how the tropical fauna, even amphibians (Hippopotamus), could move very quickly up to the Southern Hoggar during pluvial periods such as the Early Holocene.

SANDSEAS (ERGS) ACCENTUATE WETNESS OR DRYNESS

ARID: UNFAVOURABLE TO LIFE

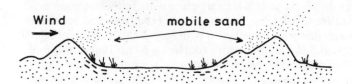

TRANSITIONAL: RISING WATER-TABLE

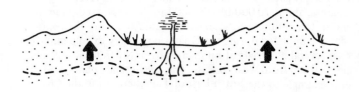

PLUVIAL: THRESHOLD EFFECT WHEN WATER-
-TABLE APPROACHES SURFACE

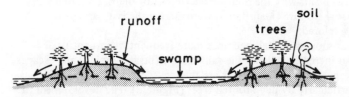

Figure 2. Sandseas (ergs) accentuate wetness or dryness:
1. Arid: unfavourable to life;
2. Transitional: rising water-table;
3. Pluvial: threshold effect when water-table approaches surface.

Since trees find conditions in the sandy soils very favourable for growth, plant life in the sandseas during the pluvial periods probably included some exceptionally wet facies which must be taken into account particularly when considering the ratio AP/ANP in the pollen analyses.

In the same way, the abundant remains of lakes or swamps in the sand-seas should not be interpreted too hastily as evidence of pluvial periods. As

48

soon as the rising water-table reaches the threshold mentioned above, deposits of carbonates rich in Mollusc shells (North Sahara Sandseas), or of diatomites (in the Centre and South), can form. Because of the permeability of dune sands, the drainage system remains totally disorganized. Thus increase of rainfall makes very many ponds appear as all the available waters are concentrated into the limited area of interdune hollows.

Therefore, sandseas accentuate effects of aridity during arid phases, but when the water-table approaches the surface (threshold effect) during pluvial phases, they accentuate wet conditions, much more so than do other environments.

2. THE INCREASE OF AVAILABLE WATER AT THE FOOT OF HILLS AND SCARPS

In arid countries, the upper steep rocky slopes rise above regularly inclined planes (glacis, pediments) whose gradient does not exceed a few degrees. The most typical example of this type of landform are the *inselbergs,* chiefly occurring in crystalline rocks and particularly spectacular in the Hoggar or Aïr massifs. Rainwater runs very quickly down the steep rocky flanks of the inselberg, which constitutes an 'impluvium', 100 to more than 1 000 m high. At the floot, the flow divides into tiny unstable rivulets or spreads as a sheet flow. It infiltrates or evaporates before reaching the wide plains where the hydrographic network becomes disorganized. So, as a result of runoff, one can sometimes find around the inselbergs vegetation and soils indicating wetter conditions than could be explained by local rainfall alone (Bocquier et al. 1978), particularly in Sahelian or Sudanian regions. In the Hoggar on the other hand the downpours are at present too small and sporadic to maintain such rings of vegetation. Here, too, there is a threshold, below which rainwater is not conserved and therefore not actually available.

Non-climatic factors were certainly effective during the Quaternary Pluvials, since the borders of many inselbergs still show the results of deep weathering, partly truncated and protected by more recent debris. Some inselbergs are surrounded by a ridge of dunes which acts as a dam, so that swamps or small lakes formed around the borders of the inselberg during the pluvials. For instance, in the North of Aïr, water running down the slopes of the 1 000 m high inselberg Adrar Bous during the Early Holocene fed such a lake of area 5 km^2 (Clarke et al. 1973). In the same way, the large inselbergs in the Hoggar (In Ekker, In Tounine . . .) were surrounded by a swampy ring during the pluvial phases. The amount of available water is closely connected with the importance of impluvium. Thus the small inselbergs in the Eglab, located on the old West African craton, rise to only 100-150 m above the pediplain and cannot act in the same way as those of the Central Sahara.

Generally speaking, the steep slopes of the inselbergs always concentrate

ACCENTUATION OF PLUVIAL CONDITIONS AT FOOT OF INSELBERGS
e.g : Adrar Bous

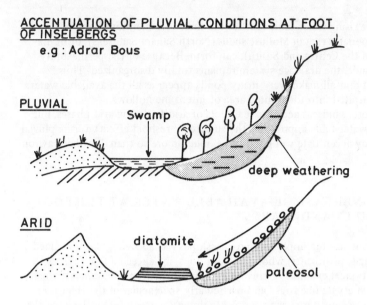

AREIC PLAINS ACCENTUATE EFFECTS OF ARIDITY

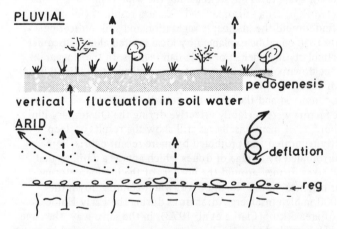

Figure 3. Accentuation of pluvial conditions at foot of inselbergs (e.g. Adrar Bous):
1. Pluvial; 2. Arid.
Areic plains accentuate effects of aridity:
1. Pluvial; 2. Arid.

whatever water is available and thus, during the arid as well as during the pluvial phases, accentuated wetter conditions at their base.

On the other hand, there are in the Sahara *large and almost perfectly flat plains*. The most famous of these is the Tanezrouft that spreads westward from the Hoggar and extends about 900 km from North to South, on the part of the precambrian basement belonging to the old West African craton. Other plains as regular as this, but smaller, surround the Central Hoggar (Amadror, Tafassasset) and the Tibesti (Serir Tibesti). These plains, now exposed to an intense deflation, are covered with sand or gravels (regs, serirs), and are wholly areic.

In Libya, Meckelein (1959) used these areic plains to demonstrate the constancy of Quaternary aridity in these 'Kernwuste'. But such reasoning seems to be questionable, because there were no inselbergs to concentrate water by runoff or into lakes. Pachur (1979) recently showed, from remains of paleosols and swamps, that pluvial periods did occur in the Serir Tibesti. But, on these perfectly flat plains, soils developed only by vertical fluctuations of soil water. No lake could be formed because of loss by evaporation over such an extensive surface, and no drainage network could be organized due to lack of gradient.

On the contrary, during the arid phases, intense deflation removed the thin unprotected soil. This is why some authors have spoken of constant aridity during the Quaternary. Every arid period accentuated the perfect horizontality of these aeric plains.

So that, while the inselbergs always accentuated the wet conditions of the environment, the vast areic plains allowed no concentration of water, giving a false impression of constant aridity.

3. THE PREDOMINANT ROLE OF DRAINAGE NETWORKS

In some regions of the Sahara, short-lived floods frequently still exploit drainage networks, thus helping to prevent the disintegration of hydrography that is one of the results of aridity. In the Atakor, Tibesti, Northern Tassili or Mzab, for instance, a dendritic drainage network rapidly concentrates rainwater, thus preserving it from immediate evaporation. Rainwater runs off along the thalwegs, resulting in much moister conditions than the average yearly rainfall would indicate. The seepage of this water into the alluvium of the valleys feeds the permanent shallow phreatic water table.

A dense plant cover of bushes and graminaceae, often with riparian vegetation of Tamarisks or Acacias, grows along the beds of the watercourses. Sometimes underground water emerges, or water remains in potholes (gueltas) along the wadis in the Atakor or in the Tassili, where animal life can subsist. Finally, in the endoreic basins which are subject to floods, one finds

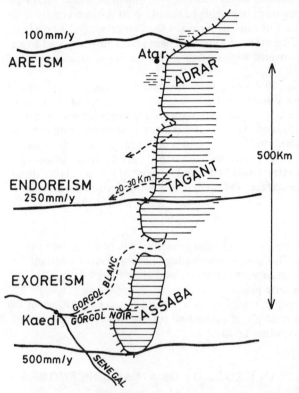

PROGRESSIVE DISINTEGRATION OF DRAINAGE
NETWORK AS FUNCTION OF DIMINISHING MEAN
ANNUAL RAINFALL (e.g Mauritania)

Figure 4. Progressive disintegration of drainage network as function of diminishing mean annual rainfall (e.g. Mauritania).

temporary and generally brackish pools (sebkhas). When the salt percentage is low (maaders), these concentrations of water constitute an environment favourable to life.

3.1 *Why some well-organized drainage networks are found in a few Saharan regions?*

This chiefly depends upon present or Quaternary climatic factors. In Mauritania for instance, along the Paleozoïc sandstone escarpment that runs about

52

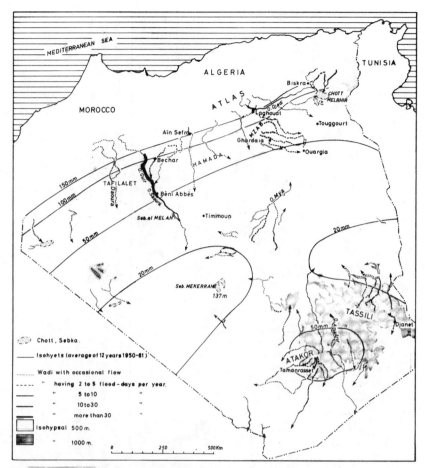

Figure 5. Map showing the mean discharge of wadis of the Algerian Sahara from 1950 to 1961 (from Tessier 1965). This map illustrates the relationship between the frequency of floods, and annual rainfall and altitude. But other factors intervene: geomorphic heritage, lithology, etc.

600 km from North to South (Figure 4), we observe initially a complete disintegration of the drainage network in the Adrar (<100 mm/y rainfall). Even the ephemeral Wadi Seguelil has been blocked by growing dunes and flows into the Sebkha Chemchane where water evaporates: the downstream sections of the valley are completely disrupted. Further south, in the Tagant (about 250 mm/y rainfall, the disorganization of the drainage network probably dates only from Ogolian time (20 000 to 10 000 years B.P.), and it is not complete. Even further to the South, in the Assaba, the drainage patterns

53

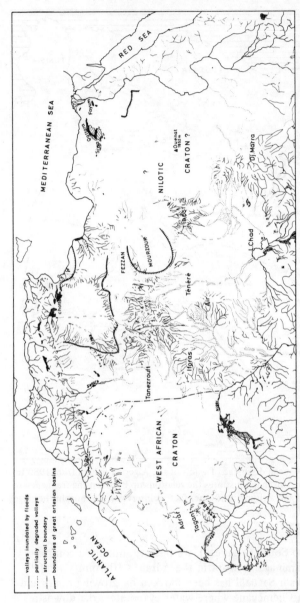

Figure 6. Map showing the hydrographic basins of the Sahara. Valleys which are still well-defined are represented by continuous lines, more degraded valleys by broken lines. Note the contrast between the still active networks of the central Sahara and the almost areic regions corresponding to the oldest cratons to the west and east. The boundaries of the artesian basins mentioned in the text are indicated.

remain well-ordered, and are still active as far as the Senegal river, in a country where the average rainfall is 300 to 500 mm/y.

However, the distribution of well-organized hydrographic patterns does not depend only upon present rainfall. On a map showing the frequency of present floods in the Algerian Sahara (Figure 5), Tessier (1965) pointed out that some drainage networks are still active, both in the Atakor and the Atlasic piedmont (more than 100 mm/y), as well as in the Tassilis, in the Wadi Mya and in the M'zab (20 to 50 mm/y). A finely textured drainage network facilitates water concentration, and thus maintains the channel system. Developed during pluvial periods, such networks persist even during arid periods. Thus the wadis of the Tassilis have annual flood frequencies almost equal to those of the Atakor, even though they receive three or four times less rainfall.

Non-climatic factors, specifically lithology and tectonics, exert a strong influence upon the characteristics of drainage networks:

3.1.1 *Connection between lithology and drainage density.* When climate is arid, lithology is very important since it influences drainage density. High drainage density permits concentration of rainfall; water very quickly runs off towards the higher order stream segments where it escapes intense evaporation. In the Northern Tibesti, N.W.Roland (1974) showed that drainage density in the shales of the precambrian basement is twice that in the paleozoic sandstones. On calcareous rocks, hydrographic networks are often very poorly developed. In the thin and very karstified limestone in the South of Laghouat, the valleys of the Hamada become series of dolines. On the Mesozoic limestone of the Tademaït, all water infiltrates or evaporates and there are no traces of a drainage pattern. On the contrary, in the Senonian limestone of the M'zab, the hydrographic network is very dense and branching. These differences are related to the more or less argillaceous facies, and the degree of faulting of the various limestones.

Owing to the high permeability of the sandstones in the Tassili n'Ajjer, there is perennial flow in the Wadi Imirhou, 22 km long, below a drainage basin whose area is only 900 km^2 and which receives a rainfall of 30-50 mm/y.

3.1.2 *Tectonics.* Figure 6 shows an obvious contrast between the Western Sahara, which is almost devoid of organized networks, and the Central Sahara where valleys are numerous. This contrast is related to small differences in rainfall, but chiefly to topographical differences related to the regional tectonics. The Western Sahara shield is a very old craton that became completely immobile about 2×10^9 years ago, which explains its flat landscapes and moderate elevations (generally between 100 and 500 m). On the contrary, the Central Sahara remained mobile until very late, as this part of the craton was formed about $500\text{-}550 \times 10^6$ years ago as part of an 'African' orogenic belt, and its sedimentary basins were deformed again during the Tertiary. The best developed hydrographic networks are situated chiefly in the great steep

55

volcanic massives which formed between the Oligo-miocene and the Quaternary. The Eastern Sahara is almost devoid of hydrographic networks, perhaps because it forms part of another old craton (the Nilotic craton) whose existence is disputed (Vail 1976).

Tectonics have two effects:
— they increase the gradients of longitudinal wadi profiles, and thus the velocity of the floods and the distances they can cover;
— they extend the area of the drainage basin which depends on the entire volume of the mountain. For instance, the Tefedest (Hoggar) has a similar drainage density and only slightly lower rainfall than the Atakor. The floods of the Tefedest wadis, however, hardly flow beyond the mountain itself, while those of the Atakor cover distances of 300 to 400 km to the west and north of the massif. The average area of the drainage basins in the Tefedest is 104 km^2, in the Atakor 388 km^2.

Thus non climatic factors are very important. They may promote or impede the formation or the expansion of dense and well-defined drainage networks. These provide the most important means of conserving available water and result in abnormal wetness over small areas. Favourable conditions for water conservation were particularly intensified during the alternations of Quaternary Pluvial and Arid periods.

3.2 Changes in the drainage networks in relation to Quaternary climatic oscillations

Changes in the Saharan drainage networks accentuated the effects of the Quaternary climatic oscillations.

3.2.1 *When aridity increases, networks are prone to disintegration.* Decrease in rainfall results in decrease in the discharge of the wadis. Their capacity to transport load is also reduced, resulting in massive deposition of alluvium, and a greater instability of the stream channels.

The permeable alluvial material absorbs increasing amounts of water. Because of this seepage, the floods travel increasingly less far downstream, and the lower parts of the valley are abandoned. This can be clearly observed over the whole alluvial piedmont of the Atlas between Laghouat and Bechar, where the wadis flow over great alluvial fans whose size is out of proportion to the present wadi discharge. A considerable part of the water that infiltrates into this alluvium reaches the deep water-table, and is entirely lost to superficial flow.

On these gently sloping fans, the braided channels shift constantly, particularly at the canyon mouths (foums). The streamlets divide and spread out, and thus water evaporates more easily, decreasing still further the quantity of available water downstream.

Thus the dynamic mechanisms of the fluvial processes increase the effects of an arid period. This disorganization of hydrographic networks occurs par-

56

ticularly in enclosed basins and alluvial piedmonts. But eolian processes can also be very effective. Their effect is unimportant when the valleys are frequently flooded; but if the floods are smaller or more scarce, wind has time to build large dune dams that impede flow. So, as a result of its strong and frequent floods, the Wadi Saoura reaches the Sebkha el Melah, 600 km from the Atlas, every year. Every tenth year, it reaches 180 km further south; in 1959, it even reached the Touat, and according to the narratives of the local people, it sometimes ran as far as its terminal basin (Sebkha Mekerrane) during the 19th century. At the present time, 30 m high dunes may be established across the channel of this watercourse. The dune dams, when strongly established, increase the effects of climatic desiccation. The efficacy of such eolian accumulations during Quaternary times is well-known: they obstructed the Senegal river 500 km upstream of its mouth, and made the Niger river an endoreic stream between 20 000 and 10 000 years B.P. Thus aridification increased in the downstream parts of these valleys, owing to the hydrographic disorganization by factors other than climatic ones. Even the White Nile, during this arid phase, was not immune to obstruction by dunes; it was probably interrupted below the Bahr el Ghazal basin which was almost closed by dunes (Williams & Adamson 1974).

In rare instances, as to the north and east of Tibesti, the valleys were cut again by elongated ridges (yardangs) aligned parallel to the continental trade-wind. The hydrographic network is then disintegrated to such an extent that it is hard to imagine how, even during pluvial periods, a drainage network could take shape again.

During the Quaternary arid phases, progressive and often irreversible network disintegration occurred, so that during the successive Pluvials, similar quantities of rainfall could not produce the same effects because flowing water was blocked by the disorganized drainage pattern. Thus some effects of partially non-climatic origin must be added to those of climatic aridity.

3.2.2 *At the commencement of a Pluvial, some hydrographic networks could however be reintegrated.* Stronger and more frequent floods can break the dune dams and restore stable channels on alluvial fans, so that greater amounts of water can concentrate in the thalwegs and feed lakes in the terminal endoreic basins. Available water increases much more quickly than does the pluviosity.

This explains why the lower parts of some endoreic basins, such as those of the Chad and Central Afar, could be filled by paleolakes, the extent of which seems out of proportion to the present sheets of water. For instance, during the early Holocene, Lake Chad (presently 10 to 20 000 km²) covered about 360 000 km², almost the area of the present Caspian Sea. The catchment basin also was much more extensive (Figure 7) since a great part of the network that is now inactive had been reintegrated, in spite of obstruction by the sand-seas. Thus rain falling on the Aïr or Tibesti massifs could reach Lake Chad. Small lakes, fed by drainage basins a few tens or hundreds of square kilometres

57

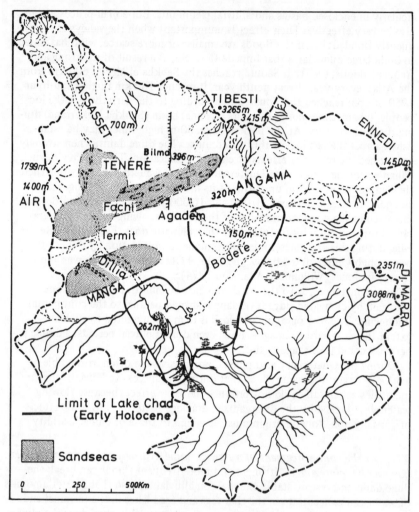

Figure 7. The theoretical hydrographic basin of the Chad depression.

in area, lay at the foot of the escarpments of Termit, Agadem, Fachi or Bilma. When the Pluvial was at a maximum, these lakes could overflow into the general network, for instance at Termit.

The revival of previously disintegrated networks, added to the increase in rainfall, explains the great extent of the lakes.

4 ROLE OF SUBSURFACE WATER

When attempting to reconstitute the succession of former climates, paleolakes are considered as the most convincing evidence in favour of Pluvial. But in the Sahara, most of the paleolakes are situated in favourable localities where artesian ground-water frequently emerges. These major artesian aquifers concentrate, into small areas, huge volumes of water which have fallen over hundreds of thousands of square kilometres, often beyond the desert itself. As the movement of ground-water is very slow (for instance about 3×10^{-7} m/sec in the 'Continental intercalaire' aquifer in Algeria), this water emerges with a considerable time lag after infiltration.

The geological structure of the Sahara explains the existence of wide artesian basins. The Precambrian basement reaches an elevation of 2 600 m in the Hoggar, about 2 000 m in the Tibesti, and 1 800 m at Auenat, in the central Sahara. Sedimentary formations of porous sandstone (Tassilis, 'Continental intercalaire') or permeable limestone (Upper Cretaceous and Eocene) overlie this basement; they are very thick, uniform and extensive. Since these strata are inclined in vast intracratonic basins, a favourable situation may exist for the development of artesian springs, in which the water flows upward toward the surface under its own pressure. This is particularly noticeable in the part of the craton dating from the 'African' orogeny.

Figure 8 shows some examples of the ground-water movements. It can be seen that the ground water of the 'Continental intercalaire' aquifer originates in an outcrop of sandstone, about 10 000 km² wide, located at about 1 000 m above sea level in the Saharan Atlas. The ground-water flows 800 km as far as Touat and Tidikelt, at about 400 m above sea level: there, in the Central Sahara, the overflow of the aquifer probably emerged during pluvial periods. During these periods, rainfall probably increased not only upon the Saharan Atlas (where the present rainfall average is 250-300 mm/y), but also upon the karstified calcareous formations of the 'hamadas' (75 000 km² wide in the South-West of the Atlas, and 13 000 km² in the Arbaa plateau).

In the North-Eastern part of the Algerian Sahara, the water of this aquifer joins that of Cretaceous and Tertiary limestones, and feeds the Algero-Tunisian 'chotts' in the centre of the synclinal basin. The rainwater that infiltrates in the Atlas, particularly in the Aures mountains, flows upward to the surface of the chotts under its own pressure. During the Pluvials, the area of the watershed was much greater, including some Saharan plateaux such as the Tademaït. Therefore, the chotts are only the residues of former large lakes that left many traces of *Cardium* bearing deposits. These deposits show that water was brackist, probably partly because of the salinity of the ground-water, but also owing to the very flat topography. These lakes extended over vast areas and were very shallow, resulting in a very significant loss by evaporation.

The lacustrine deposits of the Wadi Chatti (Fezzan), at about 400 m above sea level, also represent former lakes where *Cardium* lived. They were probably fed by ground-water moving from the Paleozoic sandstones (J.Acacus), and

59

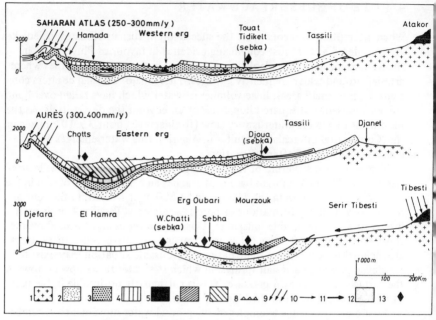

Figure 8. Schematic sections of the three great artesian basins of the Sahara:
1. Crystalline basement;
2. Aquifers of Palaeozoic sandstones;
3. Aquifers of Mesozoic sandstones;
4. Upper Cretaceous limestones;
5. Tertiary and Quaternary volcanic massifs;
6. Hamada limestones;
7. Tertiary sedimentary basin;
8. Dunefields (ergs);
9. Principle regions of provenance for the aquifers;
10. Surface flow;
11. Water circulation in the aquifers;
12. Aquifer outcropping in a syncline;
13. Possible sites of lakes fed by the aquifers.

even perhaps from the Tibesti massif, situated many hundreds of kilometres away. Radiocarbon dating of some samples of W.Chatti ground-water indicates ages between 30 000 and 50 000 years B.P. (Klitzsch et al. 1976).

All these paleolakes were thus ultimately fed by rain water which had fallen over vast areas, parts of which might be beyond the Sahara. This is also true for the paleolakes of the Afar or Chad where both local rainfall and allogenous ground-water should be taken into account. It should not be difficult to estimate the water contribution from the Sahara itself in hydrologic basins which do not have any outside supply, for instance in the syneclise of Taoudeni (Western Sahara) or in the Kufra basin (Libya). The latter has recently been studied, from the point of view of paleoclimates, by Edmunds & Wright (1979) but unfortunately, no precise data were published on the variations of paleolake levels in the ground-water discharge area (Siwa Oasis, Qattara depression in the Western Desert). Emergence of the Taoudeni or Kufra aquifers,

60

fed in Saharan regions where present rainfall is 10-15 mm/y, would provide more unambiguous arguments for climatic oscillations in the Sahara itself.

5 CONCLUSION

For the last few decades, our knowledge of Saharan paleoenvironments has made great strides. Regional differences became apparent, particularly between the northern and southern borders of the desert. It then became possible to put forward some hypotheses about the mechanisms by which the variations in climate could be explained (Rognon 1976). A parallel was drawn with the climatic oscillations in another great desert, the Australian one (Rognon & Williams 1977). But it is now necessary to take account of more topographic and geologic factors. Ideally, only environments of similar geology and geomorphology should be compared, The role of non-climatic factors has already been recognized in research on former glaciations and paleoclimatic interpretations have accordingly been modified. For instance in North America, the Cochrane ice advance is no longer explained by colder temperature, but by huge ice surges connected with the glacio-eustatic transgression. The same care must be taken in reconstituting Saharan paleoclimates. In this desert soils differ greatly in their field-capacity and in their texture, topography is varied, hydrographic organization often differs, and the artesian basins are exceptionally wide. It is necessary to bear in mind that, in the Sahara, lakes require particular conditions in which non-climatic factors may be very important. Thus their climatic interpretation is a delicate question.

REFERENCES

Bocquier, G., P.Rognon, H.Paquet & G.Millot 1978. Interprètation pédologique de dépressions annulaires entourant certains inselbergs. *Sci. Geol. Strasbourg Bull. 30*(4):245-253.

Clark, J.D., M.A.J.Williams & A.B.Smith 1973. The geomorphology and archaeology of Adrar Bous, Central Sahara: A preliminary report. *Quaternaria 17:*245-297.

Daveau, S. 1965. Dunes ravinées et dépôts du Quaternaire récent dans le Sahel de Mauritanie. *Rev. Géogr. Afr. Occid., Dakar 1*(2):7-48.

Dubief, J. 1953. *Essai sur l'hydrologie superficielle au Sahara.* Direct. Serv. Colonis. et Hydraul., Algiers. 458 p.

Edmunds, W.M. & E.P.Wright 1979. Groundwater recharge and palaeoclimate in the Sirte and Kufra basins, Libya. *J. Hydrol. 40:*215-241.

Klitzch, E., C.Sonntag, K.Weistroffer & E.M.El Shazly 1976. Grundwasser der Zentral Sahara: Fossile Vorräte, Sonderd. *Geol. Rundschau 65*(1):264-287.

Meckelein, W. 1959. *Forschungen in der zentralen Sahara-thesis.* Braunschweig.

Pachur, H.-J. 1979. Late Quaternary of Southern Libya, climatic history. In: *Second Symposium of the Geology of Libya,* Tripoli (in press).

Rognon, P. 1979. Comparison between the Late Quaternary terraces around Atakor and Tibesti. In: *Second Symposium of the Geology of Libya.* Tripoli (in press).

Rognon, P. & M.A.J.Williams 1977. Late Quaternary climatic changes in Australia and the Sahara: A preliminary interpretation. *Palaeogeogr. Palaeoclim., Palaeoecol. 21:*285-327.

Roland, N.W. 1974, Geomorphologisches Undersuchungen im Raum Serir Tibesti (Zentral Sahara). *Berliner Geogr. Abh. 17.*

Smith, J. 1949. *Distribution of tree species in the Sudan in relation to rainfall and soils texture.* Minist. Agric. Sudan Gvt., 4, Khartoum.

Tessier, M. 1965. Les crues d'oueds au Sahara algérien entre 1950 et 1961. *Trav. Inst. rech. Sah. Algiers 24:*7-29.

Vail, J.R. 1976. Outline of the geochronology and tectonics units of NE Africa. *R. Soc. London A 355:*127-141.

Williams, M.A.J. & D.A.Adamson 1974. Late Pleistocene dessication along the White Nile. *Nature 248:*584-586.

SOIL ACCUMULATIONS AND CLIMATIC VARIATIONS IN WESTERN SAHARA

DANIEL NAHON

Laboratoire de pétrologie de la Surface, Université de Poitiers

INTRODUCTION

In Western Mauritania soils, catenas have been studied in different areas with yearly rainfalls varying from 20 to 60 mm. On some sequences, several types of accumulations have been recognized in soils. These are as follows:
 – Iron crusts (ferricretes) which cap the sequences
 – Calcareous crusts (calcretes) which run through the iron crusts
 – Salt crusts (sylvite and halite) which eventually run through the calcretes.
The iron crusts are coupled with kaolinic weathering profiles. They are, on basement, the oldest pedological vestige offered by the landscapes. It is known that these iron crust profiles are inherited from older humid tropical periods, which could be Pliocene period (Elouard 1959, Monod 1962-1963, Hebrard 1973, Nahon & Ruellan 1972).

Only the calcretes and salt crusts, according to the area and the parent rock, keep their balance with the actual arid conditions. Nevertheless, the calcretes appear as being the most frequent type of accumulation in soils. Calcretes are connected with montmorillonitic weathering profiles.

Two sequences, in which calcrete settles down in the voids of an iron crust, have been studied in detail.

INKEBDENE AREA (Figure 1)

1. Well-preserved iron crusts (20°30″05 N and 15°32′ W). The highest iron capped hills (upslope) permit observation of complete weathered profiles whose main horizons are from top to bottom as follows:
 – a thick iron crust (2 to 3 m) that is dark brown and very indurated and the texture of which is most often coarse to finely nodular. Lithorelics are recognized at the purple red core of the nodules;
 – a soft iron crust, brown to purple red with a massive structure in which the texture of the original parent rock is preserved (2 m);
 – a ferruginous nodular horizon whose indurated nodules are purple red.

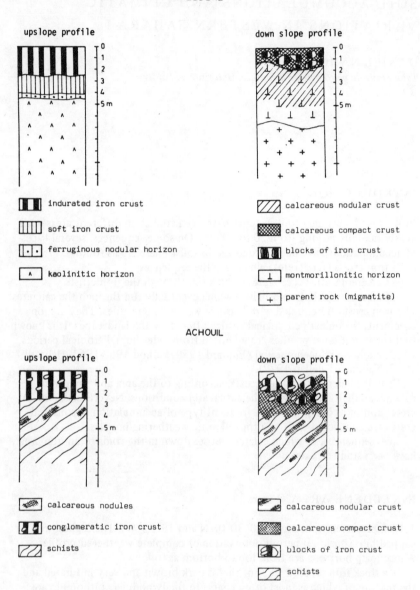

Figure 1. Upslope and downslope profiles at Inkebdene and Achouil

Between the nodules a mottled kaolinic material is developed. We can recognize in the clay and in the nodules the original texture of parent rock (10-20 m);
— a thick kaolinic saprolite, dark red with ochreous stains (6 m).

2. Fragmented iron crusts (20°29'30" N and 15°36' W). Downslope, the iron crust profiles change. The main horizons are as follows:
— a horizon (1 to 2 m thick) consisting of blocks, pebbles, grits of an iron crust with a similar facies as the former one, set in a compact calcareous crust;
— a thick greenish montmorillonitic horizon (3 to 4 m) with grey and ochreous stains. The structure is polyedric. Calcareous nodules (nodular crust) are abundant in this horizon. Some unweathered micas and feldspars can be found;
— the parent rock (migmatite), very hard and a little weathered.

ACHOUIL VALLEY (22°41'W, 12°30'N)

The different terrace or pediment levels which are seen in Achouil Valley are capped by iron crusts, but also downslope by calcareous crusts. Sometimes, on the same profile, we can recognize both of these types of accumulation. Thus from upslope to downslope we can observe the following profiles:

1. Upslope
— conglomeratic, indurated iron crust (2 to 3 m thick). The lower limit of the horizon is clear;
— yellow, soft schists with rare calcareous nodules in which we can recognize the original texture of the parent rock.

2. Downslope
— iron crust, dismantled into blocks and grits in which there are voids and cracks. Between the blocks and in the voids, a compact calcareous crust occurs (1 to 2 m);
— compact calcareous crust in which we can recognize the original texture of the schist (0,5 m);
— nodular calcareous crust. Here too, the original texture is preserved (0,3 m);
— yellow schists with scarce calcareous nodules.

Thus we can observe in Inkebdene area and in Achouil Valley that the calcareous crust is formed downslope in the sequences, where the iron crust is fragmented and shows the strongest porosity. These fragments give a dismantled aspect to the iron crust. And the calcareous crust seems to cement the iron crust in this stage.

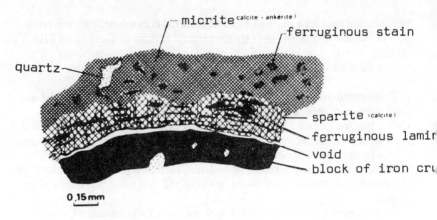

Figure 2. Petrography

PETROGRAPHIC OBSERVATIONS (Figure 2)

Under microscope the limit between the calcareous crust and the blocks of iron crust is underlined by a void (crack).

On each side of the cracks, we can observe the following data:

1. The sparitic crystals of calcite

The sparitic crystals on the edge of the voids are almost always found between the micrite which forms the core of the calcareous crust and the blocks of iron crust. The sparitic crystals are joined side by side and perpendicular to the outer shape of the ferruginous blocks. Under crossed polarizers we can observe that the sparitic crystals are monocrystals.

The sparite, near the void, shows discontinuous ferruginous laminae (films) but always parallel to each other and set concentrically around the blocks of iron crust. These laminae run through the monocrystals, but they also divide crystals with a different extinction.

The ferruginous laminae are different according to their situation near the block of iron crust or near the micrite. They can cover the end of the sparitic crystals thus forming the wall of the crack. In this case, the laminae is the result of the tearing out of the outer film of the ferruginous block. The boundary between ferruginous laminae and sparite is distinct. But the further we go from the edge of the void to the micrite, the more indistinct this boundary is. And little by little the laminae become ferruginous diffuse stains. At the contact with the micrite, the laminae have disappeared. Only the diffuse stains remain.

2. The micrite

The sparite transforms progressively, over about ten microns, to a mosaic of

66

micritic crystals of calcite. At the same time, the ferruginous stains are mixed with the micrite. Some ferruginous tiny fragments can remain in the core of the micrite. They consist of goethite.

The calcite which forms the micrite is a ferruginous calcite. The geochemical analysis made in those sequences show that this micrite contain 20 to 55 per cent of Fe_2O_3. It is obvious that the goethite which is in the micrite cannot explain such a rate. Furthermore in the Achouil sequence, we could see by X-ray diffraction that this micrite was formed with 75 per cent of calcite and 25 per cent of ankerite. Downslope in the Achouil Valley, chemical analyses of the most recent calcareous crusts show that they contain 10 to 20 per cent of Fe_2O_3. So, far from the direct influence of the iron crusts, the calcareous crusts still have a high iron content.

DISCUSSION

The crystallisation of sparitic crystals of calcite from solutions occurs easily in the voids. Each new generation of sparite protrudes from the surface of the block of iron crust a ferruginous film. With the alternation of damp-dessication stages, the phenomenon is repeated several times as we can see through the concentric ferruginous laminae shut in the sparitic crystals. Therefore, these monocrystals of calcite are not the result of a single crystallisation, but the result of a succession of 'autoepitaxic' crystallisations.

Then as the ferruginous laminae are incorporated amid the micrite, the iron is successively dissolved and precipitated giving diffuse stains. Furthermore by this method, the iron progressively penetrates into the network of calcite.

Thus when the calcareous accumulations extend along the sequence from downslope to upslope until they penetrate iron crusts, they are able to replace progressively the iron crusts by this mechanism. The iron crust is replaced from downslope and from bottom of the profile. That is why downslope of the pediments and terraces, iron crusts remain in blocks packed in the calcareous crust. It is only at the top of the highest hills that we can find complete and well-preserved profiles with iron crust and kaolinite horizon (saprolite). The ascending encroachment of carbonates has not disturbed them yet.

CONCLUDING REMARKS

In the Sahara of Mauritania, iron crusts inherited from older tropical humid periods are frequently affected by calcareous encrustation. Microscope observations show that iron crust is replaced by carbonates which means mechanical stripping of thin ferruginous films and then reduction of iron which enters the carbonate structure. In this way, the iron crusts, which are residues of

67

ancient iron-capped glacis, are dismantled, obliterated and incorporated in the calcareous crusts of the new peneplanes.

Thus, the iron crusts which are the best sign of the old humid periods are geochemically 'digested' by the calcretes if arid conditions remain long enough.

SAND CIRCULATION IN THE SAHARA:
GEOMORPHOLOGICAL RELATIONS BETWEEN
THE SAHARA DESERT AND ITS MARGINS

M. MAINGUET / L. COSSUS
Laboratoire de Géographie Physique Zonale, Université Reims

Winds transporting sand across the Sahara leave behind traces at the surface allowing reconstruction of a global wind system. Within this system, this desert and its northern and southern margins form a great functional eolian unit.

This contribution will not demonstrate the existence of this global atmospheric pattern that disintegrates into trajectories as treated principally in Mainguet (1972) and Mainguet & Canon (1975) (Figure 1), but will

— show the complex origin and termination of this functional unit on the continent, analysing some examples from the northern and southern margins of the Sahara;

— demonstrate how these eolian trajectories relate the northern margins to the arid center of the Sahara and this center to the southern margin.

To avoid confusion, we would like to insist on the difference between the transport vehicle — the wind — and the transport of material, which is a function of wind velocity or, more precisely, wind competence. The transport vehicle follows distinct and continuous pathways but the material transport is a discontinuous process. It would be erroneous to imagine that a sand grain could be transported directly across the Sahara because the transport along the trajectories is not homogeneous: the streamlines consist of currents with dominant transport sectors, interrupted by dominant deposition sectors (erg). Each of these sectors are accompanied by specific dunes related to these dominant processes.

The eolian trajectories governing the sand transport correspond with the general trade wind (Harmattan) pattern complicated in detail by the topography responsible for the localisation of the ergs and the areas of transport.

1 NORTHERN SAHARA MARGIN

1.1 *Double circulation in southern Morocco* (Figure 2)

In the Central Sahara, with the rare exceptions of the double circulation NE-

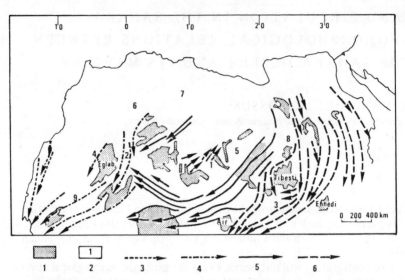

Figure 1. Trajects of sand transport across the Sahara (after NOAA observations):
1. Obstacle for winds or zones without sand deposits; 2. Ergs or surfaces with sand cover;
3. Atlantic sand transport area; 4. Western sand transport area; 5. Central transport cell;
6. Eastern sand transport area.
Ergs: 1. Admer; 2. Chech; 3. Fachi Bilma; 4. Iguidi; 5. Murzug; 6. Western Erg;
7. Eastern Erg; 8. Rebiana; 9. Makteir.

SW and SW-NE that exist at Tassili N'Ajjer (Canon-Cossus & Galichet 1975) and in the Hoggar, the sand transport is mono-directional and corresponds with the Harmattan streamlines including deviations by obstacles (Mainguet 1976).

On the contrary, sand transport patterns are much more complicated at the northern margins, i.e. in Morocco, Algeria and Tunisia.

We have investigated the configuration of the sand supply trajectories in southern Morocco where the oasis of the Draa, Ziz and Rheris wadis are at present suffering from dramatic sand deposition. South of a line from Erfoud-Alnif-Zagora, two supply trajectories were observed:

– Trajectory SW-NE as a result of the 'Sahel' wind, coming in southern Morocco from a 210-240° SW sector and from long distances because erosion and transport effects can be observed as far as west of Foum Zguid, across Jebel Bani, the Kem Kem plateau and the Ziz and Rheris planes.

– Trajectory NE-SW as a result of the 'Chergui' wind from 45° NE to 100° ESE, with a sand supply restricted to eastern Morocco between the meridian of the Chebbi Berg and the palmtree areas of the Rheris Oued where the Chergui deposits all its sand load and again becomes able to transport new sands.

70

Therefore in Morocco the winds export sand to the Sahara and, even more, import sand from this desert towards the palmtree area.

1.2 *Hydro-eolian depressions of the Maghreb and of Egypt*

Water and wind activities formed the depressions of table 1.

Table 1. Depressions at the Northern Sahara margin

	Latitude	Longitude
El Fayum	29°10' to 29°30'N	30°20' to 31°10'E
El Qattâra	29° to 30°25'N	26°30' to 29° E
Sebkhra Sidi en Nouak	34°20' to 34°23'N	10°10' to 10°20'E
Chott el Fedjadi	33°46' to 34° N	8°50' to 9°50'E
Chott el Djerid	33°20' to 34°05'N	7°45' to 8°49'E
Chott el Rharsa	34° to 34°11'N	7°30' to 8°11'E
Chott Melrhir and Merouane	33° to 34° N	6° to 7° E
Chott el Hodna	35°15' to 35°35'N	4°15' to 5°10'E
Zahrez Chergui	35°08' to 35°20'N	3°20' to 3°45'E
Zahrez Rharbi	34°50' to 35° N	2°40' to 3° E

They originate from both the disintegration of materials by decementation due to alternating wettening, drying, precipitation of evaporites and 'Salz-sprengung', and from the subsequent erosion and exportation of the separated individual grains by the wind. These processes explain the relations between
— some of these depressions such as Chott Melrhir, Rharsa, Djerid and Fedjadj and leeside ergs and dunes which are situated upwind of the 'Erg Oriental',
— the El Fayum, El Qattara and Zahrez Chergui depressions and the leeside transport dunes clearly related to these depressions.
The latitudes of these depressions run from West to East diagonally from about 35° to 29° and may indicate the northern boundary of the functional and therefore dynamical eolian unit of the Sahara.

However, the monodirectional sand transport as a result of the Harmattan activity affecting the Sahara, and originating in the Chotts, and the closed depressions in the Northern Sahara do not allow us to reject the existence of sand currents from the Sahara toward the northern Sahelian boundary. This is demonstrated by the important sand circulation as observed all over southern Morocco and in the northern Sahara in Algeria around Djelfa on Landsat images.

2 SOUTHERN SAHARA MARGIN

The complexity of the geomorphological wind effects at the Southern Sahara margin results from the latitudinal variations of the present Sahara (<150 mm precipitation) and former more expanded stages. Therefore, the southern

71

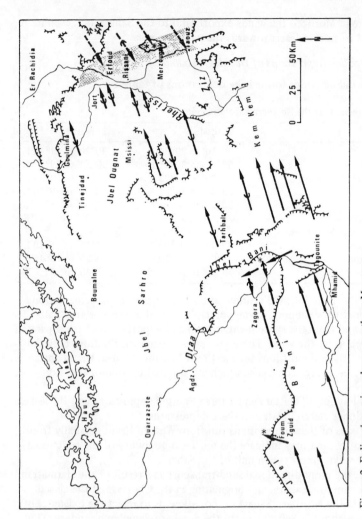

Figure 2. Eolian dynamics scheme, southern Morocco.
Key: * Ghourd; C Barkhanes; Sahel = wind from WSW; Chergui = wind from NE
Boundary region between the two sand transport systems

72

boundary of the functional Sahara unit can be discussed from three aspects:
- The actual southern boundary of wind effects (wind erosion).
- The southern boundary of the area where the wind is the only mechanism of external dynamic. This zone is situated between the active ergs of the southern Sahara and the dunes covered with an open steppe vegetation in the Sahel area. The wind there is responsible of a sand transport illustrated by barchan dunes. At present there is no deposition in this zone because the wind capacities and competences are too high.
- The areas with eolian paleo-deposits which at present include dunes and ergs of the Sahel that are covered with vegetation.

2.1 Zone with fixed dunes

The thickest continuous sand covers are not found in the Sahara but within the Sahel zone where they form ergs with vegetation. As an example, the erg Hausa, Republic of Niger, will be discussed.

Niger has a northern arid margin, the prolongation of the erg of Bilma, belonging in its central parts to the Sahelian zone and in the southern parts to the Sudanian zone. The sahelian zone may be defined by rainfalls between 150 mm (actually active dunes) and 550 mm or even 625 mm. The specific conditions of the erg Hausa is comparable with most of the southern Sahara ergs that are covered with vegetation. There are stable or vegetated dunes of three categories:
- Periodic systems, common for active ergs of the arid or Sahel area with vegetation: 'Longitudinal' undulations, i.e. alternating rises and depressions with a limited extension in the erg Hausa and situated at the leeside of the Termet massif (15°50' to 16°N / 10 °30' to 11°E) and of the Damagaram massif (13°50' to 14° / 8°50' to 9°E and 15° to 16°N / 6° to 7°E). They are replaced by a system of aperiodic sand dune chains as on the Tegama plateau in the North and the Damergou and Ader Doutchi massifs in the South, extending from Niger to Upper Volta and therefore crossing the Niger river. These large asymmetric undulations with a steeper southeastern side running ENE-WSW are 10-25 km long and up to 20 m high.
- 'Transverse' undulations, mostly near negative or positive obstacles, at leeside of the Tanout massif (14°19' to 14°24'N / 8 ° to 8°12'E) and surrounding the Koutous massif (13°50' to 14°30'N / 9°35' to 10°E). They run perpendicular to the predominant wind direction during the dry season and are orientated NNW-SSE (between 153° and 167°). They are 5-10 km long and 500-1 000 m wide, with an asymmetric profile, steeper towards SW, the leeside, and with sandy interdune depressions.
- Hilly accumulations, arranged in coalescent patterns are restricted to the ergs covered with vegetation. Between 12° and 14°N, in Niger, the hilly region ('monticulaire') is characterised by small sand domes with convex profiles separated by round depressions. These accumulations occur with three arrangements: they may be strewn without pattern, arranged transversally

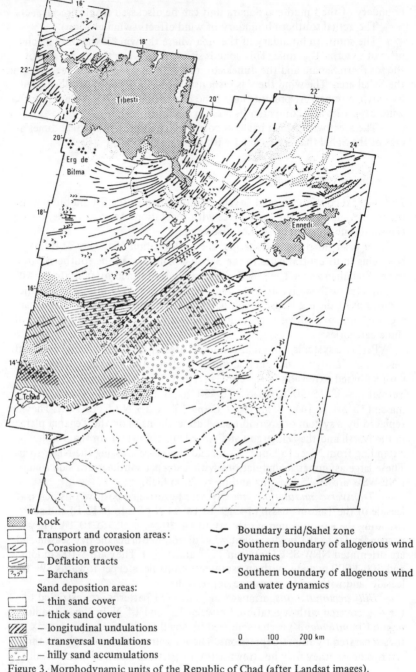

Rock

Transport and corasion areas:
  – Corasion grooves
  – Deflation tracts
  – Barchans

Sand deposition areas:
  – thin sand cover
  – thick sand cover
  – longitudinal undulations
  – transversal undulations
  – hilly sand accumulations

Boundary arid/Sahel zone

Southern boundary of allogenous wind dynamics

Southern boundary of allogenous wind and water dynamics

0   100   200 km

Figure 3. Morphodynamic units of the Republic of Chad (after Landsat images).

74

(NNW–SSE), or arranged longitudinally WSW-ENE. The sand cover may reach thicknesses of 60 m. Around Ader Doutchi the landscape consists of hills several tens of metres high.

— Following the wind, from the arid zone towards the Sudan zone, we encounter the classical scheme of 'longitudinal' dunes — transversal dunes — hilly accumulations.

2.2 Southernmost area of wind influenced landforms in the Sahel, south of the Central Sahara

2.2.1 Wind activities in the Chad and the northern Central African Republic.

The effects of wind activities as observed in Chad are also seen in the northern Central African Republic, at about 11° to 20-21°N (Figure 3). In northern Chad, around 17°N, wind activity consists of deflation, transport and corasion (sculpturing of the rocks by wind). The wind currents are split by topographic obstacles, into distinct streamlines: the first is situated west of the Tibesti, the second — and the most important because of the large sand masses there — is between Tibesti and Ennedi, and the third, the easternmost current, surrounds Ennedi in the south.

The sand transport is demonstrated by series of barkhans sometimes arranged like the flight patterns of ducks and by 'silk dunes' north of Erdi Ma and south of Ennedi. The southernmost silk dunes can be observed at 13°30'N, at the western foot region of the Ouaddai massif, the southernmost barchans at 16°N, south of the lacustrine sediments of the Bodele region, and in eastern Chad, between 21° and 21°30'E / 14°30'N. Undifferentiated sand covers occur, too, particularly in northeastern Chad, on the westernside of the Ennedi-Ouaddai area and on the leeside of the Tibesti massif, where the first 'Ghourd' pyramids of the Grand Erg of Bilma can be seen. Towards the south of 17°N, in the direction of the wind, this transport unit is taken over by a continuous tract of sand deposits beginning on the pediment of Ennedi and Ouaddai around 20°30'E, continuing westward to Niger and ending southward around 13°N at the latitude of Lake Chad. From North to South, the sand cover is fashioned by four dune types, arranged from North to South:

— Longitudinal dunes between 16°30'N and 14°30'N about 4°30 wide from west to east and running roughly NE-SW,
— A group of periodic diagonal dunes, around 16°N, 15°45' to 17°40'E,
— Two types of transverse dunes with large or smaller wavelengths.

Their northern boundary coincides exactly with the beginning Saharan-sahelian vegetation. The transverse dunes with the largest wavelengths correspond with the former Bar El Ghazal and the northern peripheres of Lake Chad.

— Finally a hilly region, receiving the deposits of temporary oueds coming from the Ouaddai massif.

2.2.2 Wind effects in Niger (Cossus 1979). At the western foothill region of

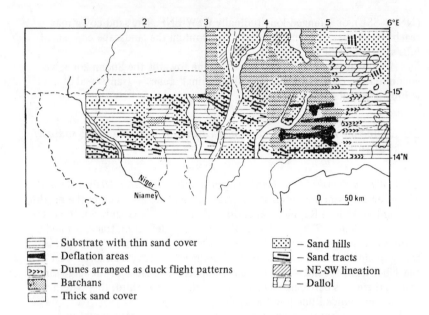

Figure 4. Eolian transport forms between Ader Doutchi and the Niger river

Legend:

- Substrate with thin sand cover
- Deflation areas
- Dunes arranged as duck flight patterns
- Barchans
- Thick sand cover
- Sand hills
- Sand tracts
- NE-SW lineation
- Dallol

Ader Doutchi (Niger) wind acts predominantly to transport material (Figure 4).

On the western rocky foothill region, gently inclined to the west, barchans occur, arranged as flight patterns of ducks, indicating eolian sand transport. The 'flight pattern' is directed to the East and opens to the West, with roughly an East-West axis. These dunes are accompanied by corasion forms *(yardangs)*. Near 5°15′E, they combine in a discontinuous, undifferentiated sand cover. Locally hilly regions occur, divided by tracts with less sand deposition, caused by deflation, and shaped like elongated triangles pointing eastward. At the leeside of Ader Doutchi they reach a length of 100 km. Between these tracts, the very thick sand cover forms hills or small coalescent barchan dunes with a 400 m long axis.

Two facts indicate that at 4°E, i.e. 120 km from the foot of the massif, the effects of the obstacle become indistinct:

- The deflation bands disappear.

- A new direction of wind activity, ESE-WNW, can be observed from the reversal of the direction of sand transport to the NW, surrounding Ader Doutchi in the south and resembling, at a smaller scale, the situation SW of Tibesti

76

(Mainguet 1972). The massif forms an obstacle that divides the sand transport streams some of which incurvate their direction behind the obstacle. On the leeside the streamlines meet and form the Grand Erg of Bilma (Mainguet & Callot 1978). The Ader Doutchi area is not totally impenetrable to wind and forms therefore a less effective obstacle, with persistent E-W directions and NE-SW lineations NW of the massif and ESE-WNW on the SW side.

2.2.3 *Wind dynamics in Upper Volta and Eastern Mali.* In Nigeria and Upper Volta eolian erosion reaches 11°30'N latitude. The examination of Landsat images demonstrates that deflation was more accentuated in 1973, i.e. at the end of the climatic crisis, than in 1976. In Mali these processes end at about 12°N, the latitude of the Bandiagara plateau.

Therefore, between the Tibesti massif in the Northeast and the Bandiagara massif in the West, present and paleoclimatic wind activities belonging to the same functional unit furnished a sand cover about 2 000 km long and 1 100 km wide and only interrupted by some rocky massifs (Termit, Air, Ader doutchi, Ifoghas Adrar). West of 4°E, this cover becomes discontinuous. Wind erosion is demonstrated by deflation tracts or, on the rocky massifs, by corasion grooves together forming a great trajectory system, corresponding with the NE-SW trade winds west of Tibesti, then bending to 8°E and finally to ENE-WSW approaching the Sahel zone. Perhaps these direction changes are the result of a drag effect from the roughness due to vegetation.

3 CONCLUSION

The Sahara is a desert closely related to its northern and southern margins. However, at the northern margin a double circulation to and from the Sahara is detectable, in contrast to the southern margin, where sand always is exported to the Sahel zone, explaining its abundance of sand. The Qattara depression and the Algero-Moroccan Chotts are depressions formed by the interactions of humidity and wind whereby sand is escaping. Perhaps they are a source of the great sand streams crossing the Sahara from NE to SW before they are diverted in E-W direction in the southern Sahel zone. This curvature explains the export of Saharan sand and dust from the continent to the Atlantic.

REFERENCES

Canon, L. & Galichet, M. 1975. Etude de télédétection de la dynamique éolienne de transport, d'accumulation et de corrasion entre l'erg de Mourzouk et l'erg d'Admer à travers les Tassili N'Ajjer. M.de Maîtrise, Univ. de Reims.
Cossus, L. 1979. Recherches de géomorphologie dynamique sur les confins nigéro-maliens au nord de Niamey: la dynamique éolienne. Thèse 3e cycle (in press).

Mainguet, M. 1972. Etude d'un erg (Fachi-Bilma, Sahara central). Son alimentation sableuse et son insertion dans le paysage d'après les photographies prises par satellites. *C.R. Acad. Sci. Paris 274. sér D:* 1633-1636.

Mainguet, M. 1972. *Le modelé des grès. Problèmes généraux.* Inst. Géogr. Nat. Etudes de photo-interprétation. Thèse de Doctorat d'Etat.

Mainguet, M. 1976. Etude géodynamique de l'effet d'obstacle topographique dans le transport éolien. Résultats obtenus à l'aide des images aériennes et satellites au Sahara. *Soc. Fr. Photogram. B. 61:* 31-38.

Mainguet, M. & L.Canon 1976. Vents et paléovents du Sahara. Tentative d'approche paléoclimatique. *Rev. Géogr. phys. Géol. dyn. 18*(2-3):241-250.

Mainguet, M. & Y.Callot 1978. *L'erg de Fachi Bilma (Tchad-Niger).* Contribution à la connaissance de la dynamique des ergs et des dunes des zones arides chaudes. CNRS, Paris.

DUNE FORMS AND WIND REGIME, MAURITANIA, WEST AFRICA: IMPLICATIONS FOR PAST CLIMATE

STEVEN G. FRYBERGER
Research Institute, University of Petroleum and Minerals, Dhahran, Saudi Arabia

INTRODUCTION

The Mainz Symposium brought together scientists who have used differing techniques to interpret climatic change in the Sahara, particularly in the Western Sahara and Mauritania. In the offshore and onshore Western Sahara, Quaternary climate has been reconstructed using detrital clays (Chamley 1979); biological results of upwelling (Diester-Haass 1979); stratigraphic techniques in dune and other deposits (Michel 1979; Rhodenburg & Sabelberg 1979); biogeography (Petit-Marie & Rosso 1979); palynology (Rossignol-Strick & Duzer 1979), and distribution of dust and sand of desert origin in marine sediments (Schutz & Jaenicke 1979; Sarnthein & Koopman 1979).

These techniques have provided evidence of past transport of sediments by wind from land to sea, or variously on land, mainly during arid glacial times. A continuing question in this work is, to what extent might present-day surface wind circulation patterns serve as a model for a particular glacial/interglacial interval? Further, is it possible to describe these winds in detail in terms of sediment transporting ability, particularly direction and amount of potential sand and nearsurface dust transport, seasonally and geographically?*

This report describes techniques for analysis of surface wind data in terms of eolian sand drift, and the application of these techniques to present-day wind systems in Mauritania, West Africa. Using the record of ancient winds embodied in inactive longitudinal dunes of Mauritania and the Sahelian Zone, it is possible with available data to reconstruct certain aspects of ancient wind regimes and to directly compare present-day effective wind systems to those of earlier dry glacial periods.

---

* The assumption underlying this approach is that the analysis of wind energy in terms of sand-sized quartz particle transport is likely to be more meaningful than analysis of wind energy in terms of watts, or percentage histograms, which may show different geographical and seasonal patterns.

THE SAND ROSE TECHNIQUE AND THE CLASSIFICATION OF WIND REGIMES

The sand rose technique

Detailed surface wind records can be analyzed systematically in terms of potential sand transport. The essentials of the technique used by the author (Fryberger 1978, 1979) consist of the following steps:

1. An equation for rate of sand drift by wind is selected. I have used the Lettau equation (Lettau 1978) which is

$$\frac{qg}{C''p} = V_*^2 \, (V_* - V_{*t}) \tag{1}$$

where: q = rate of sand drift

g = gravitational constant

C'' = empirical constant based on grain diameter

p = (greek letter rho) density of air

V_* = shear velocity of wind

V_{*t} = impact threshold shear velocity

additionally,

$$C'' = C' \, (d/d^*)^n \tag{2}$$

where: C' = universal constant for quartz sand (approximately 6.7)

d = mean diameter of sand moved

d* = 0.25 mm (standard diameter)

n = empirical constant approximately equal to 0.5

2. The above equation (1) is simplified by eliminating constants to

$$q \, \alpha \, V_*^2 \, (V_* - V_{*t}) \tag{3}$$

Wind velocity at a given height is proportional to drag (shear) velocity (Belly 1964) and thus equation (3) becomes

$$q \, \alpha \, V^2 \, (V - V_t) \tag{4}$$

where V = wind velocity at 10 m height, and

V_t = impact threshold wind velocity at 10 m height.

Equation (4) is a weighting equation. It can be used to evaluate the relative power of wind of a given velocity to move sand, if the value of V_t is either determined or assumed. In general, velocities (values for V, in knots) chosen for substitution into equation (4) will be mid-points of velocity categories of surface wind summaries.

3. Finally, since the weighting equation (4) represents an instantaneous rate, and percentages of a typical (direction x speed) wind summary represent time (usually annual or monthly periods) then

$$Q \, \alpha \, V^2 \, (V - V_t) \, t \tag{5}$$

where Q = annual (or monthly) rate of sand drift

t = length of time wind blew, expressed as a percentage of a year's observations on a wind summary.

When multiplication is carried out in the described manner for an entire wind summary, using $V_t = 12$ knots, and dividing by 100 as a scaling factor,

Table 1. Average annual drift potentials for 13 desert regions, based on data from selected stations

Desert region	No. of stations	Drift potential	
		Vector units	m³/m-w yr
High energy wind environments			
Saudi Arabia, Kuwait	10	489	34
(An Nefud desert)			
Libya (N.W. desert)	7	431	30
Intermediate energy wind environments			
Australia (Simpson desert south)	1	391	27
Mauritania (W. desert)	10	384	$\simeq$27
U.S.S.R. (Peski Karakumy,	15	366	25
Pesky Kyzylkum)			
Algeria (Erg Oriental, Erg Occidental)	21	293	20
South West Africa (Namib desert)	5	237	17
Saudi Arabia (Rub al Khali, north)	1	201	14
Low energy wind environments			
South West Africa (Kalahari)	7	191	13
Mali (Sahel, Niger R.)	8	139	10
China (Gobi desert)*	5	127	8.8
India (Thar desert)*	7	82	5.7
China (Takla Makan)*	11	81	5.6

* DP's estimated

the results are in 'vector units' (Fryberger 1979). The vector unit total for any wind summary is proportional to the sand-moving power of the wind at the station of record and is known as the drift potential (DP) at the station (Fryberger 1979). This method was used to evaluate relative sand-moving power of winds in various deserts around the world for certain chapters of the U.S. Geological Survey Professional Paper 1052 entitled 'A Study of Global Sand Seas', edited by Edwin D.McKee. For that study, it was felt that direct estimates of drift rates on the world's deserts would be misleading and possibly inaccurate, since so little is known about drift rates in deserts. Thus the 'vector unit' approach seemed a satisfactory way of comparing relative sand-moving power of wind in various deserts (Table 1).

Nevertheless, in discussing a particular desert, it is perhaps more meaningful to use drift potentials not in terms of vector units, but in terms of predicted amounts of sand transport. The Lettau equation for sand transport was evaluated quantitatively using two presumed threshold drag velocities to produce a graph showing the relationship of drift potentials to rate of sand transport in m³/m-w yr (meter³ per meter-width per year) assuming a bulk

81

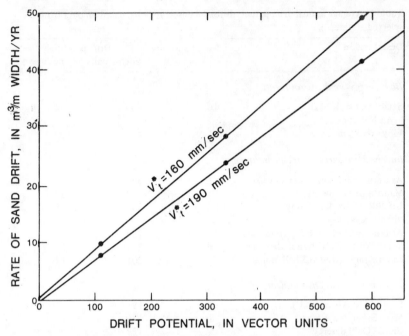

Figure 1. Graph showing comparison of annual drift potential in vector unit notation (used by the author in some reports cited) to annual DP in m^3/m-w year, for two presumed threshold drag velocities. Based on evaluation of Lettau (1978) equation (#1 in text) for two presumed threshold drag velocities.

density of 1.34 g/cm³ for loosely packed quartz sand (Figure 1). For this report then, drift potentials, which are predictions of the total sand-moving power of the wind at a given place, are expressed in m^3/m-w yr. The reader is cautioned that the values presented in this report are estimates requiring a number of assumptions which are not described here due to limitations of space. The reader is referred to Fryberger (1979) for details of these assumptions. The technique is essentially a manual integration.

The directional nature of wind data makes feasible the construction of circular histograms known as sand roses (Figure 2). In these diagrams arms are proportional in length to potential sand transport from the direction indicated by the arm toward the center of the rose. If the quantities (vector units, m^3/m-w yr, or other units) represented by the arms are treated as vectors and resolved, the resultant points in the net direction of sand drift and is known as the resultant drift direction (RDD). The magnitude of this resultant is known as the resultant drift potential (RDP). At stations where wind often comes from several directions during the year, RDP's are much lower than DP's. This relationship can be conveniently expressed by the RDP/DP (ratio).

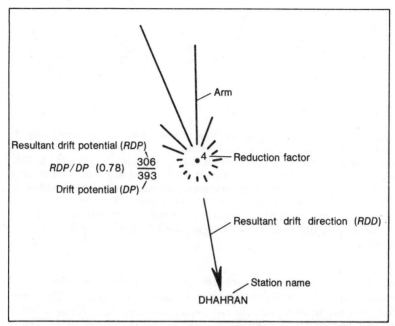

Figure 2. A sand rose. Arm of the sand rose is proportional to potential amount of sand drift along the arm toward the center of the rose, in vector units or other units. This sand rose has 16 arms, but sand roses can be plotted with as many arms as desired. Other terms explained in text. This sand rose was plotted by a programmable calculator-plotter from data entered directly into a program from a wind summary.

Wind energies in terms of sand transport range widely from place to place. Thus, if sand roses were all plotted at the same scale on a given map, a wide range of sizes would occur. In order to correct this, sand rose arm lengths can be divided by two until a convenient plotting size is reached. The number by which the arm lengths are divided is known as the reduction factor (Figure 2). Naturally, reduction factors are not required with every sand rose, provided that drift potential is indicated with the rose. This is the procedure followed for this report.

Classification of wind regimes

Wind regimes can be classified in terms of directional properties, as expressed in sand roses, and in terms of drift potentials (vector units) (Fryberger 1979). In general, five types of wind regimes are recognized, based on relationships of modes (groups of arms) on sand roses (Figure 3). *Narrow unimodal* wind regimes are those with 90 per cent or more of the drift potential at a station within a 45° arc of the compass, i.e. included in two adjacent directional cate-

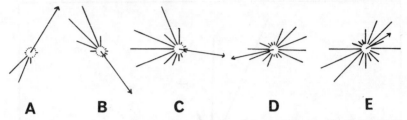

A	**B**	**C**	**D**	**E**

Figure 3. Sand roses illustrating the five types of wind regimes distinguishable using 16-point directional data from surface wind summaries. A. Narrow unimodal wind regime, Walvis Bay, South West Africa (DP = 518 vu; RDP = 448 vu; RDP/DP = 0.86); B. Wide unimodal wind regime, Bahrain, Arabian Gulf (DP = 540 vu; RDP = 435 vu; RDP/DP = 0.81); C. Acute bimodal wind regime, Badanah, Saudi Arabia (DP = 528 vu; RDP = 327 vu; RDP/DP = 0.62); D. Obtuse bimodal wind regime at Timimoun, Algeria (DP = 240 vu; RDP = 46 vu; RDP/DP = 0.19); and E. Complex (polymodal) wind regime at Ghudames, Libya (DP = 662 vu; RDP = 60 vu; RDP/DP = 0.09). North is towards top of the page.

gories on 16-point wind summaries (Figure 3A). A *wide unimodal* wind regime is any other wind distribution with a single mode (Figure 3B). *Acute bimodal* wind distributions have two modes, in which the peak directions of the two modes form an acute angle, including also the right angle (90°) (Figure 3C). *Obtuse bimodal* wind distributions have two modes in which peak directions form an obtuse angle (Figure 3D). *Complex* wind distributions include polymodal wind distributions such as trimodal, or wind distributions with poorly defined modes (Figure 3D). The 16 point directional data commonly available in wind summaries usually will not show more than three modes clearly. Modes observed on sand roses are not considered significant for purposes of classifying a wind regime if the modal direction and two adjacent directional categories (on 16-point data) constitute less than about 15 per cent of the drift potential at the station.

Wind regimes are classified in terms of vector units as high energy (400 vu or greater), intermediate energy (399-200 vu) and low energy (less than 200 vu) (Table 1). Using the $V_t^* = 19$ cm/sec in Figure 1, this same scale indicates that for purposes of this report high-energy wind regimes have drift potentials (approximate) of 28 m^3/m-w yr; intermediate energy wind regimes have DP's of 28 to 14 m^3/m-w yr and low-energy wind regimes have DP's of less than 14 m^3/m-w yr.

Large dune forms visible on Landsat imagery of modern deserts have certain characteristic relationships to wind regimes as defined using sand roses (particularly in high-energy wind regimes) (Fryberger 1979, Breed, Fryberger et al. 1979). This is also true if wind regimes are evaluated by plotting drift potentials (wind energy, y-axis) against RDP/DP ratio (wind variability, x-axis) (Fryberger 1979), although use of such a plotting technique must be tempered by certain reservations.

Large linear dunes can occur in wind regimes ranging from complex through bimodal to unimodal, but preliminary data suggest that as wind energy increases, RDP/DP ratios must also increase (i.e. wind variability decrease) for the linear dune type to occur (Fryberger 1979). Thus, in high energy wind regimes, unimodal and bimodal distributions of effective winds are commonly associated with linear dunes. As energy decreases, complex wind regimes can be associated with linear dunes.

Where linear dunes are the principal type present in a high-energy wind regime, they are aligned with the resultant drift direction of modern day surface winds in most instances studied by this worker (Fryberger 1979, Breed, Fryberger et al. 1979). The relationships just described are used later in this report to assess the fit of modern winds to ancient dunes in Mauritania.

The analysis of surface winds presented here uses surface wind summaries supplied by the National Climatic Center, Asheville, North Carolina, U.S.A. on the form called 'N-summary'. These N-summaries contain quantitative data from stations operated in accordance with World Meteorological Organization standards and procedures, and each contains roughly 10 years data. This data is summarized as percentages in 16 compass directions and 9 velocity categories from 0 to 40 knots. The Landsat image analysis is based on single Landsat images and on slightly reduced photocopies of 1:1 000 000 scale Landsat mosaics of Mauritania prepared while the author was with the U.S.Geological Survey, Flagstaff, Arizona, U.S.A.

MODERN EFFECTIVE WIND REGIMES IN MAURITANIA

Sand rose classification

Wind regimes in Mauritania are unimodal (Port Etienne) or acute bimodal (Fort Gouraud, Atar, Boutilimit) and become more directionally variable southward (Figure 4). Resultant drift directions in the northern inland areas, for example at Fort Gouraud, are from northeast to southwest, parallel to the approximate boundaries of major sand seas, as sketched in Figure 4. Resultant drift directions swing southward near the coast, in Erg Trarza and near the Senegal River. Thus, present-day resultant drift directions are most directly offshore near Port Etienne, becoming subparallel to parallel to the coast southward toward Nouakchott. Resultant drift directions vary considerably during the year at southerly stations of Nouakchott, Boutilimit, and Rosso. During winter, the easterly winds are dominant; during summer, northerly and northwesterly winds are stronger in this area, which perhaps accounts for the southward (landward) turn of annual RDD's in this area (Breed, Fryberger et al. 1979). The bimodal wind regimes at Fort Gouraud and Fort Trinquet also result from two effective wind seasons. For example, at Fort Gouraud, the northeasterly winds are strongest in summer, whereas the southeasterly winds are strongest in winter (Breed, Fryberger et al. 1979). Shifting surface wind

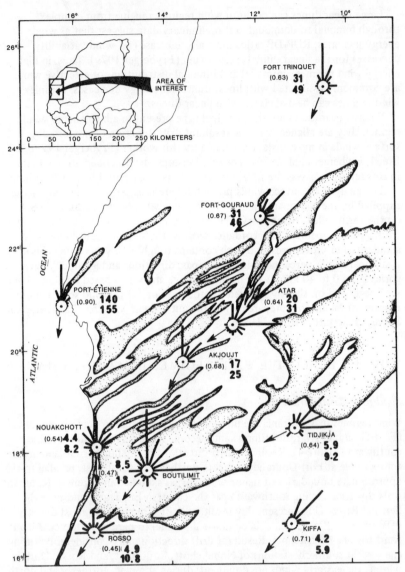

Figure 4. Map of sand sea areas of Mauritania (after *Times Atlas of the World* 1959) with annual sand roses for meteorological stations in Mauritania. Upper boldface numbers are resultant drift potentials; lower boldface numbers are drift potentials; light type indicates RDP/DP. DP and RDP values are in m³.

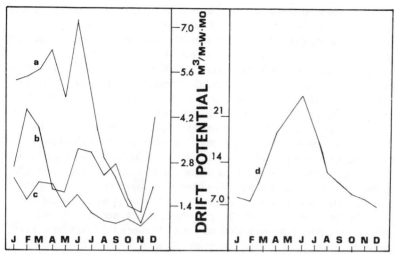

(a) Fort Trinquet;
(b) Atar;
(c) Boutilimit;
(d) Port Etienne.
June peaks in drift potential are less pronounced at southern stations, and annual drift potentials are also lower at southern stations.

Figure 5. Graphs showing variation of drift potential during the year at four stations in Mauritania (station locations shown in Figure 4).

circulation patterns, reflected in the bimodal wind regimes of Mauritania, indicate the possibility of different source areas and distribution patterns for sand and other particles brought toward the ocean during different seasons.

Drift potentials

Mauritania is sharply zoned in terms of wind energy, with high annual drift potentials in the north along the coast (Port Etienne, 155 m^3/m-w yr) and low drift potentials in the south near the Senegal River (Rosso, 10.8 m^3/m-w yr; Kiffa, 5.2 m^3/m-w yr) as shown in Figure 4. Contours of equal drift potential extend roughly northeast to southwest across most of the region shown in Figure 4.

It is apparent that even using the conservative estimates of this report (based on $V_t^* = 19$ cm/sec) the present day surface winds can potentially move enormous quantities of sand in northern Mauritania. Further south drift potentials of winds blowing across inactive sand dunes between Boutilimit and Rosso are lower, but still capable of producing geologically significant deposition. For example, decrease in RDP from 8.5 m /m-w yr at Boutilimit

87

to 4.9 m³/m-w yr at Rosso is 3.6 m³/m-w yr. If the distance between them is approximately 175 km, then a rate of deposition of 0.020 6 mm/yr can be expected between the stations if winds at Boutilimit were to be fully charged with sand. This rate is not great, but is geologically significant, since it amounts to 20.6 mm/1 000 yrs, or about 0.206 m in 10 000 years. This may not be enough to account for the sand deposition of the ancient Ogolien erg of this region, but suffices to indicate the power of the wind even in the low to intermediate energy wind regimes of the area.

Wind energy release is highly seasonal in Mauritania, with much of the effective wind energy occurring during either June, and/or winter-spring, depending upon location (Figure 5). In general, areas further north and closer to the coast have the strongest June peaks in drift potentials (Figure 5). During this time of year, anticyclonicity of the Azores high pressure cell is well-developed, and thermal low pressure is strongly developed over the Sahara (Breed, Fryberger et al. 1979). This produces a strong, narrow, northeast-southwest extended zone of high wind energy between the two pressure cells. The winter-spring peak of drift potentials (earlier but weaker at Atar as opposed to Akjoujt) reflects outflow from the wintertime Saharan high pressure cell, or a tongue of high pressure which sometimes extends from the Azores High across North Africa. In general, this circulation produces weaker and more easterly winds than the summertime circulation.

The broad pattern sketched above, of high drift potentials at latitudes above 20°N in Mauritania, and low DP's at lower latitudes, is also typical of a number of areas in the Sahelian zone, as well as Mauritania (Breed, Fryberger et al. 1979). For example, in Mali, drift potentials at Gao (16°16'N), and Tessalit (20°12'N) are 27 m³/m-w yr and 31 m³/m-w yr respectively, whereas at Segou (13°24'N), Mopti (14°31'N) and Tillaberi, Niger (14°12'N) drift potentials are 1.5, 11 and 5.7 m³/m-w yr respectively.

COMPARISON OF SAND FEATURES TO WIND REGIMES IN SOUTHWEST MAURITANIA

Shape and extent of major sand seas

The shapes of active sand seas in Mauritania are parallel or subparallel to resultant drift directions of present-day surface winds (Figure 4). For example, the RDD at Fort Gouraud is parallel to edges of the nearby sand seas, but at Akjoujt the RDD is subparallel to the edges of nearby sand seas (Figure 4). Some linear dunes visible on Landsat imagery near Akjoujt are parallel to the RDD at Akjoujt, but oblique to the edge of the sand sea, as noted elsewhere (Mckee, Breed & Fryberger 1977). A rough parallelism of major active sand sea borders with resultant drift directions of present-day high energy wind regimes also occurs near Bilma, Niger and the An Nefud of Saudi Arabia (Breed, Fryberger et al. 1979, Fryberger & Ahlbrandt 1979).

Borders of inactive (vegetated) sand seas in Mauritania can be either parallel or strongly oblique to RDD's of present-day winds, depending upon locality. For example, the RDD at Kiffa is roughly parallel to the southern edge of Erg Trarza. Yet the present-day RDD at Nouakchott is strongly oblique to the northern edge of Erg Trarza. This situation may result from a restricted coastal wind regime. Such coastal wind regimes are present elsewhere, for example such a region occurs along the coast of South West Africa (Namib desert) (Breed, Fryberger et al. 1979). Such restricted coastal wind regimes can sometimes result when offshore winds skip over coastal inversions caused by the presence of cold waters. Perhaps the lack of close fit or parallelism of inactive sand sea edges and modern-day winds at stations such as Nouakchott in Mauritania occurs because the sand seas (south of about 19°N) are the product of multiple depositional events.

Active linear dunes. The large, active linear dunes of Northern Mauritania are aligned parallel to the RDD's of the modern high-energy wind regimes of the area (Figure 6, McKee, Breed & Fryberger 1977). This alignment of large linear dunes with RDD's of high-energy wind regimes is typical of many active dunefields around the world (Fryberger 1979). Moreover, active linear dunes of this size do not form crosscutting sets of linear dunes in a single wind environment (Fryberger 1979). Thus, overlapping or crosscutting of long linear dunes (50 km or more) over wide (100 x 100 km) areas implies long-term changes in effective wind regime.

Inactive linear dunes. Inactive linear dunes visible on Landsat imagery record two major eolian depositional events, plus a reworking of the youngest dunes in southern Mauritania (Figure 6). These depositional events, as recorded by linear dunes, may correlate with the chronology suggested by Michel (1979).

The remains of the oldest linear dunes extend east-northeast to west-southwest and are now almost destroyed by processes of denudation, at least in terms of their visibility on Landsat imagery. These dunes are visible on Landsat imagery as dark bands, aligned northern and southern ends of (white reflectance) interdunes of a younger set of dunes, and southwest-trending blurring of younger dunes (Figure 6, pattern marked by # 2). The older dunes cannot be traced to the sea using Landsat imagery, but probably did reach the sea, based on evidence provided by Sarnthein & Diester-Haass (1979). These dunes are aligned parallel to the boundaries of active ergs to the north (example is marked by # 1 in Figure 6) and also to the southern boundary of the inactive Erg Trarza, to the south (Figure 4, Figure 6). These dunes, or more exactly, remnants of dunes, align with the remnants of linear dunes visible in Ferlo, Senegal, and thus may be equivalent to the Ferlo erg of Michel (1979). Hence, the dune remnants may be over 40 000 years old. They must certainly represent a result of a pre-Würm desertification. These inactive dunes are conspicuously out of alignment with the resultant drift direction of present-day winds in the area of Boutilimit where they are most clearly visible on Landsat imagery.

The next youngest set of dunes (marked with a # 3 on Figure 6), is the most conspicuous group of dunes in Mauritania as viewed on Landsat imagery. These are the Ogolien dunes of about 15 000 to 20 000 years ago (Michel 1979). These dunes are clearly visible as dark lines on the Landsat imagery. They are aligned parallel to the resultant drift direction of present-day winds at every station in Mauritania except Nouakchott. These inactive dunes can be traced northward aligned with local RDD's step by step into regions of high drift potentials and active sand movement.

The acute bimodal and wide unimodal wind regimes typical of the southern, inactive dune fields are compatible with the linear dune type, although drift potentials are somewhat low (Fryberger 1979). In summary, the Ogolien dune system built by winds during the Würm glacial period are compatible with present-day wind regimes in terms of both RDD's and sand roses.

The apparent fit of both active and inactive (Ogolien) linear dunes in Mauritania to the modern-day wind regime, with no break in continuity from inactive to active areas, is not a solitary occurrence. A similar situation occurs in the Kalahari desert, South Africa. In the Kalahari desert, simple linear dunes extend from relatively low-energy wind regimes and higher rainfall near Mariental southward to drier, higher-energy wind regimes near Upington without a break in morphology. Within the Kalahari desert as a whole, these dunes align closely with the resultant drift directions of present-day winds (Breed, Fryberger et al. 1979), but the age of the dune system is not known to the author.

The youngest linear dunes in the study area visible from the Landsat imagery are too closely spaced to be marked on Figure 6. They are visible on the imagery mainly near Rosso, north of the Senegal River. These dunes are very short (1 km or so in length), and are oriented slightly oblique to the Ogolien dunes, and usually developed along the Ogolien dune crests. These dunes seem to have resulted from reworking of the Ogolien dunes, possibly prior to the maximum Nouakchottian transgression of 5 500 years ago (Michel 1979).

Crossed trends of relatively inactive linear dunes occur elsewhere in the Sahelian zone in a pattern similar to that of Mauritania (Figure 7). These

(a) solid lines with dots: tracings of linear dune crests visible on the imagery. Dots occur where the dunes can no longer be traced.
(b) solid lines with no dots: barchanoid dunes (three areas only).
(c) dashed lines: approximate boundaries of dunefields, clouds and other features visible on the imagery.
(d) C: cloud cover.
(e) EI: edge of image.
(f) # 1: edge of sand sea.
(g) # 2: pattern of older east-northeast to west-southwest oriented dunes.
(h) # 3: Ogolien (?) dunes.
(i) Arrows: RDD's of modern day wind regimes at the towns indicated by dots. Town of Boutilimit is indicated by large 'B' in right center of map.

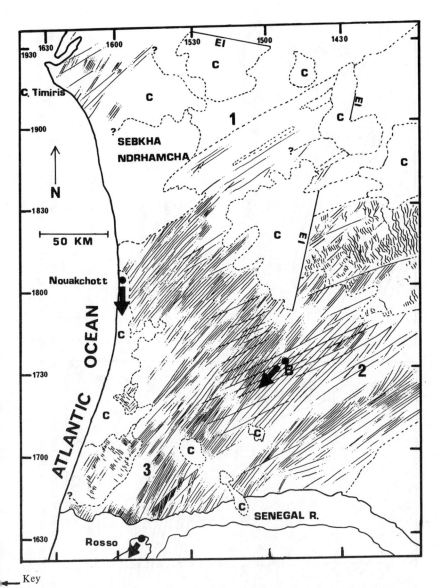

Figure 6. Sketch map showing selected eolian and other features of the southwest portion of Mauritania. Map is based on analysis of a slightly reduced black and white photocopy of a 1:1 000 000 mosaic of bulk processed, false-color Landsat imagery of the region. Dune crests and other features were traced from the photomosaic using a light table.

91

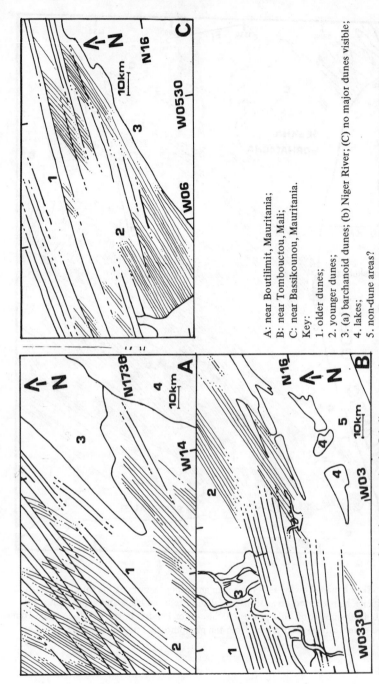

Figure 7. Sketch maps of the three areas in the Sahelian zone where two eolian dune systems cross. All maps produced by tracing of dune crests using black and white or false-color bulk processed Landsat imagery and a light table. Heavier lines represent author's interpretation of older dune pattern in each area.

A: near Boutilimit, Mauritania;
B: near Tombouctou, Mali;
C: near Bassikounou, Mauritania.
Key:
1. older dunes;
2. younger dunes;
3. (a) barchanoid dunes; (b) Niger River; (C) no major dunes visible;
4. lakes;
5. non-dune areas?

92

patterns consist of older, broader, widely spaced dunes crossed by younger, narrower, more closely spaced dunes. In every instance observed by this worker, the older dunes are aligned in a more east to west direction than the younger dunes (Figure 7).

The similarity of the older and younger dune trends in Mali and Niger to those in Mauritania suggests a correlation may exist between them, but no data is presently available to substantiate this idea. The conspicuous crossing of two sets of mainly inactive linear dunes at about 16°N in the Sahelian zone does, however, confirm the past occurrence of at least two major eolian depositional episodes (though not necessarily synchronous between areas) over a wide geographic area.

IMPLICATIONS OF COMPARISON OF SAND FEATURES TO WIND REGIME

The present analysis of modern surface wind data and study of dunes of Mauritania using Landsat imagery has perhaps complicated the job of reconstructing climatic history in this area. Certain tentative conclusions seem possible for Mauritania, however, as follows.

The two major crossed sets of dunes in this region, which in all probability represent the effects of wind regimes of different arid glacial episodes, indicate that these different arid episodes (Würm versus pre-Würm, Riss?) were associated with different patterns of surface wind circulation. Put another way, this implies that different surface wind circulation models may be required for at least two glacial episodes.

The alignment of the oldest set of inactive dunes (pre-Würm) is parallel to the RDD's of the northern Mauritania wind regimes, and thus would fit well a southward shift of this belt of high-energy winds (which extends roughly northeast to southwest through Atar to Akjoujt and includes Port Etienne). Merely increasing the wind energy of the present surface wind circulation pattern, with dessication, is not sufficient to account for the orientation of this older set of dunes. Certainly a stronger westward flow during a pre-Würm arid episode is indicated by these dunes.

The high degree of compatibility of the inactive Ogolien dunes with present-day wind regimes indicates that effective surface wind circulation during the Würm glacial resembled present-day effective surface winds. Thus, only dessication is required — not a southward shift of wind belts — to explain the arid Ogolien dune system in Mauritania.

The present geographic zonation of drift potential from northwest to southeast over Mauritania, as well as seasonal changes in drift potential and resultant drift direction in Mauritania should be reflected in offshore eolo-marine sediments of the Würm arid episode. This is primarily because these changes control intensity of sedimentation and wind scour on land, and source areas for eolian transported sand and dust.

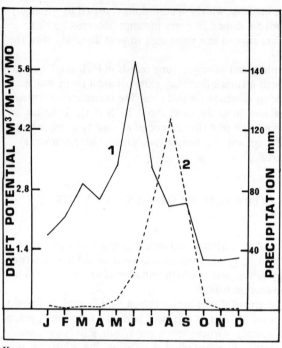

Key:
1. drift potential;
2. precipitation.
Precipitation data from Griffiths & Soliman (1972).

Figure 8. Graph comparing seasonal occurrence of drift potential and precipitation at Gao, Mali, 16°16′N, 00°03′W. Annual drift potential at Gao is 420 vector units (or about 29 m³/m-w year).

COMMENTS ON DESERTIFICATION

Presently there is considerable discussion in the literature regarding the origins and significance of Sahelian desertification. In particular, the question seems to be, to what extent is desertification man-caused (directly or indirectly) versus the result of natural processes not affected by man (Mensching 1979, Sec. of U.N. Committee on Desertification 1977).

The data presented in this report, plus data presented in earlier reports (McKee, Breed & Fryberger 1978, Breed, Fryberger et al. 1979) have some bearing on this question. First, it is apparent from the foregoing analysis that even by conservative estimates, the sand-transporting power of winds at stations in the Sahel, where many soils are developed on old dune sand, is enormous. Thus, the drift potential of 18 m³/m-w yr at Boutilimit amounts

94

to potential downcutting of 18 vertical meters in a disturbed patch of ground 1 m².

Second, in the Sahelian zone, two natural, continuing processes operate in many areas to intensify any disturbances caused by either man, lack of precipitation, or other agents. These processes are first the concentration of much effective wind energy of the area into intense sand-moving seasons, or at least potential sand moving seasons in vegetated areas. Second, maximum drift potentials usually occur after long dry spells, i.e. just prior to the summer monsoon (Figure 8). Thus, effects of a missed season of rainfall or other disturbance are enhanced.

Third, it has been noted above that there exists a certain compatibility of present-day wind circulation patterns (analyzed in terms of sand drift) to ancient Ogolien dunes in Mauritania, and to other relatively inactive dune systems, which in all probability are products in part of the most recent glacial arid episode (Würm). This may indicate that drastic changes in surface wind regime are not required for return to more arid conditions associated with dune construction in the Sahelian zone. These considerations, in turn, indicate that a more delicate balance of active desert versus inactive desert in terms of overall climate may exist in the Sahel than previously recognized. This proportionately increases the potential upsetting effects of human activities on this balance in the desert system.

REFERENCES

Belly, Pierre-Yves 1964. Sand movement by wind, with addendum 2, by Abdel-Latif Kadib. US Army Corps of Engineers, Coastal Eng. Research Center Tech. Memo. 1.

Breed, C.S., S.G.Fryberger, Sarah Andrews, Camilla McCauley, Frances Lennartz, Dana Gebel & Kevin Horstman 1979. Regional studies of sand seas using Landsat (ERTS) images. In: E.D.McKee (ed.), *A study of global sand seas*. US Geological Survey Prof. Paper 1052 (in press).

Chamley, H. 1979. Marine clay sedimentation and climate in the late Quaternary off NW Africa. *Palaeoecology of Africa 12*.

Diester-Haass, L. 1979. Upwelling and climate off NW Africa during the Late Quaternary. *Palaeoecology of Africa 12*.

Fryberger, S.G. 1978. Techniques for the evaluation of surface wind data in terms of eolian sand drift. US Geological Survey Open-file Report 78-405.

Fryberger, S.G. 1979. Dune forms and wind regime. In: E.D.McKee (ed.), *A study of global sand seas*. US Geological Survey Prof. Paper 1052 (in press).

Fryberger, S.G. & T.S.Ahlbrandt 1979. Mechanisms for the formation of eolian sand seas. *Z. Geomorphol. Stuttgart* (in press).

Griffiths, J.S. & K.H.Soliman 1972. The Northern Desert (Sahara). In: H.E.Lansburg (ed.), *World survey of climatology, climates of Africa*. Elsevier, Amsterdam, New York, London.

Jaenicke, R. & L.Schütz 1979, Eolian dust from the Sahara desert. *Palaeoecology of Africa 12*.

Lettau, K. & H.Lettau 1978. Experimental and micrometeorological studies of dune migration. In: H.Lettau & K.Lettau (eds.), *Exploring the world's driest climate*. Center for Climatic Research, Inst. for Environmental Studies. EIS Report 101, pp.110-147.

McKee, E.D., C.S.Breed & S.G.Fryberger 1977. Desert sand seas. In: *Skylab explores the earth*. NASA SP-380, pp.5-48.

Mensching, H.G. 1979. The Sahelian zone and the problems of desertification. Climatic and anthropogenic causes of desert encroachment. *Palaeoecology of Africa 12*.

Michel, P. 1979. The SW Sahara margin. Sediments and climate changes during the Recent Quaternary. *Palaeoecology of Africa 12*.

Petit-Maire, N. & J.C.Rosso 1979. Recent biogeographical variations along the NW African coast (28-19°N). Paleoclimatic implications. *Palaeoecology of Africa 12*.

Rohdenburg, H. & U.Sabelberg. 1979. Northwestern Sahara margin. Terrestrial stratigraphy of the Upper Quaternary and some paleoclimatic implications. *Palaeoecology of Africa 12*.

Rossignol-Strick, M. & D.Duzer 1979. Late Quaternary West African climate inferred from palynology of Atlantic deep-sea cores. *Palaeoecology of Africa 12*.

Sarnthein, M. & L.Diester-Haass 1977. Eolian-sand turbidites: *J. Sed. Petrol. 47:*868-890.

Sarnthein, M. & B.Koopmann 1979. Late Quaternary deep-sea record on the NW African dust supply and wind circulation. *Palaeoecology of Africa 12*.

Secretariate of the UN Conference on Desertification 1977. Desertification: An overview. In: *Desertification: Its causes and consequences*. UN Conference on Desertification, Nairobi, Kenya, 29 August-9 September 1977. Pergamon Press.

The Times Atlas of the World 1959. Mid-Century Edition. Bartholomew, ed. Houghton Mifflin Co., Boston, 5 vols.

EOLIAN DUST FROM THE SAHARA DESERT

L. SCHÜTZ / R. JAENICKE
Max-Planck-Institut für Chemie, Mainz; Institut für Meteorologie, Université Mainz

A major source for mineral dust are the deserts and the semiarid regions of the Earth which cover roughly one third of the continental surfaces. A rather continuous horizontal flux of dust particles is created in areas where the dust is fed into a steady atmospheric circulation system like the Hadley cell of the subtropics. This desert dust transport is studied best for the Sahara, one of the largest deserts on Earth. Over its western margins across the north Atlantic ocean, almost continuously mineral dust is transported towards the Caribbean Sea. Only the airmass above the trade-wind inversion is supplied with desert mineral dust directly from the Sahara. The airmass below the inversion – the trade wind layer – is originally free of dust and is filled with dust only from above. The strong trade wind inversion hinders the vertical turbulent exchange and thus influences the vertical flux of mineral particles. Thus the submicron particles are penetrating very slowly the inversion, whereas the larger particles – mainly transported vertically by sedimentation – reach the trade wind layer without delay. This is causing a significant accumulation of the concentration of micron mineral particles close to the ocean surface within the first couple of thousand kilometers of transport. That could be confirmed by measurements and model calculations. Because of the rather rapid depletion of micron particles due to sedimentation within the first part of the transport, long range transport can be observed only for submicron particles, causing haze in the atmosphere and significant contribution to the formation of deep-sea sediments.

Based on the transport model for the Saharan dust, a budget was calculated and the source strength estimated to be of the order of 260 MT (million tons) annually. Compared to this transport towards the west, the transport into other directions is of minor importance. Roughly two thirds of the mineral dust mass is deposited within 2 000 km and after a distance of 5 000 km, about 50 MT per year reach the Caribbean Sea. Because the deep-sea sediment mainly comes through eolian dust transport, the model predicts the accumulation of sediments reasonably well.

Dust transport from other deserts can be estimated from distribution patterns of frequency of dry haze, and the deposition pattern of clay minerals in deep-sea sediments. In addition, the distribution of quartz particles and isotopic data from sedimentary deposits indicate that other deserts contribute to the atmospheric mineral dust burden too. An estimate of the global production rate of mineral dust was obtained by using data of mean sedimentary accumulation, vertical fluxes of dust particles in sandstorms and mean mineral dust concentration in the atmosphere. The figure shows the considerable variation in the data and indicates a global production of mineral dust of the order of 1 000 MT per year.

REFERENCES

Jaenicke, R. 1979. Monitoring and critical review of the estimated source strength of mineral dust from the Sahara. In: *Saharan Dust: Mobilization, transport and deposition.* Scope 14. Ch. Morales (ed.). John Wiley, Chichester, pp.233-242.

Jaenicke, R. & L.Schütz 1978. Comprehensive study of physical and chemical properties of the surface aerosols in the Cape Verde Islands region. *J. Geophys. Res. 83:*3585-3599.

Junge, C. 1979. The importance of mineral dust as an atmospheric constituent. In: *Saharan dust: Mobilization, transport, deposition.* Scope 14, Ch.Morales (ed.). John Wiley, Chichester, pp.49-60.

Rahn, K.A., R.D.Borys, G.E.Shaw, L.Schütz & R.Jaenicke 1979. Long-range impact of desert aerosol on atmospheric chemistry. In: *Saharan dust: Mobilization, transport, deposition.* Scope 14, Ch.Morales (ed.). John Wiley, Chichester, pp.243-266.

Schütz, L. 1980. Saharan dust transport over the North Atlantic Ocean – Model calculations and measurements. In: Ch.Morales (ed.), *Saharan dust: Mobilization, transport, deposition.* Scope 14, John Wiley, Chichester, pp.267-278.

Schütz, L. 1979. *Long-range transport of desert dust with special emphasis on the Sahara.* New York Academy of Sciences (in press).

Schütz, L. & R.Jaenicke 1974. Particle number and mass distribution above 10^{-4} cm radius in sand and aerosol of the Saharan desert. *J. Appl. Meteor. 13:*863-870.

Schütz, L., R.Jaenicke & H.Pietrek 1980. *Saharan dust transport over the North Atlantic Ocean.* Proc. Conf. Desert Aerosols, Denver. Univ. of Arizona Press (in press).

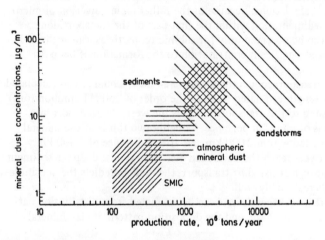

YOUNG PLEISTOCENE LOESS AS AN INDICATOR FOR THE CLIMATE IN THE MEDITERRANEAN AREA

KARL BRUNNACKER
Geologisches Institut der Universität zu Köln

1. PRELIMINARY REMARKS

The studies on Mediterranean loess were mainly carried out in SE Europe, Greece, Spain, Turkey and Tunesia during the period 1965 to 1971 (Figure 1). On the one hand, they were intended to continue the research on the expanse of loess in the Mediterranean area (e.g. Range 1925, Rathjens 1928, Schwegler 1944, and summarized by Scheidig 1934), and on the other hand to contribute climate history during the young Pleistocene in this area. The results were particularly surprising as the theory of the last ice age being a pluvial period in what are now arid regions (refugium for the pretentious Central European flora and fauna), became invalid (summarized by Brunnacker 1974). The available data about the vegetational history fit very well into this new picture (summarized by Frenzel 1967). This is further confirmed by more recent discoveries in the coastal plain of Israel (Brunnacker, Schütt & Brunnacker 1979) and on Crete.

For better understanding of the Mediterranean loess, its northern extension, the periglacial loess, also is mentioned in this paper as far as necessary.

2. THE EXTENT OF LOESS

During the last ice age in Central Europe, the northern loess boundary lay north of the Central mountains. The zone on periglacial congelifraction follows further north. The northern loess boundary is a consequence of a too-cold climate, because the glacial congelifraction-zone lacks suitable vegetation to allow the final dust deposit. The southern loess boundary is expected to be in North Africa and in the Near East, where the climate was too dry for a suitable vegetation (today about 150(-200) mm precipitation per year). This loess zone can be subdivided into two regions: a northern one periglacial loess and a southern one with Mediterranean loess. The periglacial loess is found together with appearances of solifluction and partly shows ice wedges and cryoturbations. Furthermore, it contains the typical periglacial mollusc

99

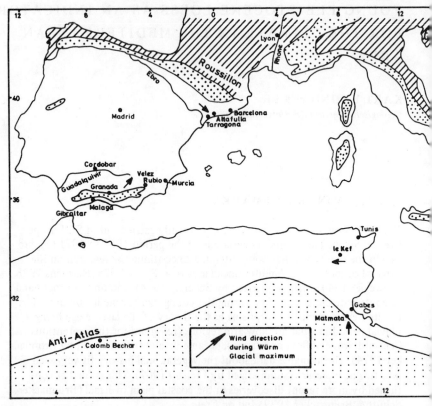

Figure 1. Regional review of the Mediterranean area based on present classification of the climate according to Walter & Lieth (1960). Furthermore, the wind direction during the glacial maximum, as it can be deducted from the eolian sediments, has been included.

fauna. 'Typical' loess is characteristic for this part of the zone. Dust loam, free of carbonate, is found in glacial as well as in today's humid areas. Deluvial loess, marked by reworked deposits, is increasingly found toward the maritime NW Europe, mainly in certain fine stratigraphic layers (Brunnacker & Hahn 1978). The southern boundary extends through the lowlands across the southern Balkans, northern Italy and SE Spain. The Mediterranean loess, to the south, is marked by a characteristic highly tolerant Mediterranean mollusc fauna. In more humid regions, it still shows the properties of typical loess. Toward more arid regions (the number of arid months and not the absolute amount of precipitation is decisive for comparison), it is more and more replaced by deluvial loess with fluviatile reworked sediments (Brunnacker 1974).

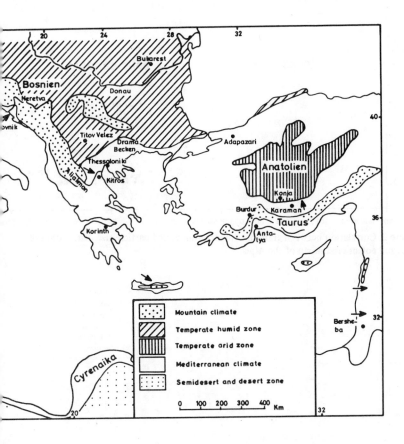

For various reasons, the Mediterranean loess is more patchily distributed than the periglacial loess. The reason for this lies in the limitation of the area supplying the loess particles; for example, massifs, broader valleys and coastal zones. Increasingly the loess was reworked, mainly from arid regions and was removed by fluviatile transport over large distances from the original place of sedimentation (high flood deposits). For this reason, the identification of the loess material becomes considerably more difficult. In addition, there have been redepositions due to soil erosion caused by man. This again adds to the difficulties in recognizing loess material from a geological mapping especially as loess in this region is too often rated as a subordinate geological

101

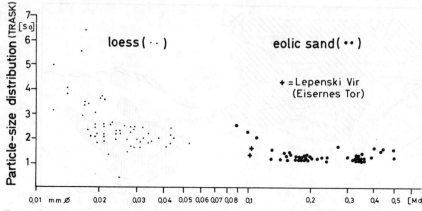

Figure 2. Correlation between grain-size (median-value) and particle-size distribution of loess and eolian sand north of the Alpes.

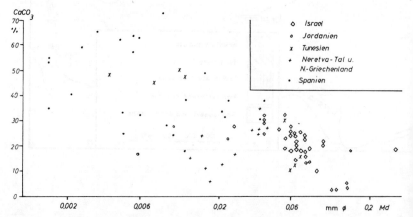

Figure 3. Correlation between grain-size (median-value) and carbonate content of the Mediterranean loess.

element. The glacial refugium of the less tolerant Central Europe flora and fauna was restricted to areas in SE Europe, favoured by better climate, e.g. in the Neretva Valley, south of Mostar (Brunnacker et al. 1969).

3. LITHOLOGY OF LOESSES

The grain size and the carbonate content are generally significant for the loess, in particular their combination (Table 1). 'Typical' loess shows a uniform grain size with median values of about 30 μm (Figure 2). Its carbonate con-

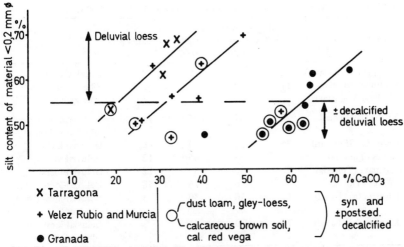

Figure 4. Correlation between silt and carbonate content of the deluvial loess in Southeast Spain with regard to the syn and postsedimentary decalcification.

tent lies between 5 and 40 per cent according to the source of origin. As a rule, the loess is of a finer grain the higher the carbonate content is (Figure 3). Besides that, a strong local influence of heavy minerals can be detected (Tillmanns & Windheuser 1979).

But only in the primary dust from which the loess is finally derived, is there a direct correlation between the carbonate content and the source of origin, because a larger or smaller portion of the carbonate is dissolved during its genesis. Occasional washouts have occurred (brown loess, dust loam). In part, the dissociated carbonate was precipitated in a relatively fine fraction without larger vertical transport. Therefore, loess that has been partly decarbonized during and after sedimentation, shows a higher clay content with a simultaneous reduction of the silt content (Figure 4).

The carbonate of the loess, which has developed through the carbonate reaction during the process of as well as later final disposition, generally shows a new distribution of stable C^{12}/C^{13} and O^{16}/O^{18} values (Figure 5) (Manze et al. 1974). Among the Mediterranean loesses the results are similar, even though partly not quite as distinct, e.g. also in the extreme form of deluvial facies: the deluvial loess shows an extraordinarily large variance of the median value, due to fluviatile influence, which can become dominant in suitable morphological locations (Figure 4). Furthermore, there is a more poorly sorted particle-size distribution (it has to be taken into consideration, that the standard method for determining grain size of typical loess, i.e. without prior decalcification, yields more or less false results).

The Mediterranean loess shows a clear correlation between the carbonate

Table 1. Examples of grain size and carbonate content of loess sediments

Locality	Grain size mm ϕ (%)			$CaCO_3(\%)$	
	<0,002	0,002-0,06	0,06-0,02		
Kitzingen (Germany)	22	69	9	17	Periglacial-loess
Hodbina (Neretva-Valley, Yugoslavia)	11	79	10	35	S-boundary of periglacial-loess
Kitros (N-Greece)	28	67	5	11	Mediterranean-loess, relatively humid
Granada (SE-Spain)	36	61	3	66	Deluvial-loess
Gabes (Tunesia)	20	34	46	12	Deluvial-loess
Matmata (Tunesia)	18	30	52	12	Deluvial-loess
Le Kef (Tunesia)	38	36	26	48	Deluvial-loess
Kisufim (a) (Israel)	11	50	39	4	Deluvial-loess (shelf-derived)
Kisufim (b) (Israel)	29	30	41	8	Deluvial-loess (shelf-derived)

content and the source of origin (Figure 3 and Table 1). In the vicinity of the Sierra Nevada, the carbonate character is as a result of the carbonate reaction during sedimentation, which is documented by the relation between grain size and the carbonate content. Accordingly, the stable isotopes of the carbonate do not show the typical distribution of a marine carbonate, which surely was the main source for the primary carbonate in the original loess dust. Instead, we find a completely different isotope distribution (Figure 5). It should be noted that the deviation of O^{18} in the Mediterranean loess is not so negative as compared to a standard such as the periglacial loess (Manze et al. 1976). The C^{13} deviation in particular shows a complete change during the loess immature soil genesis.

4. INTERSTADIAL SOIL WITHIN THE LOESS

Intercalated in the loess are soils, which partially have retained a carbonate content. They can be correlated with the interstadial of the Upper Würm. In the periglacial loess, they carry names like 'Tundra wet soil' and brown horizon (Brunnacker 1974). Pararendzina also occurs in a weak form. In the Mediterranean loess, the names calcareous brown soil, calcareous pseudogley and relatively well-developed 'Pararendzina' have been applied (Brunnacker 1974). In so far as these soils have a carbonate content, the deviation in the carbonate isotopes is clearer than in the loess (Manze & Brunnacker 1977).

The negative C^{13} deviation clearly increases in the interstadial soils of the Periglacial loess (Figure 5). The content reaches up to 60 per cent. According to the high clay and silt content, the carbonate shows signs of transformation, which was confirmed through isotopic measurements (Manze et al.

1974). In Northern Greece, loesses with a high clay content (5 to 12 per cent) are present. In Middle Tunesia (Matmata) the carbonate content lies between 12 and 17 per cent in relatively fine sandy loess. The relative fine sandy and at the same time clayey loesses in the surroundings of Bersheba occupy a special position with their carbonate content of 4 to 10 per cent.

In both the deluvial loess and in the periglacial loess, no transitions to aeolian sands occurs. Only loess deposits containing sandy layers in the vicinity of its origin indicate a transition zone. An exception is the late glacial periglacial loess from the 'Eisernen Tor' (Lepenski Vir) (Figure 2). The place of origin, the river bed of the Danube, was only a mere 100 m away. According to the grain size, it is an aeolian sand; from its structure, carbonate content and mollusc fauna, it has to be regarded as loess.

5. LOESS AS IMMATURE SOIL

Paleopedologically, loess is seen as an immature soil (Brunnacker, Urban & Schnitzer 1977), as there are signs of life during its genesis (Rhizosolenien as a cover of roots in periglacial loess, opal phytolith as skeletal elements of gramineen, mollusc shells and traces of organic derivates).

An additional important criterium for its immature soil character lies in the dampening of the soil by thawing during the warm seasons. During the winter months, the exchange of air in the soil with the atmosphere was prevented due to frost and crusted snow. But in the deeper region of the thawed soil, the biogenic CO_2 reaction could continue. The enrichment of biogenic CO_2 might be the reason for the larger negative deviation of the C^{13} isotope from the immature soil loess.

In the Mediterranean loess, on the other hand, there is a quite different distribution of the stable isotopes (Figure 5), i.e. a lower negative O^{18} deviation from the standard. Therefore, the difference found between the Mediterranean and periglacial loess is kept. This difference, considering the Mediterranean loess only, cannot be understood easily as many factors can influence the behaviour of O^{18}. In view of the results from the poorly developed interstadial soils in the Mediterranean loess, the following interpretation is obvious (Manze & Brunnacker 1977).

Carbonate precipitations from remaining solutions alter the O^{18} content as described above. This would mean for the Mediterranean loess and especially for the intercalated poorly developed soils that relatively moist winters must have followed very aried summers. The Mediterranean molluscan fauna showed by its tolerance the same trend as does the increasing deluvial character of the soil, at least in the arid region (Brunnacker 1974). Therefore, a Mediterranean climate must have been present during the genesis of these Mediterranean loess having the soils intercalated in the circum-Mediterranean area. The arid season was especially well-developed. Furthermore, the yearly average temperature was relatively low. Therefore, the annual precipitation must have been largely reduced.

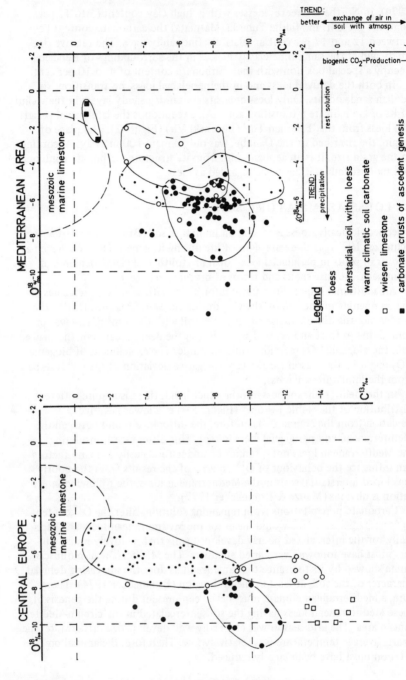

Figure 5. Distribution of the stable isotopes in the carbonate content of loess, in interstadial soils from loess and warm climatic (interglacial and postglacial) soil carbonates.

6. WARM CLIMATIC SOILS FORMED FROM LOESS

The interglacial soil relicts together with the periglacial loess are developed as parabrown soil with transitions to Pseudogley. Additionally, we observe the tirsoide soils, in SE Europe called black soil, as they are developed for example in the humus zone of the Early Würm.

Accordingly, the tirsoide soils are to be separated into those which are related chronologically, genetically and climatically with the interglacial warm periods in Central Europe, and those in the warmer Mediterranean area which also developed in suitable localities during the last glacial period. The reasoning behind this differentiation within the tirsoide soils is the relatively warmer climate in the south, even during the Upper Würm, in respect to the region north of the Alps.

Geomorphologic, geologic and recently discovered archeologic aspects show that brown loamy soils were present in the Mediterranean region during the last interglacial, although their exact stratigraphic position often is not very easy to locate. They can also be found in the early Würm (as a counterpart to the humus zones north of the Alps) as an equivalent to the Ammersfoort, Broerup and probably also to the Odderade interstadial (Brunnacker 1979). It is characteristic for these soils to be very closely linked with the records of the last interglacial, due to a slow rate of sedimentation between the pedogenic time marks. It should be noted that they also are characterized by a layered profile (Schichtprofil), in which a thin layer of sediment that has undergone a pedogenesis lies upon a foreign base. This is especially the case for vega (meadow soil) type soils. Furthermore, these types of brown loam soils and vega occur in the whole circum-Mediterranean area.

In the coastal plain of Israel, these Paleosoils are documented partly as a soil complex, in which fluvial sediments with red and brown autochthon vega, a type of brown loam, and tirsoide soils lie one on top of the other (middle-paleolithic soil and soil complex). This form is especially clear during the late glacial in the whole Mediterranean region. One can conclude, that in comparison to the periglacial area, the Mediterranean climate of the Middle Würm was again slightly warmer, with well-developed humid phases during the winter months. Similar conditions naturally were also valid for the Early Würm. Also the molluscan fauna within the last glacial soils again show signs of milder climate (Brunnacker et al. 1978). In the periglacial loess, on the other hand, the transition is slow. The typical loess fauna is replaced by one preferring slightly more humid and warmer climate.

Whether similar soils have developed in the Mediterranean area during the Holocene is the object of our present investigations. The answer to this question is quite difficult, as human activities tend to change the true picture, and exact datings of soils become very difficult.

At the base of the warm climatic parabrown soil directly on top of periglacial loess, we find, especially in arid regions, Cc-horizons as carbonate

107

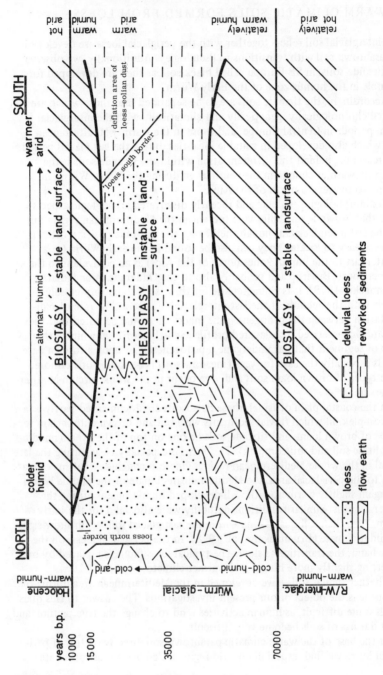

Figure 6. North-south differentiation from Central Europe to North Africa during the last glacial, regarding loess and its solifluctional and fluviatile redepositions.

108

secretion in the form of loess concretions.

Intercalated with the corresponding soils of the Mediterranean area and very often as a relict at the base of an eroded soil, carbonaceous crusts occur in the arid districts of the Mediterranean region. Additionally gypsum can also be found in even dryer areas, depending on rock materials and water supply. As they are pedologically unstable, their interpretation is nearly impossible.

The Mediterranean carbonaceous crusts belong mainly to the Cc-horizon type when using the O and C isotopes for interpretation (Manze & Brunnacker 1977). Furthermore, there are crusts which developed through ascendent water (Figure 5). Significant isotope values are also present for this case. From field work we can separate two different types of crusts.

7. STRATIGRAPHY

In the absolute timetable the boundary of the Riss Würm interglacial (Eemium s.l.) with the cold periods (Weichselium s.l.) lies at 70 000 to 75 000 years B.P. But older dates are also used for this boundary. According to widely accepted opinion, the boundary from the Weichselium to the Holocene lies at 10 000 years B.P. North of the Alps, the change in climate occurred at this point. This is documented by intensive periglacial redeposition and through loess genesis. Finally, we naturally find a change in the flora and fauna, and the start of a warm climatic pedogenesis. In the Mediterranean area, we do not observe such a conformity regarding the soil-stratigraphy (Figure 6), since the highly developed warm climatic soils (brown loamy soils and vega) also occur in the Early Würm which underlies the Middlepaleolithicum. The intensive pedogenesis also begins earlier than north of the Alps.

The warm period before the last glacial, as well as the post-glacial warm period (without the effect of mankind) were at a time of biostasy, as defined by Erhard (1965), that is a period during which the land surface was covered with vegetation. According to Erhard the last glacial was a period of rhexistasy, i.e. a period with different types of redepositions (unstable land surface). But the boundary between biostasy and rhexistasy is not a stratigraphic but one of facies. The reason for this is the generally warmer climate towards the equator. Chonostratigraphy and pedostratigraphy are therefore not identical towards the south.

8. RELATIONSHIP BETWEEN THE GLACIAL EFFECTS AS OBSERVED ON LAND AND IN THE SEA

The build-up of great masses of inland ice caused marine regressions. For the last glacial, the time of the largest ice expansion is placed about 20 000 years B.P. This is in agreement with the findings in the periglacial area. Additionally, another climax in the Early Würm has been suggested. But this first climax (or several ones) cannot be clearly proven in glacial areas. Also in the periglacial area, little evidence for that can be found. On the other hand, the evidence is concentrated in the Mediterranean area. The young beach terraces give evidence for this. Besides this, the paleolithic stations in the coastal plain of Israel give further indications. But here the following problem is encountered. In the southern part of the coastal plain of Israel, approximately at the height of Gaza-Bersheba, lie several aeolianite ridges (kurkar), which are comprised of fine sand. The sand was deflated from the shelf area during times of marine regression. On the other hand, more inland, in the region of Bersheba, loess type deposits occur, which are interpreted as desert loess (Yaalon 1969, Bruins 1976). The interfingering of both is present at Kissufim. Therefore, the winds that transported the material must have partly originated from the west and partly from the east. According to our investigation (not yet completed), it is possible that the aeolian sand and the not-too-well graded loess (with high clay and fine sand content, but relatively little silt, Table 1), were derived from the same dried-up marine area.

This is confirmed by the uniformity of the heavy mineral association (investigated by Tillmanns). According to the profile near Kissufim, this would mean that the sea level dropped quickly several times during the last glacial. Especially during the early sea-level reductions, fine sand together with silt and clay lumps on the dry shelf area were available for aeolitic redeposition. But it must be stressed that it is not yet proven that this type of fine sediment had its origin in the marine area. Only then can it be said if it is possible to subdivide the last glacial according to the different types of sediments from the shelf area.

9. THE CLIMATE DEVELOPMENT IN TIME AND SPACE DURING THE LATE PLEISTOCENE IN THE CIRCUM-MEDITERRANEAN AREA

Contrary to prior belief that the temperate climate zone was shifted towards the equator, our investigations (Brunnacker 1974) show that the northern boundary of the Mediterranean Late Glacial climate lay more or less in the same position as today. Therefore, the flora and fauna of the last glacial is of a Mediterranean type. Adjoining to the north, was the periglacial area with a typical flora and fauna. The warm-climate flora and fauna found refuge in local areas with better climate. The boundary zone lay mostly in SE Europe. To a certain degree, from this position, organisms could advance further to the south particularly in regions with an ascending land surface. Towards higher attitudes, the Mediterranean environment also was confined by periglacial conditions. East of Granada for example one can find at a height of 800 m a periglacial loess mollusc fauna, but which already contains some tolerant Mediterranean species.

The facies classification of the Mediterranean loess and the interbedded poorly developed soil horizons yield a regional subdivision into humid and arid regions, which are characterized by the present number of dry months as defined by Walter. On the other hand, it is hard to correlate the amount of recent precipitation with this subdivision.

The objections to these results, as raised by Sabelberg (1978), miss the mark in so far as they ignore, among other things, the experience of many years underlying our investigations. Furthermore, they do not take into consideration the limitations arising from the dominance of fluviatile sediment occurrences in the still arid regions. Therefore, it is surprising that Sabelberg (1978) came practically to the same result in arid regions of West Morocco.

At least along the northern boundary of the Mediterranean area, the temperature (according to mollusc fauna found near Mostar and Tarragona) during the glacial maximum was similar to that of the periglacial area (i.e. $10°$ to $15°C$ lower than today). This is opposed to the concept that e.g. in SE Spain, the temperature was only $4°C$ lower than today.

A somewhat slighter temperature depression can be assumed for the beginning and end of the last glacial. But one must bear in mind that the intensive soils that can be found from this period are primarily a result of higher winter humidity and summer aridity. The precipitation during the winter was in a form of a Mediterranean shower during the whole glacial. By this, the deluvial character was mainly caused in areas with only slight vegetation. Regarding pedogenesis, it must be added that threshold values rarely became effective here. This takes into consideration that the changes between very intensive soil and less intensive are barely perceivable. Also the time for its development was not necessarily so long as for example the epipaleolithic soil complex (several thousand years).

The glacial maximum was characterized, according to the findings, as a savanna to prairie at least in the lowlands. Instead of the currently generally accepted postulate of a pluvial climate during the last glacial in the circum-Mediterranean area, a very arid climate with an open vegetation was also effective here during the high glacial just as it was in the adjacent northern and southern regions (Figure 6).

The concept of a pluvial climate can therefore only be applied to the beginning and end of the last glacial. A further subdivision of the climate of the last glacial (sensu lato) may be possible as probably several marine ingressions took place.

Indications of the wind direction for the transportation of dust are scarce. The final place of deposition, which is marked by relief and base, suitable for loess sedimentation (clay and sand), is more important than the place of origin. Of course, some particularly suitable sites for the origin of the loess (i.e. broad valleys) must have been present. Furthermore, a transport distance of only 10 to 20 km is quite normal. The Mediterranean loess, mostly of a deluvial character, could have been transported somewhat further. The broad

111

zones with little vegetation such as the congelifraction zone in the north and in the mountainous regions as well as the more arid Sahara might have formed additional sources of loess.

Therefore, it is only possible to give few data on the wind direction during the climax of the last glacial (Figure 1). Southerly winds brought the loess dust from the Sahara and deposited it in Middle Tunisia, in the vicinity of Matmata, mountainous regions were in general the source for such dust. But due to poor mapping, the extent of these loess deposits is not known, which is an inescapable condition for further investigations. The Ebro valley must have played a role in the loess deposits near Tarragona; the Aliakmon valley for the loess near Kitros. This indicates winds from a northerly direction. In the Neretva valley, below Mostar, the loess generally lies only a few hundred meters on both sides of the valley floor. From this, it is inferred that west wind was dominant, whereby unevenness in the relief caused very local deposition of the dust. In the coastal plain of Israel, west wind must be responsible for the aeolinites. If the loess in the southern region, which lies inland, should have come from the shelf area, then the mentioned variations of the wind directions from west to east is superfluous. Also the young aeolinites from the north coast of Crete, which can be compared with the ones from Israel, were mainly derived from north or northwest.

REFERENCES

Brunnacker, K. 1957. Die Geschichte der Böden im jüngeren Pleistozän in Bayern. *Geol. Bavarica, München, 34.*

Brunnacker, K. 1960. Zur Kenntnis des Spät-und Postglazials in Bayern. *Geol. Bavarica 43:*74-150.

Brunnacker, K. 1971. Geologisch-pedologische Untersuchungen in Lepenski Vir am Eisernen Tor. *Fundamenta, Köln/Wien, A3:*20-32.

Brunnacker, K. 1973. Einiges über Lös-Vorkommen in Tunesien. *Eiszeitalter u. Gegenwart, Ohringen, 23/24:*89-99.

Brunnacker, K. 1974. Lösse und Paläböden der letzten Kaltzeit im mediterranen Raum. *Eiszeitalter u. Gegenwart, Ohringen, 25:*62-95.

Brunnacker, K. 1977. Beiträge zum 'epipaläolithischen Boden' bei Tel Aviv (in press).

Brunnacker, K. 1979. Zur Bodengeschichte des Jungquartärs im mediterranen Raum. *Z. Geomorphol. Suppl., Stuttgart* (in press).

Brunnacker, K., H.-J.Altemüller & H.J.Beug 1969. Das Profil von Kitros in Nord-Griechenland als Typus-Profil einer mediterranen Lösprovinz. *Eiszeitalter u. Gegenwart, Ohringen, 20:*90-110.

Brunnacker, K., Dj.Basler, V.Lozek, H.-J.Beug & H.-J.Altemüller 1969. Zur Kenntnise der Lösse im Neretva-Tal. *N.Jb. Geol. Paläont. Abh., Stuttgart, 132:*127-154.

Brunnacker, K. & H.Hahn 1978. Der jungpleistozäne Lös samt seinen paläolithischen Kulturen in den Rheinlanden als Glied einer zeitlichen und räumlichen Faziesänderung. *Beiträge zur Quartär- und Landschaftsforschung, Wien:*37-51.

Brunnacker, K. & V.Lozek 1969. Lös-Vorkommen in Südostspanien. *Z. Geomorph. Berlin/ Stuttgart, N.F.13:*297-316.

Brunnacker, K. & A.Ronen 1977. *Zur Geologie und Paläopedologie der Paläolith-Station Evron/Israel* (in press).

Brunnacker, K., H.Schütt & M.Brunnacker 1979. *Uber das Hoch- und Spätglazial in der Küstenebene von Israel* (in press).

Brunnacker, K., B.Urban & W.Schnitzer 1977. Der jungpleistozäne Lös am Mittel- und Niederrhein anhand neuer Untersuchungsmethoden. *N. Jb. Geol. Paläont. Abh., Stuttgart, 155:*253-275.

Bruins, H.J. 1976. The Origin, Nature and Stratigraphy of Paleosols in the Loessical Deposits of the NW Negev (Netivot, Israel). Thesis, Hebrew Univ. of Jerusalem.

Erhart, H. 1965. Le témoignage paléoclimatique de quelque formations paléopédiques dans leur rapport avec la sédimentologie. *Geol. Rundschau, Stuttgart 54:*15-23.

Frenzel, B. 1967. *Die Klimaschwankungen des Eiszeitalters.* F. Vieweg, Braunschweig.

Manze, U. & K.Brunnacker 1977. Uber das Verhalten der Sauerstoff- und Kohlenstoff-Isotope in Kalkkrusten und Kalktuffen des mediterranen Raumes und der Sahara. *Z. Geomorph, Berlin/Stuttgart, N.F.21:*343-353.

Manze, U. & K.Brunnacker 1977. Der kalkumsatz in würmeiszeitlichen Lös-Interstadialböden am Rhein auf grund der Kohlenstoff- und sauerstoff-Isotopenverhältnisse. *Kölner Jb. Vor. u. Frühgeschichte 15.*

Manze, U., I.C.Vogel, R.Streit & K.Brunnacker 1974. Isotopenuntersuchungen zum Kalkumsatz im Lös. *Geol. Rundschau, Stuttgart, 63:*885-897.

Range, P. 1925. Das Diluvium Palästinas. *Die Eiszeit, Leipzig, 2:*116-118.

Rathjens, C. 1928. Lös in Tripolotanien. *Z. Ges. Erdk., Berlin, 1928:*211-228.

Sabelberg, U. 1978. Jungquartäre Relief- und Bodenentwicklung im Küstenbereich Südwestmarokkos. *Landschaftsgenese u. Landschaftsökologie, Cremlingen-Destedt, H.1:* 171.

Scheidig, A. 1934. *Der Lös und seine geotechnischen Eigenschaften.* Dresden/Leipzig.

Schwegler, E. 1944. Bemerkungen zum Vorkommen von Lös im lybischen und tunesischen Gebiet. *N. Jb. Min. Mh., Stuttgart, Abt. B.:*10-17.

Tillmanns, W. & H.Windheuser 1979. Der quartäre OsteifelVulkanismus im Rahmen der Lösbildung. Ein Beitrag zur Lösgenese. *Eiszeitalter u. Gegenwart, Ohringen, 29.*

Walter, H. & H.Lieth 1960. *Klimadiagramm-Weltatlas.* Jena.

Yaalon, D.H. 1969. Origin of Desert Loess. *Etudes sur le Quaternaire dans le monde, Paris, 2:*755.

FLUVIAL PROCESSES IN THE SAHARA

HORST HAGEDORN
Geographisches Institut der Universität Würzburg

When looking at a satellite image of the Sahara, one can recognize drainage networks everywhere with the exception of the large ergs (dune areas), even though resolution is limited on those images. They show disturbances of these networks only in the regions of eolian relief, as can be seen better on aerial photographs of rather large scale. There the mostly dendritic fluvial pattern is superimposed by the striped arrangement of wind-eroded bedrock (Hagedorn 1971), often to the extent that major portions have become inactivated or that occasional runoff utilizes only certain sections of the network.

Where major structural elements, such as basin rims or mountain areas, are present, they largely determine the drainage pattern. In the open tableland regions, the pattern is often rather chaotic, or, in other words, it shows a high degree of variegation. The pattern is highly dependent on the major types of relief of the Sahara, such as serir, hamada, erg and mountains. In addition, there are peculiarities that pertain to fluvial systems in arid regions.

1. *Endorheic drainage* dominates in many parts of the Sahara. This has an important geomorphic consequence, because it is the processes themselves at work in the river channels which change the height of the base level of erosion. These self-induced changes then in turn affect the work of the processes.

2. Another peculiarity is the type of runoff under arid conditions, which is characterized by high runoff peaks in response to short-term high precipitation intensity. Special properties of the topsoil such as the existence of a vesicular layer (Schaumboden) or of various types of crusts, or in many cases also bare rock surfaces are responsible for low infiltration rates, which further increase the runoff. Of course, there are variations with the type of land surface (Rognon, this volume).

There are indications that the highest runoff peaks may occur in arid regions. Runoff does not set in unless there is a short-term precipitation of 5 to 10 mm depending on the type of surface. Episodic severe rainstorms well above this threshold together with the high runoff rate are the reasons for the often-described flash-floods of major wadis (Klitzsch 1966).

115

During flash-floods in fully arid regions only a relatively small amount of bedload is transported. This can be explained by the low weathering rate on the slopes and by the shortness of a runoff event. Hence, fluvial relief formation is presently not very effective in the arid to hyperarid central Sahara. This has been corroborated by field-work (Hagedorn et al. 1978).

Taking together field observations and the general considerations regarding arid regions given above, it appears that there is some vertical erosion today in the upper courses of wadis originating in mountains and uplands, along escarpments and in the surroundings of major depressions. The eroding reaches are followed downstream by those where through-transport prevails. Deposition takes place in the lower reaches, the point of furthest deposition gradually moving upstream from the respective base-level into the higher relief. Deposition near the head may have come to a halt under present conditions. It will always do so whenever most of the sediment that has been deposited in the end-basins by fluvial activity is exported from the area by deflation. This is presently the case for the rivers Bardagué and Yebigué on the north slope of the Tibesti (Jäkel 1979).

If individual rivers are studied, it shows that fluvial terraces exist in major valleys everywhere in the Sahara. They have been studied and described as tokens of climatic change, but their interpretation with respect to climatic history meets with a number of problems. Before discussing these, a few general remarks have to be made:

In order to derive climatic information from river terraces, one must be sure that they were not formed due to the influence of other agents such as tectonics. This is a very trivial statement in theory, but in the field, it is often quite difficult to prove that a certain terrace has been formed in response to climatic causes alone. For example, the most conspicuous terrace in the Bargagué valley of the Tibesti Mountains has most likely come into existence because of an abundant supply of volcanic ashes choking the river.

This leads to the second remark: for the growth of sediment bodies and for their dissection into terraces, it is not only important what happens in the river-bed, but the processes of weathering and denudation on adjacent slopes, i.e. the nature and supply rate of debris entering the river, are of similar importance.

For the model of terrace development to be presented here, evidence is taken only from those terraces that can be followed over long distances of the major Saharan river-beds, especially of the Tibesti Mountains. Isolated relics of older and frequently higher terraces — up to 100 m above the present river-beds — that occur in many places have to be neglected in the following discussion. They indicate, however, that the history of erosion and deposition in a given river system has certainly been more complicated than that which is to be outlined here. They also show that the climatic history behind these events becomes increasingly difficult

116

to grasp with increasing age and status of preservation of the terrace remnants.

With these constraints in mind, four terrace bodies, which also show as distinctive terrace levels, can be taken into consideration. The valleys of both the Hoggar and the Tibesti Mountains generally show the same number of three to four morphologically significant terraces. The sequence from the oldest to the youngest is as such:

Upper terrace (*Oberterrasse*, OT), made up of mostly sandy to coarse gravelly sediments;

Middle terrace (*Mittelterrasse*, MT), made up of silt, clay and fine sand with occasional gravel lenses, in places passing into lake deposits;

Lower terrace (*Niederterrasse*, NT), somewhat similar to OT, but generally more coarse-grained;

Oasis terrace (*terrace de palmeraies*), made up of silts, clays and fine sands, often the substratum of oasis gardens. This terrace, which seems to be more important in the Hoggar (Rognon 1967) than in the Tibesti, is only found to exist intermittently.

Whereas this is the sequence that has been described from the Bardagué-Zoumri system, there are differences even in neighboring systems that do not seem to be due to less careful work in one area as compared to the other. For instance, Gabriel (1977) found five distinct terraces in Enneri (= river) Dirennao, a large tributary to the Bardagué, and Ergenzinger (1969) and Grunert (1975) could prove the existence of only three terraces (or rather: two, as the oasis terrace is largely absent) in the Enneris Misky and Yebigué respectively. OT and MT are there regarded as parts of one body that was called *Hauptterrasse*. The sequence on the Messak/Magueni escarpment NW of the Tibesti resembles that of the Yebigué, with modifications due to eolian activity at altitudes of only 600 to 700 m (Busche et al. 1979, Hagedorn et al. 1978). In the Hoggar Mountains, a seeming similarity of four terraces is complicated by − as it seems − a different age relationship (Rognon) 1967).

Regardless of the differences of number, the terraces fall into two groups, which may be called coarse-sediment and fine-sediment terraces. This basic difference of material has to be explained as well as the development of the bodies, in order to arrive at palaeo-climatic conclusions.

1. COARSE-SEDIMENT TERRACES

The Upper Terrace (OT)

This sandy to coarse-gravelly terrace laterally interfingers with blankets of subangular slope debris. The latter were a prerequisite for the development of the terrace body as suppliers of debris to the river. The debris that makes up these blankets − often preserved between slope runnels to a thickness of

117

several meters — show evidence of having been frost-weathered under the influence of repeated freeze-and-thaw cycles (cf. the descriptions by Pachur 1970 or Hövermann 1972), although such subangular debris, when found in deserts, is generally attributed to the specifically arid processes of insolation weathering or salt-wedging. Both of these processes, however, do not seem to have been significant.

The appearance of these debris blankets permits the conclusion that the climate must then have had moist and cool winters. Vegetation cannot have been of much importance on slopes of slowly moving debris. There are neither fossil remains nor signs of pronounced soil formation. As for the river-beds, precipitation must have been high enough and evaporation rates low enough to allow fluvial debris transport all along the rivers and into the mountain forelands. In the central-Saharan mountains the debris supply from the slopes — roughly simultaneous with the aggradation of the river-beds — occurred down to altitudes of 800 to 1 000 m. It should be noted here that different models have been proposed for the extension of the OT into the foreland by Jäkel (1971) and Ergenzinger (1969), in which the terrace is viewed as the result of upstream migration of the depositional area with increasing aridity. This does not account for the simultaneous growth of slope debris blankets, however.

The course of events in a climatic oscillation leading to the growth of a terrace like the OT may be envisaged as follows: With increasing humidity at the end of hyper-arid phase there first will be some incision and/or through-transport of older valley fill, until sufficient humidity and time have started debris production on the slopes. This will become most effective with further winter cooling and the formation of frost debris. The debris entering the river is moved downstream, the length of transport increasing with runoff intensity. At the height of the humid phase, debris supply will be somewhat slowed by vegetation. When the latter is reduced again with increasing dryness, this will be reflected in the terrace first by deposition of finer sediment from soils and chemical weathering, and then by coarse grain-sizes again. Further decrease of humidity going hand in hand with decreasing weathering on the slopes will then, as floods are still frequent and vigorous enough, result in the concentration of runoff in a narrow talweg and in erosion (cf. Mortensen 1930).

In the longitudinal profile, this dissection will grade downstream in a reach of through-transport and then of deposition. With further aridification, deposition will move upstream, as the floods terminate more and more upstream, until the turning point is reached at the height of the arid phase. Except for the opposite direction of depositional growth, another important difference to the growth of the OT-type body lies in the fact that the amount of sediment involved is much less. Thus headward deposition of the Bardagué under the present hyper-arid conditions could be detected by Jäkel (1971) only with the aid of precision leveling. At the point of maximum aridity, material transport in the foreland and lower mountains takes place only by deflation. In the Tibesti Mountains, the upper limit of deflation presently lies at 1 000 m altitude.

The body of the upper terrace thus records a winter-cool climate probably of a Mediterranean type with a rainy season of two to three months in winter and a precipitation rate of about 200 to 300 mm at altitudes below 1 000 m. This is ten times more than today. In the mountains above 2 000 m precipitation must have been as much as 600 to 1 000 m, at least some of it falling in winter as snow (cf. Messerli 1972).

The Lower Terrace (UT)

Another type of coarse-sediment terrace is presented by the Niederterrasse of the Tibesti rivers. For the accumulation of this terrace body increasing humidity does not seem to have coincided with increased debris production on the slopes, but a sufficient amount of debris was available from remnants of the OT time slope deposits that could be reworked and transported by increasingly frequent rainstorms. In other respects the coarse of events will have been similar to that of the OT. Development of this terrace took place over a much shorter period of time, however. Evidence places it in the Neolithic (Gabriel 1977).

2. FINE-SEDIMENT TERRACES

The Middle Terrace

The body of this terrace contains sediment such as fine rock debris, fine sand, silt and clay which are the products of predominantly chemical weathering. In the Messak/Mangueni region it could be shown that at least some of the fine sediment derives from the erosion of thick soil profiles. As soil profiles take some time to develop, it has to be assumed that accumulation was preceded by a time of slope stability under a rather complete vegetation cover. Only when deterioration of the climate toward the arid side set in, decreasing surface protection by vegetation allowed sedimentation to start. In this model, the terrace would not stand for the height of a humid phase, but rather for the time of its decline, although humidity must still have been considerable by present-day standards.

It can be assumed that with an increase of summer rains and the onset of a periodic instead of an episodic rainy season (such as the *Tchadien* of the French authors, cf. Michel, 1973) first led to a short-term increase of erosion, which soon came to an end with the development of a savanna-type vegetation perhaps similar to that of the southern Aïr region today. For some time then, downwash of chemically weathered sediment from the slopes must have been in balance with the transport capacity of the rivers. Conditions gradually must have been shifted toward the depositional side as renewed increase of aridity led to progressive destruction of the vegetation cover and thus to increased slope-wash. Fossil remains in the terrace indicate that the rivers must have had the appearance of reed-swamp ribbons, locally widening into shallow lakes that in the case of Enneri Bardagué were ponded by calc-sinter barriers.

The Sudanic type vegetation prevailing at the maximum of the humid phase under consideration may have succumbed to occasional winter frosts, when, with a retreat of monsoonal summer rains, winterly intrusions of Mediterranean cold air became more frequent. Because of reduced vegetation cover the winter rains, although less intense than summer monsoon rains, will have been more efficient than the latter in slope denudation. There is evidence for such a Mediterranean-type climatic regime in the middle terrace of the Tibesti, as pollen of Mediterranean species and — perhaps more stringent proof because of the problems with pollen long-distance transport — shells of holarctic molluscs were found in its sediments (Jäkel & Schulz 1972, Böttcher et al. 1972).

Other evidence for the development of this terrace body under Mediterranean conditions, i.e. by those governed from the north, comes from the continuous existence of this terrace from Tibesti to Serir Canalchio across the presently hyperarid Serir Tibesti and the Libyan desert (Pachur 1974). The rivers would hardly have managed to flow so far north under a monsoonal regime of exclusively summer rains governed from the south.

With increasing aridity, the dissection of the accumulation body must have resembled that described above for the OT, with erosion upstream and headward deposition extending from the foreland. The MT thus represents a climate that — at least in the times of principal sediment preparation — must have been warm and relatively humid with a precipitation of about 300 to 400 mm in the lowlands and 800 to 1 200 mm in the higher mountains, with a rainy season three to four months long.

The oasis Terrace

This loamy terrace mostly consists of soils washed from adjacent slopes. It is so young — late to post-neolithic — that human interference such as soil erosion in the aftermath of savanna burning by nomadic herdsmen, has to be reckoned with.

Dating the Terraces

As said before, similar terrace sequences have been found in all parts of the Sahara but the ages assigned to them differ considerably, especially between Hoggar and Tibesti. When research started on the terraces of the Tibesti, the terraces were first held to represent most of the Pleistocene. With the first radiocarbon dates for the middle terrace, however, the morphologically most significant development became rather young. The middle terrace appears to have been deposited between 16 000 and 7 000 y. B.P. (Geyh & Jäkel 1974). In the northern Tibesti foreland, the base of the corresponding terrace is considerably younger, however, around 10 000 y. B.P. (unpublished data presented by Pachur at the Dt. Geographentag, June 7, 1979). Even younger is the base of the corresponding terrace in a tributary to Enneri Achelouma at the south rim of the Plateau du Mangueni, which was dated as about 8 000 y. B.P. (unpublished data). In all cases, however, the dates for the top of the terrace have yielded an age of ca. 7 000 y. B.P., indicating a synchronous development of the upper portion of the terrace bodies in the various regions.

The lower terrace is placed between 6 000 and 3 000 y. B.P. Data were obtained from the study of sediments and tamarisk hills from the end-basin of Enneri Bardagué by Jäkel (1979) and from prehistoric findings by Gabriel (1977).

Extrapolating from the rather safe C^{14} data, the upper terrace would then have to be placed at the time of the Würm maximum, but actually evidence is so feeble that the Riss glacial also cannot be excluded as the possible time of origin.

As said before, the terrace sequence in the Hoggar Mountains is similar in appearance, but the dates affixed to the individual terraces comprise all of the Pleistocene (cf. Rognon 1967 *inter alia*). Unfortunately, very few absolute age determinations have been made so far, so that the correlation between the two mountain areas is still open to discussion and to further research.

SUMMARY

One can learn from these discrepancies that terrace sequences of similar appearance cannot be paralleled over large distances without conclusive evidence.

The problems of age are perhaps less pressing, though, than those of terrace formation. Thus the model presented here with the help of evidence from the Tibesti and environs — downstream progressive accumulation beginning in an arid-to-humid transition phase and lasting beyond the peak of a humid phase — stands in striking contrast to that by Chavaillon (1964). According to him, accumulation takes place with increasing aridity and erosion with increasing humidity. As mentioned above, the same idea is held by Jäkel (1971) and others for the erosion/deposition mechanism in the Tibesti. However, as far as Chavaillon is concerned, this seeming contradiction can be resolved when one takes into account that he developed his cycle for Wadi Saoura in the northwestern Sahara, away from the effects of mountain areas. His model may thus be correct for the lower courses of large table-land wadis, but not for the mountains where, according to the model outlined above, river behavior is different.

Regardless of the differences of interpretation there remain a number of points that have to be kept in mind when working with desert river terraces which are summarized here:

1. Conclusions as to climate history can be drawn from terrace bodies provided their origin by climatic control is securely established.

2. For the deposition of terrace bodies, the transitional phases between two climatic regimes are the most important ones. These also seem to be the longest ones.

3. The sediment found in a terrace body may reflect climatic conditions predating its accumulation by a shorter or longer period of time, as has been shown for the MT and NT respectively. Additional evidence, from the surroundings of rivers, such as from slope debris blankets, from gravel pediments, or from glacis formation phases, as well as from fossils within the terraces — if there are any, also must be taken into account.

4. When discussing terrace sequences, one must always look at the whole length of a terrace system from the headwaters to the foreland, as conditions are likely to have varied along the longitudinal profile. Furthermore, the influence of older climatic phases has to be reckoned with, as in the case of the resumed destruction of older slope debris sheets for the building of the lower terrace.

5. Terrace sequences must not be paralleled over long distances without really conclusive evidence. This is especially true when sequences from the northern and the southern parts of the Sahara are to be compared.

6. The role of vegetation as a stabilizing or destabilizing factor on adjacent slopes has to be taken into account, but one must also keep in mind that different types of vegetation — in the present case Mediterranean or Sudanic — react differently to climatic changes.

Only with these precautions in mind somewhat reliable information concerning climatic history can be gathered from central Saharan river terraces. But it has to be seen that at the present moment our knowledge of the mechanisms which have actually governed the development of individual river terraces, and of their position in time are still too incomplete to write a climatic history of the central Sahara from them.

REFERENCES

As this paper was not intended to be a review of existing literature on Saharan river terraces, the bibliography only contains titles referred to in the text. Easy access to further literature, especially on the Tibesti, can be gained from the volumes of 'Berliner Geographische Abhandlungen'.

REFERENCES

Böttcher, U., P.Ergenzinger, S.Jaeckel & K.Kaiser 1972. Quartäre Seebildungen und ihre Molluskeninhalte im Tibestigebirge und seinen Rahmenbereichen der zentralen Ostsahara. *Z. Geomorph. N.F.16:*182-234.

Busche, D., J.Grunert & H.Hagedorn 1979. Der westliche Schichtstufenrand des Murzukbeckens (Zentral-Sahara) als Beispiel für das Gefügemuster des ariden Formenschatzes. *Festschr. dt. Geographentag Göttingen 1979*, pp.43-65.

Chavaillon, J. 1964. *Les formations quaternaires du Sahara Nord-Occidental.* CNRS, Sér. Géologie.

Ergenzinger, P. 1969. Rumpfflächen, Terrassen und See-ablagerungen im Süden des Tibestigebirges. *Tagungsber. u. wiss. Abh. dt. Geographentag Bad Godesberg 1967*, pp.412-427.

Gabriel, B. 1977. Zum ökologischen Wandel im Neolithikum der östlichen Zentralsahara. *Berliner Geogr. Abh. 27.*

Geyh, M.A. & D.Jäkel 1974. Spätpleistozäne und holozäne Klimageschichte der Sahara aufgrund zugänglicher 14-C-Daten. *Z. Geomorph. N.F.18:*82-98.

Grunert, J. 1975. Beiträge zum Problem der Talbildung in ariden Gebieten am Beispiel des zentralen Tibesti-Gebirges (Rép. du Tchad). *Berliner Geogr. Abh. 22.*

Hagedorn, H. 1971. Untersuchungen an Relieftypen arider Räume an Beispielen aus dem Tibestigebirge und seiner Umgebung. *Z. Geomorph. N.F. Suppl. Bed. 11.*

Hagedorn, H., D.Busche, J.Grunert, K.Schäfer, E.Schulz & A.Skowronek 1978. Bericht über geowissenschaftliche Untersuchungen am Westrand des Murzuk-Beckens (zentrale Sahara). *Z. Geomorph. N.F. Suppl. Bed. 30:*20-38.

Hagedorn, H. & D.Jäkel 1969. Bemerkungen zur quartären Entwicklung des Reliefs im Tibestigebirge (Rép. du Tchad). *Bull. Ass. Sénégal. Quatern. Ouest. Afr., Dakar, 23:* 25-41.

Hövermann, J. 1972. Die periglaziale Region des Tibesti und ihr Verhältnis zu angrenzenden Formungsregionen. *Gött. Geogr. Abh. 60* (Festschr. H.Poser):261-284.

Jäkel, D. 1971. Erosion und Akkumulation im Enneri Bardagué-Arayé des Tibestigebirges (zentrale Sahara) während des Pleistozäns und Holozäns. *Berliner Geogr. Abh. 10.*

Jäkel, D. 1979. Run-off and fluvial formation processes in the Tibesti Mountains as indicators of climatic history in the central Sahara during the late Pleistocene and Holocene. *Palaeoecology of Africa 11:*13-44.

Jäkel, D. & E.Schulz 1972. Spezielle Untersuchungen an der Mittelterrasse im Enneri Tabi (Tibestigebirge). *Z. Geomorph. N.F. Suppl. Bed. 15:*129-143.

Klitzsch, E. 1966. Bericht über starke Niederschläge in der Zentralsahara (Herbst 1963). *Z. Geomorph. N.F.10:*161-168.

Messerli, B. 1972. Formen und Formungsprozesse in der Hochgebirgsregion des Tibesti. *Hochgebirgsforschung 2:*23-81.

Michel, P. 1973. *Les bassins des fleuves Sénégal et Gambie. Etude géomorphologique. Mém. ORSTOM,* Bondy, France.

Mortensen, H. 1930. Scheinbare Wiederbelebung der Erosion. *Peterm. Geogr. Mitt. 76:*15-16.

Pachur, H.-J. 1970. Zur Hangformung im Tibestigebirge (Rép. du Tchad). *Die Erde 101:* 41-54.

Pachur, H.-J. 1974. Geomorphologische Untersuchungen im Raum der Serir Tibesti. *Berliner Geogr. Abh. 17.*

Rognon, P. 1967. *Le massif de l'Atakor et ses bordures (Sahara Central). Etude géomorphologique.* CNRS, Paris. Sér. géologie.

PRESENT AND PAST GEOMORPHIC EVIDENCES IN THE DEVELOPMENT OF A BADLANDS LANDSCAPE: ZIN VALLEY, NORTHERN NEGEV, ISRAEL

A. YAIR / H. LAVEE
Department of Geography, Hebrew University, Jerusalem

P. GOLDBERG   R. B. BRYAN
Institute of Archaeology, Hebrew University   *Scarborough College, University of Toronto*

1. INTRODUCTION

Studies on palaeoclimatic changes are usually based on the analysis of sedimentological, prehistoric, botanical, isotopic and other evidence derived from deposits laid down in the geological past. On the basis of data obtained, a given period is characterized as 'wet' or 'dry', referring thus to the general climatic conditions prevailing in the study area. It is tacitly assumed that the factors controlling sedimentological processes, such as spatial distribution of rainfall, runoff and erosion processes are quite uniform, at least over small watersheds of uniform lithology. But recent studies conducted in humid, as well as in arid regions show clearly that such an assumption is far from real.

Sharon (1970) in a study conducted in the extremely arid part of Israel observed that, at almost every storm, the spatial distribution of rainfall amounts in a small watershed extending over 0.7 km^2 may vary up to 100 per cent. Similar trends were obtained by Shanan (1975), Yair et al. (1978) and Sharon (in press), in other locations in Israel and by Cappus (1958) in France. Such differences in the spatial distribution of rainfall must affect runoff erosion and deposition processes which are controlled in addition by the spatial variability in surface properties.

Recent detailed field studies dealing with the problem of rainfall-runoff relationship (Betson 1964, Dunne & Black 1970, Dunne et al. 1975, Freeze, 1974; Yair & Lavee 1974, Yair et al. 1978, Bryan et al. 1978, Yair et al. in press) led to the concept of 'Partial area contribution to storm runoff'. The concept stresses the great spatial variability observed in runoff processes in very small watersheds carved in quite uniform lithological units. All studies quoted above show that the spatial variability in runoff processes is by far more pronounced than that of rainfall, over the same areas, being mainly controlled by the spatial variability in the surface properties. Under extreme conditions, the spatial distribution in runoff yield may be negatively correlated with that of rainfall amounts (Yair & Danin, manuscript).

Under such conditions, the interpretation of past events may depend greatly on our understanding of present day processes, especially the range

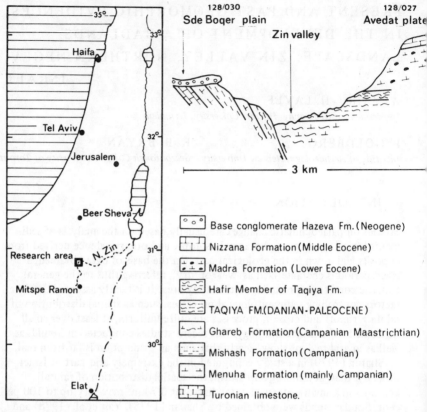

Figure 1. Location map.　　　　Figure 2. Geological cross-section through the Zin valley

of spatial variability in the frequency and magnitude of these processes. On the other hand, the type and magnitude of past geomorphic changes may provide a yardstick against which we can evaluate the scale of present day processes.

The present paper has two main objectives:

1. To present a case study of present day processes and show how it may contribute to the problem of climatic changes.

2. To compare the conclusions reached by the study to those resulting from prehistoric evidence in the study area.

2. THE STUDY AREA

The research site is located in a small tributary of the Zin valley, one of the major arteries draining the northern Negev highlands, eastwards to the Dead

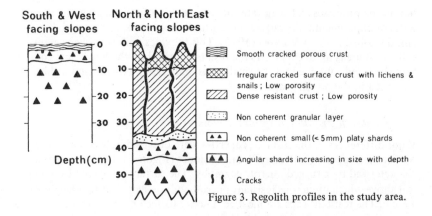

Figure 3. Regolith profiles in the study area.

Sea (Figure 1). The Zin valley is deeply incised in various geological formations of Campanian to Neogene age (Figure 2). A typical badlands topography
developed on the montmorillonitic rich shales of the Taqiya Fm., where the
present study was conducted. Average annual rainfall is about 90 mm with
extreme values of 34 and 167 mm. Only few rainstorms yield more than
20 mm/day. Characteristically, rainstorms are of short duration and consist
of several discrete showers separated by sunny and windy periods when the
surface can drain and dry (Yair et al. 1978).

3. THE STUDY OF PRESENT DAY PROCESSES

3.1 Experimental design

The study was limited to slopes of quite uniform angle and shape developed
in uniform bedrock. The slopes differed only in aspect which strongly
influenced the moisture regime of the regolith, leading to systematic differences in the properties of the upper 2-30 cm. This in turn, led to marked
variation in the nature and rate of processes active.

In order to examine present day processes active on shales slopes of varying aspect, two series of simulated rainfall experiments were carried out (Yair
et al. in press). The first series was performed with a small rainfall simulator
(Morin et al. 1967) capable of reproducing the dropside spectrum and terminal velocity of natural rainfall of varied intensity, but which can sprinkle
only 1.5 m². This unit provided the opportunity to obtain some basic and
sound data on the following aspects: precipitation threshold for runoff generation; characteristic infiltration curves; runoff and erosion rates. Two plots
representing the two basic types of slopes in the study area were sprinkled
with this unit. In order to supplement data obtained in the first series of
experiments and especially to obtain some information on the spatial varia-

tion in the processes and their rates on a slope area of a given aspect, a second series of experiments was carried out on fairly large plots, using a sprinkler unit described by Yair & Lavee (1976). This provides uniform coverage, on still air, of areas up to 90 m². As the results obtained on the large plots confirm the trends obtained on the small plots, only latter experiments will be presented in the present paper.

3.2 *Detailed description of sprinkled plots*

Major differences in the surface properties were observed between north and NE facing slopes to south and SW facing slopes (Figure 3). Former slopes are characterized by a rugged microtopography. The surface is covered by a dense seal above which numerous resistant micro-hoodoos stand out to a height of 4-8 cm. The resistant caps are most commonly formed by crustose dark brown lichens but occasionally by snail shells or gypsum crystals. The regolith profile is thick consisting of an upper 20-30 cm deep and dense crust crossed by desiccation cracks underlain by a thin non-coherent granular layer and a coarse shard layer representing the mechanically disagregated bedrock. Numerous pipes, which typically form 10-20 cm below the surface, appear. On collapse these pipes produce a discontinuous irregular rill system incised to 20 cm below the surface.

The latter group of slopes has a smooth microtopography. The surface crust is very thin, less than 1 cm, and broken by an irregular pattern of polygonal desiccation cracks. A thin fine granular shard layer underlies the crust giving way with depth to large angular shards. The surface is scored by parallel rills incised up to 8 cm below the surface with 1-1.5 m spacing.

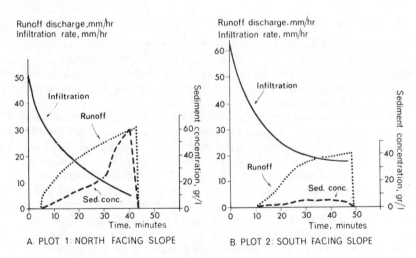

Figure 4. Infiltration, runoff and sediment concentration curves.

128

3.3 *Description of results and data analysis*

Rainfall was applied at an intensity of 36 mm/hr and lasted for 42-48 minutes. The variations in discharge, infiltration capacity and sediment concentration are given, for both plots, in Figure 4. Data obtained show a clear difference in the hydrological and erosional behaviour of the two plots. The north-facing slope responded more quickly to rainfall (lag time = 5 min) and had relatively low initial and terminal infiltration rates (51 mm and 4.1 mm respectively). Average denudation rate was 0.48 mm. The lag time for the south-facing slope was 11 minutes and the initial and terminal infiltration rates were 63 mm and 17.2 mm respectively. Average denudation rate was 0.03 mm.

In order to evaluate the geomorphic significance of the experiments performed, in terms of frequency and magnitude of processes, rainfall intensities and durations used at the small plots were compared to those prevailing in the area under natural rainfall conditions. The nearest meteorological station where rain intensities were recorded during the last 20 years is Mitspe Ramon (Figure 1). The station is located some 40 km south of the research site. Recent rainfall analysis by Kutiel (1978) are presented in Table 1. Data show very clearly that both rain intensity and duration used at the present work are rather rare. The intensity used represents less than three per cent of the average annual rainfall amount. The yearly average cumulative duration of

Table 1. Rainfall data – Mitspe Ramon Station

A. *Percentage frequency distribution of rainfall within given intensity category*

Rainfall intensity (mm/hr)	1-5	5-10	10-20	20-30	30-40	40-50	50-60	60-70	70-80	80
Percentage	42.6	16.0	10.0	4.3	3.6	0.5	0.8	2.0	0.0	1.7
% cumulative	60.6	76.6	86.6	90.9	94.5	95.5	96.3	98.3	98.3	100

B. *Cumulative duration of rainfall at a given intensity (minutes) – yearly average*

Rainfall intensity (mm/hr)	1-5	5-10	10-20	20-30	30-40	40-50	50-60	60-70	70-80	80
Duration in mins	761	95	31	7.4	4.2	0.5	0.6	1.26	–	0.8

C. *Rainfall duration at a given intensity necessary to initiate runoff (dry soil)*

Rainfall intensity (mm/hr)		5	10	15	20	25	30	35	40	45	50	55	60	65
Duration in minutes	Thick topsoil crust	40	30	22	16	11	7	4.5	3	1.5	0.5	0	0	0
	Thin topsoil crust	–	–	90	30	17	14.5	10.5	8	5.5	3.5	2	1	0

129

such intensity is less than 4 minutes and usually occurs in more than one storm.

On the basis of the infiltration curves obtained (Figure 4), the rainfall duration necessary to generate runoff at a given intensity was calculated (Table 1c). A comparison between data obtained in the present experiments with values quoted by Kutiel (1978) for natural rainfalls shows very clearly that under present day rainfall conditions surface flow over the steep slopes of the Taqiya Fm. must be extremely infrequent (Figure 5). Furthermore, considering the short duration of storms capable of generating runoff, when surface flow happens runoff rates and yields may be expected to be very low, much lower than those obtained in the tests reported here. The same applies for the sediment yields and rates of surface lowering, which are surprisingly low when compared with published data on erosion rates in badlands areas (Schumm 1956, Campbell 1974; Gerson 1977, Kin che Lam 1977).

The discrepancy between the results obtained with the characteristic high drainage density of the Zin badlands (which represents a basically fluviatile landscape), raises the problem of the origin of the badlands topography.

While experiments such as these can contribute significantly to our knowledge of present-day geomorphic processes, they remain to be evaluated and calibrated with known geomorphic changes through time. We are fortunate that in this area temporal perspective can be added by the occurrence of *in situ* prehistoric sites whose stratigraphic, geomorphic and temporal setting is well-known.

4. PREHISTORIC EVIDENCE

About 200 m east of the experimental plots, where Nahal Zin makes a sharp turn eastward (Figure 6), is an area that has been the focus of intensive Quaternary research in the fields of prehistory, geology and palaeobiology

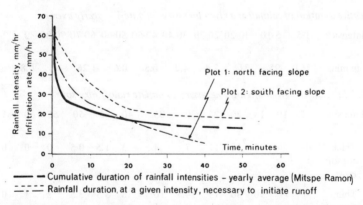

Figure 5. Frequency of flow events on badlands slopes under present day rainfall conditions.

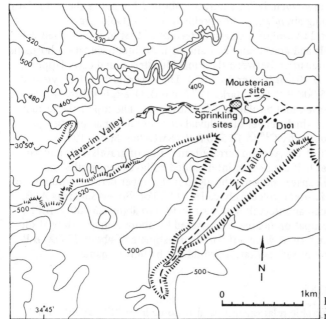

Figure 6. Map of research area.

(Marks 1976, 1977). Although the geomorphic history of the area will be a subject of a more detailed report (Goldberg, in preparation) its salient aspects will be summarized below.

The earliest phase is represented by a small hillock whose surface is strewn with massive blocks of well-indurated conglomerate (Figure 7). On the top is a large concentration of flints of an unquestionable Mousterian industry (Marks et al. 1971). Field relations show that this site (D-6 of Marks et al. 1971) was at one time more intact, and subsequent to the Mousterian (about 80 000 B.P.) has undergone considerable erosion and has been vertically lowered some 5 to 10 m to its present elevation of about 410 m. Further evidence suggests that the underlying conglomerates were formerly more widespread and filled the Zin Valley to this elevation which here, is about 35 m above the present Zin channel. These conglomerates can be traced about 3 km downstream to the east where at the exit of Nahal Aqev, they link up with similar 'Mousterian' conglomerates (Goldberg 1976)*. Although lower in absolute elevation, these conglomerates are still 35 m above the Zin.

About 150 m east of the Mousterian site D-6, and flanking both sides of

* The term Mousterian is used here to refer to a distinct lithic industry. 'Mousterian' is used in an informal sense to refer to deposits or geomorphic features associated with such an industry. A similar usage is implied for the Upper Palaeolithic and 'Upper Palaeolithic'.

131

the Nahal Zin are two well-defined terraces, about 12 m above the wadi floor and both containing abundant *in situ* prehistoric sites (Goldberg 1976, Marks 1977). The lower 2 to 3 m of the eastern terrace consist of interbedded calcareous silty sands and gravels and contains four distinct occupation horizons that represent a transition between the Final Mousterian and early Upper Palaeolithic (site D-101 or 'Boker Tachtit'). The lowermost horizon yielded a C^{14} date of ca. 45 000 B.P. (see Marks 1977, p.64). Overlying these sediments are imbricated gravels and boulders of unknown age. The top of the terrace is capped by a 1 m thick veneer of colluvial silts containing an *in situ* Epi-Palaeolithic site dated by C^{14} to 13 530 B.P. (SMU-268, Marks 1977).

The lower half of the western terrace is made up of gravels, silts and sands whereas the upper half consists of interbedded colluvial sands, silts and clays. Scattered in space and time throughout the upper part are a number of Upper Palaeolithic sites (collectively called D-100 or 'Boker' subareas A, B, C, D, E) which span an interval of at least 40 000 to about 25 000 y. B.P. This youngest occupation in area E is covered by about 3 m of colluvial silts and clays. Although these were not dated, an estimate of about 22 000 B.P. for this final phase of sedimentation would not be unreasonable.

5. DISCUSSION

The above stratigraphic relationships lead to some interesting points. First, most of the sculpting of the present landscape occurred after the Mousterian represented by site D-6 but before the deposition of the D-100 and D-101 terrace complexes. Although the deposition of these 'Upper Palaeolithic' sediments and their subsequent erosion represent significant geomorphic events, and span some 45 000 years, they are of a smaller magnitude than the 'post-Mousterian' erosional phase.

Secondly, the region was quite active geomorphologically as it experienced at least two erosive phases (ca. 80 000 to 45 000 B.P. and ca. 22 000 B.P. to present (?)) and at least one depositional phase (ca. 45 000 to ca. 22 000 B.P.) within the span of the last 80 000 years.

A similar evolution seems to fit the Havarim valley where the sprinkling experiments were performed. Remnants of at least one fluviatile terrasse were recognized at various parts of the valley.

6. CONCLUSION

Reconciling the verility of past geomorphic change with the apparent senility of present-day erosion shown by the sprinkling experiments is not easy. It seems clear, however, that present-day processes, *sensu stricto,* cannot be used to explain the geomorphic history of the area. No doubt that generalized processes such as weathering (crumb and shard formation, for example), run-off, precipitation occurred then as now. However, something in this complex system must have changed in order to effect the severe erosion in 'post-

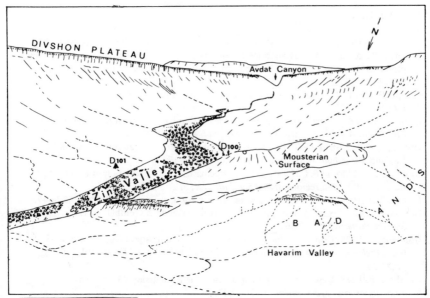

Figure 7. Zin valley: location of prehistoric sites.

Mousterian' time. Experimental data obtained clearly point out that in order to generate runoff and erosion events, the duration of high intensity storms should be increased. Analysis of rainfall data by Kutiel (1978) shows that such an increase is parallel to that of average annual rainfall amounts.

A comparison between Kutiel's data and those obtained in the present study leads to the assumption that the average annual rainfall amount under which more than a single effective storm can be expected in the badlands is in the order of 300 mm or more. This assumption is currently being evaluated through a detailed study of the prehistoric and geomorphic evolution of the Havarim valley.

7. ACKNOWLEDGEMENTS

Assistance of the Central Research Fund of the Hebrew University is kindly acknowledged. Funding for Dr Bryan's participation was provided from a National Research Council Canada grant which is gratefully acknowledged. Thanks are accorded to Mr Y.Binyamini of the Soil Erosion Research Station, Ministry of Agriculture; to Mr E.Adar and Y.Goldstein, Institute for Desert Research for their technical assistance. Thanks are also accorded to Mrs T. Sofer for drawing the illustrations.

133

REFERENCES

Betson, R.P. 1964. What is watershed runoff? *J. Geophys. Res. 69*(8):1541-1552.

Bryan, R.B., A.Yair & W.K.Hodges 1978. Factors controlling the initiation of runoff and piping in Dinosaur Provincial Park badlands, Alberta, Canada. *Z. Geomorph. Suppl. Bd. 29:*151-168.

Campbell, I.A. 1974. Measurements of erosion on badlands surfaces. *Z. Geomorph. Suppl. Bd. 21:*122-137.

Cappus, P. 1958. Répartition des précipitations sur un bassin versant de faible superficie. Proc. Gen. Ass. of Toronto, 1957. *IASH Publ. 43:*515-528.

Dunne, T. & R.B.Black 1970. Partial area contribution to storm runoff in a small New England watershed. *Water Resour. Res. 6*(5):1296-1311.

Dunne, T., T.R.Moore & C.H.Taylor 1975. Recognition and prediction of runoff producing zones in humid regions. *Hydr. Sci. Bull. XX, 3*(9):305-327.

Freeze, R.A. 1974. Streamflow generation. *Rev. Geophys. Space Phys. 12*(4):627-647.

Gerson, R. 1977. Sediment transport for desert watersheds in erodible materials. *Earth Surface Processes 2:*343-361.

Goldberg, P. 1976. Upper Pleistocene geology of the Avdat/Aqev area. In: A.E.Marks (ed.), *Prehistory and paleoenvironments in the Central Negev, Israel.* Vol. I, Part 1, pp.25-55.

Goldberg, P. 1978. Desert geomorphic processes, Field excursion y3. In: *Tenth Int. Congr. on Sedimentology, Jerusalem,* pp.100-103.

Horowitz, A. 1976. Late Quaternary paleoenvironments of prehistoric settlements. In: A.E.Marks (ed.), *Prehistory and paleoenvironments in the Central Negev, Israel,* Vol. I, Part 1, pp.57-68.

Kutiel, H. 1978. The distribution of rain intensities in Israel. M.Sc. thesis. Hebrew Univ. of Jerusalem (in Hebrew).

Lam, K.C. 1977. Patterns and rates of slopewash on the badlands of Hong Kong. *Earth Surface Processes 2:*319-332.

Marks, A.E., J.Phillips, H.Crew & R.Ferring 1971. Prehistoric sites near 'En Avdat' in the Negev. *Israel Expl. J. 21:*12-24.

Marks, A.E. (ed.) 1976. *Prehistory and paleoenvironments in the Central Negev, Israel,* Vol I, Part 1. SMU Press, Dallas.

Marks, A.E. (ed.) 1977. *Prehistory and paleoenvironments in the Central Negev, Israel,* Vol II, Part 2, SMU Press, Dallas.

Marks, A.E. 1977. The upper Paleolithic sites of Boker Tachtit and Boker: A preliminary report. In: *Prehistory and paleoenvironments in the Central Negev, Israel,* Vol II, Part 2, pp.61-79.

Morin, J., C.B.Cluff & W.R.Powers 1970. Realistic rainfall simulator for field investigations, *Eos Trans. AGU, 51*(4):292.

Shanan, L. 1975. Rainfall and runoff relationships in small watersheds in the Avdat region of the Negev desert highlands. Ph.D. Thesis. Hebrew Univ. of Jerusalem.

Schumm, S.A. 1956. The role of creep and rainwash on the retreat of slopes. *Amer. J. Sci. 254:*693-706.

Sharon, D. 1970. Topography-conditioned variations in rainfall as related to the runoff contributing areas in a small watershed. *Israel J. Earth Sci. 19:*85-89.

Sharon, D. n.d. The distribution of hydrologically effective rainfall incident on sloping ground, manuscript.

Yair, A. & H.Lavee 1974. Areal contribution to runoff on scree slopes in an arid environment: A simulated rainstorm experiment. *Z. Geomorph. Suppl. Bed. 21:*123-137.

Yair, A., D.Sharon & H.Lavee 1978. An instrumented watershed for the study of partial area contribution of runoff in the arid zone. *Z. Geomorph. Suppl. Bd. 29:*71-82.

Yair, A., R.B.Bryan, H.Lavee & E.Adar (in press). Runoff and erosion processes and rates in the Zin valley badlands, northern Negev, Israel. *Earth Surface Processes.*

Yair, A. & A.Danin, n.d. Spatial variation in vegetation species as related to the soil moisture regime over an arid limestone hillside, northern Negev, Israel. Manuscript.

135

THE RELATIVE IMPORTANCE OF CLIMATE AND LOCAL HYDROGEOLOGICAL FACTORS IN INFLUENCING LAKE-LEVEL FLUCTUATIONS

F. A. STREET
School of Geography, Oxford

INTRODUCTION

In a detailed study of Late Quaternary lake-level fluctuations in Ethiopia and Djibouti, Gasse & Street (1978) discovered characteristic differences in the number and amplitude of the fluctuations registered by individual basins during the last 30 000 years. The five lakes which we investigated are fed directly or indirectly from the high-rainfall areas on the Ethiopian and South-eastern Plateaux (Figure 1) and can be assumed to have experienced a similar regional climatic history.

The observed differences in behaviour were interpreted in terms of hydrological and topographic factors operating at the basin scale (Gasse & Street, 1978). The purpose of this paper is to examine these local factors in more detail, using modern hydrological data for eastern and central Africa, and then to attempt to establish some general principles which can be used to predict the water-level sensitivity of lakes in other regions to climatic change, for example in the Sahara.

VARIATIONS IN THE RESPONSE OF LAKE BASINS TO CLIMATIC CHANGE: EXAMPLES FROM THE HORN OF AFRICA

The differing responses of individual lake basins to climatic change are particularly well-illustrated by the large rift lakes surrounding the Ethiopian Highlands. The six basins discussed here have been independently investigated by Dr F.Gasse, Professor K.W.Butzer and the author. The modern lakes range in elevation from + 1 636 m to − 155 m a.s.l. Their catchments receive annual precipitation totals ranging from more than 1 600 mm on the highest summits to less than a few millimetres in the areas below sea level. There is accordingly a wide variation in the proportion of total inputs derived from direct rainfall, surface runoff, and groundwater resurgence.

The Ziway-Shala Basin lies near the highest part of the Main Ethiopian

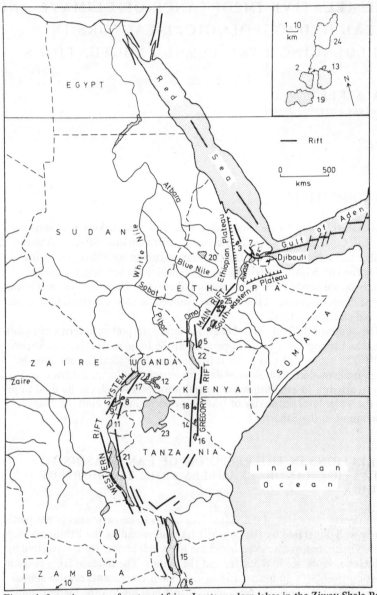

Figure 1. Location map of eastern Africa. Inset: modern lakes in the Ziway-Shala Basin.
Key to lakes: (1) Abhé; (2) Abiyata; (3) Afrera; (4) Asal; (5) Chew Bahir; (6) Chilwa;
(7) Dobi-Hanlé; (8) George; (9) Idi Amin Dada; (10) Kariba; (11) Kivu; (12) Kyoga;
(13) Langano; (14) Magadi; (15) Malawi; (16) Manyara; (17) Mobutu Sese Seko;
(18) Naivasha; (19) Shala; (20) Tana; (21) Tanganyika; (22) Turkana; (23) Victoria;
(24) Ziway; (25) Ziway-Shala Basin.

138

Rift floor (Figure 1) and contains four residual lakes (Ziway, Langano, Abiyata and Shala), situated at altitudes of 1 558-1 636 m a.s.l. These are predominantly fed by perennial rivers rising above 2 000 m on the Ethiopian and Southeastern Plateaux. At present, there is no outlet from the basin, and the lowest lake, Shala, has no surface connection with the other three. The lakes coalesced several times in the past to form a large waterbody, Lake Ziway-Shala, which overflowed into the Awash catchment to the north (Grove et al. 1975, Grove & Street 1978). Four major and six minor highstands have occurred during the past 30 000 years (Figures 2 and 3). These can be grouped into a complex Lake Pleistocene lacustral interval (ca. 30 000-21 000 B.P.) and a complex Holocene lacustral interval (ca. 11 500-4 800 B.P.), separated by a prolonged arid period (Street, in press and unpublished).

During overflow episodes, Lake Ziway-Shala became the highest in a chain of large lakes situated along the course of the Awash River, which terminated in a greatly enlarged Lake Abhé (Gasse 1975, 1977a, Gasse & Delibrias 1977, Gasse & Street 1978, Gasse et al. in press). Lake Abhé (240 m a.s.l.) is located in the central Afar, in a region of highly complex horst-and-graben topography at the junction of the Main Ethiopian and Afar Rifts (Figure 1). At the present day, it is fed primarily by the Awash River, which rises on the Ethiopian Plateau margin to the north of Addis Ababa; although a small proportion of the inflow is derived from springs (United Nations 1965, Gasse 1975, pp.36-37). The history of lake-level fluctuations is very similar to Lake Ziway-Shala (Figures 2 and 3). However, the final Late Pleistocene highstand Abhé III (30 000-17 000 B.P.) was much more stable and prolonged. There is also no sign of any lake-level rise between 12 000 and 10 000 B.P.

The palaeolakes which developed along the course of the Awash at Koka, Wonji and Gawani probably owed their existence to ponding of the river flow upstream of the gorge sections (Taieb 1974). Their fluctuations have not yet been investigated in any detail.

To the northeast of Lake Abhé, large lakes also formed in the steep-sided Dobi-Hanlé and Asal grabens (Gasse & Stieltjes 1973, Gasse et al. 1974, Gasse 1975, Gasse & Street 1978). The modern Lake Asal (−155 m) is the only surviving remnant. In this very arid area, groundwater resurgence plays a relatively more important role in the hydrology of the lowlying fault troughs. The regional water table slopes northeastwards across the central Afar from recharge zones situated along the foot of the Ethiopian and Southeastern Escarpments and along the course of the Awash (United Nations 1973). This suggests that the grabens are hydrologically interconnected. Flow rates through the fractured basalts of the Afar floor appear to be relatively rapid. The springs feeding Lake Asal, which lies only about 8 km from the sea, also receive a substantial contribution of seawater (Gasse & Stieltjes 1973). Today, direct rainfall and inputs from ephemeral streams appear to be very small, although during its maxima Lake Dobi-Hanlé was partly sustained by runoff from the Southeastern Escarpment (Gasse 1975, p.298). Although the Lake Pleistocene history of these lakes has not been dated in detail, the sequence of Holocene

139

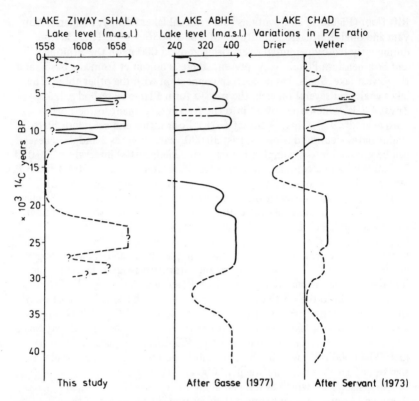

Figure 2. Water-level curves for Lakes Ziway-Shala and Abhé compared with variations in the precipitation/evaporation ratio in the Chad Basin (based on the fluctuations of Lake Chad).

fluctuations is now fairly well-established (Figure 3) (Gasse 1975, Gasse & Delibrias 1977, Gasse & Street 1978).

Lake Afrera (−80 m) is a small, hyperarid basin in the Northern Afar, surrounded by fairly gentle relief. The present salt lake is maintained by numerous springs, fed by groundwater from the northern end of the Ethiopian Escarpment. It is, therefore, hydrologically independent of the Central Afar Lakes (United Nations 1973). Direct rainfall and surface runoff are again believed to be negligible. Details of the Holocene fluctuations of the lake (Figure 3) can be found in Gasse (1975) and Gasse & Street (1978).

Finally, Lake Turkana (375 m a.s.l.) lies in a tectonically complex area between the Main Ethiopian and Gregory Rifts (Figure 1). It derives about 80-90 per cent of its inputs from the perennial Omo River, which rises on the Ethiopian Plateau just to the west of the Ziway-Shala and Upper Awash headwaters (Butzer 1971). The very large present-day lake (ca. 7 500 km²) is topo-

140

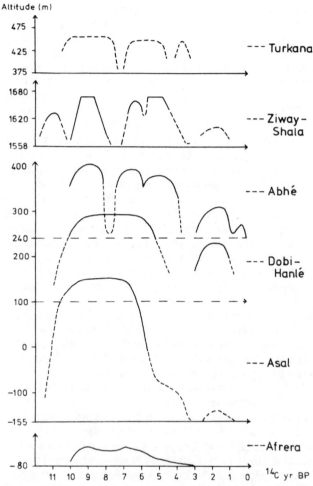

Figure 3. Water-level curves for the last 12 000 years for Lakes Turkana, Ziway-Shala, Abhé, Dobi-Hanlé, Asal and Afrera. Each curve is controlled by a total of 8 to 73 ^{14}C dates (average 31). The Turkana curve is based on data published by Butzer et al. (1969, 1972) and Butzer (1976). The other curves are based on Gasse & Street (1978) except for the Ziway-Shala Basin (Street, unpublished).

Note: 400 y has been added back onto the Turkana dates to make them compatible with the ^{14}C ages for the other five basins, which have not been corrected for dead-carbon effects. The 400 y correction originally applied by Thurber (Thurber & Broecker 1970, Thurber 1972) was based on surface-water analyses from the Great Basin, USA. Recently, however, various uncertainties in his calculations have been pointed out (Benson 1978, Peng et al. 1978), and it is clear that the appropriate dead-carbon correction depends critically on the volume, hydrology, water chemistry and circulation of the palaeo-lake in question.

graphically closed, and although its relatively low salinity has led some authors to believe that it loses significant amounts of water by influent seepage (Thurber 1972, Beadle 1974, pp.142-143), there is no general agreement on this point (Yuretich 1976). During the early and middle Holocene, Turkana probably received flows from palaeolakes Chew Bahir, Chalbi and possibly Suguta, in addition to the Omo influx (Grove et al. 1975, Bishop 1975, Truckle 1976, D.Phillipson, personal communication). It may in turn have spilled over westwards towards the Pibor-Sobat system during the periods 9 900 to 7 900 and 6 600 to 3 650 B.P.* (Butzer et al. 1969, 1972, Butzer 1976).

Figure 2 shows the water-level curves for Lakes Ziway-Shala and Abhé for the period 30 000 B.P. to present. Their close similarity to the curves for the Chad Basin (Figure 2) and the Sebkha de Chemchane (Chamard 1972) clearly demonstrates the general Sahelian character of Late Quaternary climatic change in the Horn of Africa; although certain minor differences between the Abhé and Ziway-Shala curves have already been noted. There is still too little information to draw up curves for the other Afar lakes. The history of Lake Turkana before 10 000 B.P. also remains rather uncertain. Butzer (Butzer et al. 1969, 1972, Butzer 1976) stated that it was low throughout the period 35 000 to 10 000 B.P. This conclusion, however, is based on the existence of an apparent hiatus in lacustrine deposition rather than on dates from non-lacustrine deposits. It will almost certainly be necessary to examine cores from the deeper part of the lake in order to clarify the events during this interval.

During the Holocene, however, we find that the lake-level curve for Lake Turkana bears a strong resemblance to the curves for the other river-fed basins, Ziway-Shala and Abhé; contrasting markedly with the much simpler sequence of fluctuations in the groundwater-dependent Asal and Afrera Basins. The curve for Dobi-Hanlé follows the Asal pattern, suggesting that the two lakes were hydrologically very similar.

Although the discrepancies between individual lakes (Figure 3) could be interpreted as the result of differences in intensity of investigation and in accuracy of dating, I assume here that they reflect genuine differences in lake-level sensitivity (Gasse & Street 1978, Gasse et al. in press) which will be explored in the following section. This view is supported by the much greater complexity of the stratigraphy in river-fed basins such as Ziway-Shala.

GENERAL RELATIONSHIPS BETWEEN WATER-LEVEL SENSITIVITY AND WATER BALANCE

The water balance of a lake at equilibrium is given by the following equation:
$$P_L + R + G_I = E + O + G_O \qquad (1)$$
where P_L is direct precipitation on the lake

* See caption to Figure 3.

142

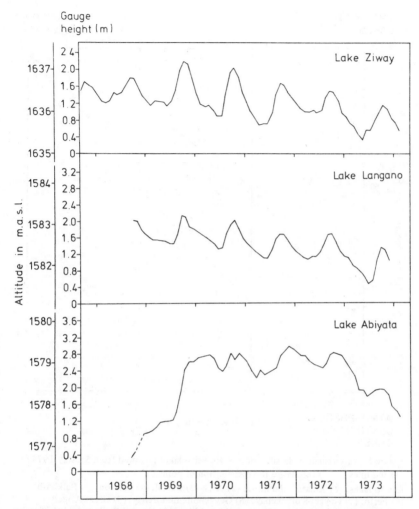

Figure 4. Water-level curves for Lakes Ziway (open), Langano (open) and Abiyata (closed) for the period 1967-1974. Data from Makin et al. (1976).

R   is surface runoff from the catchment
G_I is groundwater inflow (effluent seepage) (Ward 1967, p.291)
E   is evaporation from the lake
O   is surface outflow
G_O is groundwater outflow (influent seepage)

Szestay (1974) has outlined the extent to which the various water-balance components are dependent on climate. Precipitation on the lake surface and

143

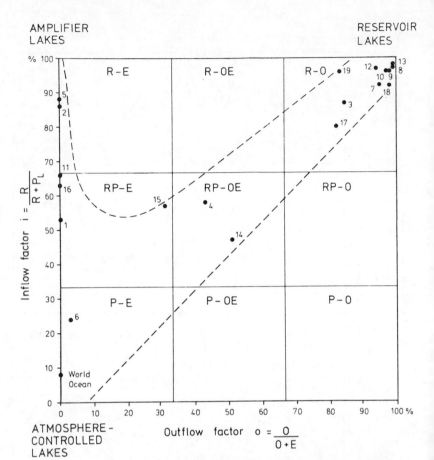

Figure 5. Water-balance classification of Eurasian lakes (adapted from Szestay 1974).
Dotted lines (.......) emphasise the distribution of points.

Key to lakes: (1) Amara – Ialomita; (2) Aral; (3) Baikal; (4) Balaton; (5) Balkhash;
(6) Fertö; (7) Geygel-Kurackehay; (8) Geygel-Shamkhorchay; (9) Great Alagel;
(10) Ilmen; (11) Issyk-Kul; (12) Karagel-Ishukhly; (13) Karagel-Perihingil; (14) Khanka;
(15) Khubsugul; (16) Kulundinskoye; (17) Peipsi-Pihkva; (18) Skadar; (19) Uzunoba.

on the catchment is controlled by macroscale atmospheric and orographic
factors; surface and groundwater inflows originate from rainfall and melting
of snow and ice and are affected by hydrogeological and orographic factors
within the drainage area. Evaporation from the lake surface represents one of
the major elements of the radiation and energy balance of the lake and its
surroundings. In contrast, however, surface and subsurface outflows are
chiefly controlled by the water level in the lake and by the hydraulics of the
outlet and underlying aquifers (Jackson 1833, Langbein 1961).

144

The most fundamental factor affecting the sensitivity of a lake to climatic fluctuations is, therefore, the presence or absence of an outlet. This was recognized more than 250 years ago (Halley 1715). Closed lakes ($O = G_O = $ zero) lose water entirely by evaporation and undergo fluctuations in level of much larger amplitude than open lakes. This behaviour is clearly seen when recent lake-level data for the open lakes Ziway and Langano are compared with the fluctuations of their terminal lake, Abiyata, during the same period (Figure 4). Open lakes are of two main kinds: drainage lakes ($O>$zero) which have surface outlets, and seepage lakes ($G_O>$zero) which lose water by percolation to groundwater although they may be topographically closed (Langbein 1961, Snyder 1962).

This simple classification has been elaborated by Szestay (1974) in an analysis of the relationships between the water balance and water-level sensitivity of modern lakes in Eastern Europe and the U.S.S.R. (Figure 5). He defined four criteria which can be used to characterize the water balance of any lake, assuming that groundwater inputs and outputs are negligible:

The inflow factor, i, is the percentage of surface runoff in the total water input of the lake:

$$i = \frac{R}{R + P_L} \cdot 100 \tag{2}$$

The outflow factor, o, is the percentage of outflow in the total water output of the lake

$$o = \frac{O}{O + E} \cdot 100 \tag{3}$$

The annual water flux, f, is most conveniently defined by:

$$f = P_L + R \tag{4}$$

Equation (4) allows for cases in which inputs and outputs do not balance, causing a change in storage. Finally, the aridity factor, a, is given by:

$$a = \frac{E}{P_L} \tag{5}$$

A further important characteristic is the turnover time, t, (the average time required for rainfall and runoff to replenish the lake waters), which is approximated by:

$$t = \frac{V}{f} \tag{6}$$

where V is the lake volume.

If i is plotted against o for individual lakes, a characteristic pattern emerges (Figure 5). In Figure 6, the limited information available for the lakes discussed in the preceding section has been supplemented with data from other modern lakes in eastern and central Africa. On both diagrams, closed lakes are located on the vertical axis. Szestay (1974) noted that the data points are unevenly distributed among the nine quadrants. Three common categories of lakes were defined with regard to lake-level sensitivity:

1. *Flow-dominated reservoirs (R−O).* P_L and E are very small compared with

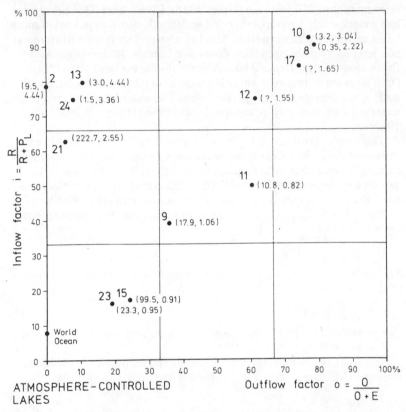

Figure 6. Water-balance classification of lakes in eastern and central Africa. In brackets (t in years, a).
Key to lakes as for Figure 1: (2) Abiyata; (8) George; (9) Idi Amin Data; (10) Kariba; (11) Kivu; (12) Kyoga; (13) Langano; (15) Malawi; (17) Mobutu Sese Seko; (21) Tanganyika; (23) Victoria; (24) Ziway. Data from Hecky & Degens 1973, Krishnamurthy & Ibrahim 1973, Viner & Smith 1973, Makin et al. 1976, Balek 1977.

R and O. These lakes tend to lie on major rivers, for example the Kariba reservoir on the Zambezi and Lake Mobutu on the White Nile (Figure 6). They are regulated primarily by the hydraulics of the outflow and are insensitive to long-term climatic fluctuations unless the inflow is dramatically reduced. Palaeolakes Koka, Wonji and Gawani can tentatively be placed in this category.

2. *Atmosphere-controlled lakes (P–E)*. P_L and E are very large compared with R and O. Lakes Victoria and Malawi are classic examples. Lake Tana may also

be of this type. These lakes are situated in relatively humid areas, and have probably overflowed quite frequently in their history. They react directly but not very sensitively to changes in the equilibrium of the water balance. The level of Lake Victoria, for example, is highly correlated with monthly mean rainfall, but has varied by only about 3 m since 1876 (Mörth 1967, Beadle 1974, p.194).

3. *Amplifier lakes (RP–E and R–E) (my own term)*. R and E dominate the water balance. This group includes the closed basins typical of arid and semi-arid areas, and others with small outflows. The modern lakes in the Ziway-Shala Basin are included in this category (Figures 4 and 6). So in all probability are Lakes Abhé and Turkana (United Nations 1965, Butzer 1971). Such lakes show a large-amplitude response to long-term climatic fluctuations, because of the numerous positive feedbacks influencing runoff generation from their tributary areas. They also tend to 'accumulate' short-term variations in precipitation, especially if there is no surface outlet.

Szestay (1974) noted that the lakes in quadrants P–E, RP–E and R–E fall on a continuum which represents increasing values of the aridity factor, a, towards the top left of the diagram. This is also clearly seen in Figure 6. Although he mentioned the apparent absence of lakes in quadrants R–OE, RP–O, P–OE and P–O, he did not comment on the tendency of the data points to cluster along the diagonal i = o. This trend represents a second continuum from lakes which are, in hydrological terms, only 'wide places in a river' (Langbein 1961), with a large annual water flux relative to their size, through to inland seas like Victoria which have water budgets resembling the ocean (Figure 6). In keeping with this hypothesis, turnover time, t, decreases by more than two orders of magnitude from lower left to top right on Figure 6.

If groundwater discharge is now considered, further important categories are introduced, although the problems of classification are greatly increased by the difficulties of making accurate estimates of groundwater flow into, or out of, lakes (Zektzer 1974):

4. *Groundwater-fed (effluent) lakes* are sustained by springs and/or by diffuse effluent seepage. Closed lakes receiving a substantial net influx of groundwater are quite common, a famous case being Great Salt Lake (Peck & Richardson 1966, Mower 1968) (Figure 7), but in eastern Africa, closed lakes fed *primarily* by subsurface inflow ($G_I - E$) are restricted to the most arid areas, in which P_L and R are both negligible. Asal, Afrera (Figure 7) and Magadi are typical examples (Gasse & Stieltjes 1973, United Nations 1973, Jones et al. 1976).

In the short term, effluent lakes are often remarkably stable, due to the damped response of aquifers to climatic fluctuations (Hughes 1974, Zemljanitzyna 1974). Lakes of category $G_I - E$ may undergo longterm fluctuations of large amplitude (Figure 3), as a consequence of slow adjustments in the

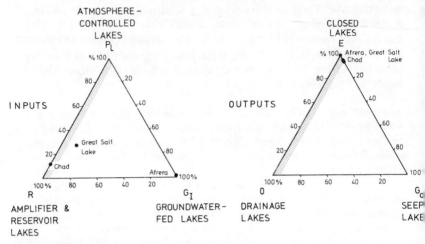

Figure 7. Water-balance classification including groundwater flows. Dotted: probable area occupied by lakes shown in Figure 6. Data for Great Salt Lake, Lake Chad and Lake Afrera (from Peck & Richardson 1966, Roche 1977, United Nations 1973).

height of the regional water table. But they may nevertheless fail to register climatic oscillations of less than 1 000 y duration which cause significant variations in the extent of amplifier lakes in the same region.

The sensitivity of effluent lakes depends on several factors: including the proportion of the total influx contributed by groundwater; the vertical fluctuations in the water table in relation to the basin relief (Zemljanitzyna 1974); the flow rate through the aquifer; and the distance of the lake from its recharge area. The influence of local topography will be explored further in the next section. One problem besetting the interpretation of the water-level record in these basins is the possibility of long-distance transfers of groundwater across surface drainage divides (Winograd 1962; Eakin 1966), which may necessitate detailed observations of the regional flow pattern in order to determine the source area of the water (United Nations 1973). It is also probable that the spatial distribution of recharge areas has fluctuated through time in response to climatic changes (Edmunds & Wright 1979).

5. *Seepage (influent) lakes* may be fed either by direct precipitation or by surface runoff, but lose water mainly by a combination of evaporation and net groundwater outflow. Lake Naivasha (Figure 1) appears to be a classic case (Richardson & Richardson 1972), although no water-balance data have so far been published. Lake Chad (Figure 7) has attracted a great deal of attention because its waters are also unusually dilute for a 'closed' lake, but in fact the contribution of seepage losses to the total efflux is quite small (Roche 1977). It is, therefore, hardly surprising that its long-term history

148

closely resembles that of amplifier lakes such as Ziway-Shala and Abhé, which are also fed primarily by large rivers (Figure 2).

The suspected seepage lakes Turkana, Naivasha and Baringo in eastern Africa appear to have experienced long-term water-level fluctuations very similar to the closed lakes in the same region (Richardson & Richardson 1972, Kamau 1977, Young & Renaut 1979). Since G_O probably remains relatively constant in relation to the other water-balance terms, the area of seepage lakes without surface outlets adjusts to balance inputs in a manner comparable with lakes which are truly closed.

CHANGES IN WATER-BALANCE STRUCTURE THROUGH TIME: THE ROLE OF TOPOGRAPHY

The long-term behaviour of any lake will be influenced not only by its present-day water balance but also by past shifts in water-balance structure resulting directly or indirectly from changes in climate. Such shifts can be viewed as thresholds or feedbacks in the hydrological system. Although it has long been recognized that many present-day closed lakes formerly had outlets (Jackson 1833), it was not until the advent of efficient lake-sediment coring techniques that the impact of climatic fluctuations on water budgets began to be recognized. The present Lake Victoria (Figure 1), for example, is a very dilute drainage lake in the atmosphere-controlled category (P–E) (Talling & Talling 1965) (Figure 6). Its Holocene gyttjas overlie an evaporite deposit which indicates that the lake was closed and accumulating salts between $\geqslant 14\,500$ and 12 000 B.P. (Kendall 1969, Livingstone 1965). During this period it was probably much more sensitive to variations in runoff and evaporation than it is today.

The most important feedback affecting the water balance results from the relationship between the basin topography and the pattern of groundwater flow. In the steep-sided rift-valley grabens of eastern Africa, the vertical range of fluctuation in the water table is often less than the altitude range from the basin floor to the lowest col. In cases where their substratum is permeable, the lakes, therefore, tend to be influent during highstands and effluent during lowstands (Gasse 1975). In areas of gentle relief, such as the Central Asian steppes (Zemljanitzyna 1974), the Texas High Plains (Brakenridge 1978) and much of the Sahara (Conrad 1969, Servant 1973), the reverse is often true, because the water table drops below the lake floors during arid phases. However, in these regions, the lakes which are fed primarily by allochthonous rivers or wadis may lose water by seepage throughout most of their history. For this reason, they are unlikely to become saline (Zemljanitzyna 1974).

In order to characterize any lake it is, therefore, necessary to take into account not only the availability of an outlet but also the interaction between the basin relief and hydrogeology.

149

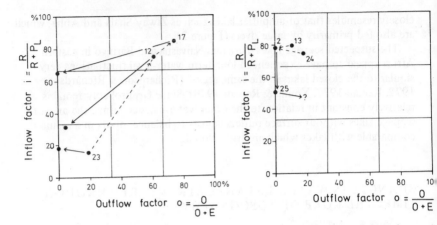

Figure 8. Reconstructed water-balance trajectories for the White Nile lakes (left) and
Lake Ziway-Shala (right). Dotted arrows (......) indicate present-day overflows. Solid arrows
(——) indicate calculated trajectories. Left diagram shows the situation from $\geq 14\,500$
to 12 000 B.P. compared with today. Right diagram shows the reconstructed water-
balance at 9 500-8 500 B.P. compared with today.
Key to lakes as for Figure 1: (2) Abiyata; (12) Kyoga; (13) Langano; (17) Mobutu Sese
Seko; (23) Victoria; (24) Ziway; (25) United Lake Ziway-Shala. Data from Harvey (1976)
and Street (1979).

RECONSTRUCTING CHANGES IN WATER-BALANCE STRUCTURE THROUGH TIME

As a result of climatic fluctuations and their direct and indirect repercussions
on the water budget, individual lakes will follow trajectories from one water-
balance category to another. These can be reconstructed using two
approaches. First, for lakes which have been closed at some period during
their history, it is possible to calculate the equilibrium water-balance at the
time of closure if the area of the water surface is known, and if a realistic
estimate of open-water evaporation can be made. From that point onwards,
the direction of change is less certain, but can sometimes be deduced from
the trends in water balance exhibited by modern lakes in the same region.
Secondly, a qualitative assessment of the nature of past changes can be made
from mineralogical, sedimentological, palaeontological and isotopic data.

    The first approach has now been applied to a number of lakes in the South-
west U.S.A., Africa, Australia and the U.S.S.R. for which G_I and G_O are
assumed to be small (Brakenridge 1978, Street 1979, Butzer et al. 1973, Ebert
& Hitchcock 1979, Coventry 1976, Schnitnikov 1974). Geomorphological
data on ancient shorelines, which are needed to establish the existence of an
outlet, enable one to calculate the catchment/lake area ratio, A_B/A_L, and its
effect on the relative magnitudes of P_L and R (Street 1979). It is more difficult

150

to assess the long-term impact of changes in open-water evaporation. All studies so far have made the rather unrealistic assumption that evaporation is solely determined by temperature, and then applied palaeotemperature estimates derived from snowline, treeline or periglacial data. In future, it may prove feasible to develop alternative methods based on radiation and energy – balance considerations, thereby eliminating this assumption (J.E.Kutzbach personal communication). Another difficult variable to estimate is the runoff coefficient for the catchment, which is usually estimated from empirical measurements of rainfall, runoff and/or evapotranspiration. Such calculations make only very crude allowances for the effects of long-term changes in vegetation cover, soils and groundwater levels (Brakenridge 1978). Other important variables which cannot at present be taken into account are rainfall intensity, seasonality, cloudiness and wind strength.

The results of two studies of this type are given in Figure 8. The Ziway-Shala example (Street 1979) shows the type of trajectory likely to be followed by closed amplifier lakes (categories R-E and RP-E) as a result of a decrease in the aridity factor, a. Changes in the water balance of reservoir lakes are potentially more dramatic. As a result of the cessation of outflow from Lake Victoria between ≥14 500 and 12 000 B.P., Lakes Kyoga and Mobutu were greatly reduced in size, so that Mobutu no longer overflowed into the White Nile. Figure 8 shows the situation with Victoria just below the lip of its outlet (Harvey 1976). If the required climatic change (a reduction of 5°C in temperature and 29 per cent in precipitation) is also applied to the downstream reservoir lakes in addition to the cessation of the Victoria Nile, then Kyoga and Mobutu collapse down towards the left-hand axis, becoming atmosphere-controlled and amplifier lakes respectively.

The second approach has not yet been widely exploited as a means of reconstructing changes in water balance, although Jackson (1833) long ago recognised that factors such as the presence or absence of an outlet, and the role of groundwater flow, exert a controlling influence on lake-water salinity and sedimentation. He also recognised that the various water-balance components can be distinguished in terms of their contributions to the sediment and salt budgets of a lake. His basic reasoning is illustrated and extended in Figure 9. This model predicts that high salinities are most likely to develop in closed lakes of categories R-E and G_I-E; and low salinities in rain-fed seepage lakes, drainage lakes of category P-E, and flow-dominated reservoirs (R-O) with rapid turnover times. (As noted earlier, category P-O is not represented in the data from eastern Africa) This model agrees well with the chemical analyses published by Talling & Talling (1965), Gasse (1975) and Jones et al. (1976). These show that Magadi, Asal, Afrera and Abhé are among the most concentrated lakes in eastern Africa, whereas Victoria, Kariba, Tana, George and Malawi are among the most dilute. Salinity can therefore be used as a gross indicator of water balance, although it is subject to such time-dependent effects as the concentration of dissolved salts during periods of lake shrinkage, losses of salts through sedimentation and

151

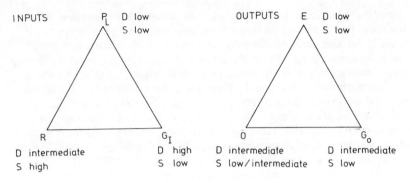

INPUTS P_L D low
        S low

OUTPUTS   E   D low
        S low

R G_I
D intermediate   D high
S high   S low

0   G_0
D intermediate   D intermediate
S low/intermediate   S low

Figure 9. Inputs and outputs of dissolved salts (D) and clastic sediment (S) into a lake as a function of water balance.

deflation (Langbein 1961, Jones & Van Denburgh 1966) and recycling from surface crusts and saline soils (Jones et al. 1976).

    In eastern Africa, clastic sediment is chiefly derived from rivers and shoreline erosion (McLachlan & McLachlan 1971). Aeolian and direct volcanic inputs are generally small (Yuretich 1976). Losses (which are essentially restricted to the suspended fraction) occur mainly by outflow. Based on modern reservoir data, the sediment-trap efficiency of drainage lakes varies from about 50 to 100 per cent (Glymph 1973, Ziminova 1974), and is directly proportional to t. The most rapid natural rates of clastic sediment influx are, therefore, to be expected in closed amplifier lakes of category R-E, especially if these are situated in mountainous or semiarid regions, where sediment production tends to reach a maximum (Gregory & Walling 1973, pp.329-337). Turkana, Abhé, Ziway-Shala are modern examples of river-fed basins accumulating deposits with a large clastic component (Yuretich 1976, Gasse 1975, Baumann et al. 1975, McLachlan & McLachlan 1969).

    Qualitative information about past fluctuations in water balance can, therefore, be derived from a wide variety of mineralogical, sedimentological, isotopic and palaeontological indicators, although only a few possibilities can be mentioned here. Calcite and aragonite marls, for example, seem to be accumulating today in a number of moderately alkaline lakes such as Balaton, Fertö, Issyk-kul and Balkhash, which are located in quadrants RP-OE, P-E, RP-E and R-E in Figure 5 (Müller et al. 1972, Kuznetsov 1975). These have low to moderate turnover times and values of a > 1, but may be either open or closed. Many closed lakes precipitate a wide variety of other authigenic minerals, especially evaporites (Reeves 1968, ch.5, Hardie & Eugster 1970), which can be used to establish the absence of outflow at times in the past. The geochemical conditions necessary for the precipitation of some of these minerals can be attributed to specific hydrological environments. Thick deposits of trona, for example, are characteristic of highly concentrated

152

alkaline lakes receiving groundwaters derived from igneous and metamorphic terrains (Eugster 1971). They may therefore serve as an indicator of group G_I-E. Hydrated sodium silicates such as magadiite and kenyaite also appear to be associated with closed basins in which silica-rich alkaline groundwaters are brought to the surface and then concentrated by evaporation (Hay 1968, Eugster 1969, Rooney et al. 1969, Maglione & Servant 1973). However, the precipitation of these metastable silicates may require a trigger such as the mixing of dilute runoff or spring waters with the alkaline brines (Jones et al. 1967).

Certain structural and microtopographic features may act as indicators of groundwater resurgence, just as deltas reveal the entry point of rivers. Chief among these are algal pinnacles or spring mounds (Scholl 1960, Scholl & Taft 1964, Fontes & Pouchan 1975) which form in the euphotic zones of lakes as a result of the precipitation of calcium carbonate by algae growing around the orifices of sublacustrine springs. In theory, ^{14}C dating of the base of such pinnacles should provide an estimate of the time of onset of groundwater effluence; although the uncertainties of dating porous algal carbonates are well known (Benson 1978). A number of minor features such as salt crusts, phreatophyte mounds and evaporite polygons are also believed to be indicative of groundwater discharge through playa surfaces (Neal 1972), but are probably more ephemeral in the geological record.

Diatoms have been widely used in eastern Africa as an indicator of lake depth and salinity (Richardson & Richardson 1972, Gasse 1975, Hecky & Kilham 1973, Harvey 1976, Holdship 1976, Gasse 1977b, Richardson et al. 1978). The percentage of alkalophilous and halophilous species in fossil assemblages is inversely related to the importance of outflow in the water budget. Certain so-called 'crenophilous' taxa are characteristically associated with spring waters and may provide an indication of the importance of groundwater inputs in the past (Gasse 1975). Other lacustrine fossils which are sometimes used to infer general levels of salinity include mollusca, ostracods and foraminifera.

Isotope studies of lacustrine deposits are still in their infancy, although stable-isotope measurements may provide a valuable indication of the degree of evaporative concentration of lake waters. So far, attention has been focused mainly on the $^{18}O/^{16}O$ ratio in calcareous tufas (Gasse et al. 1974, Fontes & Pouchan 1975) which can be used to estimate the original composition of the water from which the carbonate precipitated; provided that some estimate of palaeotemperatures is available.

CONCLUSIONS

The sensitivity of any lake to Lake Quaternary climatic fluctuations can be explained in terms of three main factors:
    (a) its modern water balance

(b) the nature of the basin relief, and its influence on long-term variations in the pattern of groundwater flow and on the availability of an outlet,

(c) the shifts in water balance occurring through time as a result of the interaction between climate and local hydrogeological factors. The most significant discontinuity in behaviour occurs at the point at which an open lake becomes closed, or *vice versa*. However, many low-latitude lakes appear to have followed trajectories from one water-balance category to another. This fact must certainly have been reflected in their sensitivity to minor climatic fluctuations. An attempt can be made to reconstruct these trajectories using a combination of geomorphological, mineralogical, sedimentological, palaeontological and isotopic evidence.

The general relationships established between water-level sensitivity, water balance and topography form a basis for predicting the value of lakes as climatic indicators in areas which have not previously been investigated.

REFERENCES

Balek, J. 1977. *Hydrology and water resources in Africa*. Elsevier, Amsterdam.

Baumann, A., U.Förstner & R.Rohde 1975. Lake Shala: Water chemistry, mineralogy and geochemistry of sediments in an Ethiopian Rift lake. *Geol. Rundschau, Stuttgart, 64:* 593-609.

Beadle, L.C. 1974. *The inland waters of tropical Africa*. Longman, London.

Benson, L.V. 1978. Fluctuation in the level of Pluvial Lake Lahontan during the past 40 000 years. *Quat. Res. 9:*300-318.

Bishop, W.W. 1975. Geological reconnaissance of the lower Suguta valley in northern Kenya. In: A.T.Grove (ed.), *Cambridge meeting on desertification*. Univ. Cambridge, Dept. Geogr.

Brakenridge, G.R. 1978. Evidence for a cold, dry full-glacial climate in the American Southwest. *Quat. Res. 9:*22-40.

Butzer, K.W. 1971. Recent history of an Ethiopian delta: The Omo River and the level of Lake Rudolf. *Univ. Chicago, Dept. Geogr., Res. Paper 136.*

Butzer, K.W. 1976. The Mursi, Nkalabong and Kibish formations, Lower Omo Basin, Ethiopia. In: Coppens, Y., F.C.Howell, G.Ll.Isaac & R.E.F.Leakey (eds.), *Earliest man and environments in the Lake Rudolf Basin*. Univ. Chicago Press, Chicago. pp.12-23.

Butzer, K.W., F.H.Brown & D.L.Thurber 1969. Horizontal sediments of the Lower Omo Valley: The Kibish formation. *Quaternaria 11:*15-30.

Butzer, K.W., G.J.Fock, R.Stuckenrath & A.Zilch 1973. Palaeohydrology of Late Pleistocene Lake Alexandersfontein, Kimberley, S.Africa. *Nature 243:*328-330.

Chamard, Ph.C. 1972. Les Lacs holocènes de l'Adrar de Mauritanie et peuplements préhistoriques. *Notes africaines 133:*1-8.

Conrad, G. 1969. *L'évolution continentale post-hercyienne du Sahara algérien*. CNRS-CRZA, Paris, Sér. géologie.

Coventry, R.J. 1976. Abandoned shorelines and the Late Quaternary history of Lake George, New South Wales. *J. Geol. Soc. Austr. 23:*249-273.

Eakin, T.E. 1966. A regional interbasin groundwater system in the White River area, Southeastern Nevada. *Water Resour. Res. 2:*251-271.

154

Ebert, J.I. & R.K.Hitchcock 1979. Ancient Lake Makgadikgadi, Botswana: Mapping, measurement and palaeoclimatic significance. *Palaeoecology of Africa 10:* 47-56.

Edmunds, W.M. & E.P.Wright 1979. Groundwater recharge and palaeoclimate in the Sirte and Kufra basins, Libya. *J. Hydrol. 40:* 215-241.

Eugster, H.P. 1969. Inorganic bedded cherts from the Magadi area, Kenya. *Contrib. Mineral. Petrol. 22:* 1-31.

Eugster, H.P. 1971. Origin and deposition of trona. *Contrib. Geol. Univ. Wyoming, Laramie 10:* 49-56.

Fontes, J.-C. & P.Pouchan 1975. Les cheminées du Lac Abhé (TFAI): Stations hydro-climatiques de l'Holocène. *Cr. Acad. Sci., Paris, 280D:* 383-386.

Gasse, F. 1975. L'évolution des lacs de l'Afar Central (Ethiopie et TFAI) du Plio-Pléisto-cène à l'Actuel: Reconstitution des paléomilieux lacustres à partir de l'étude des Diatomées. D.Sc. thesis, Univ. Paris – VI.

Gasse, F. 1977a. Evolution of Lake Abhé (Ethiopia and TFAI) from 70 000 B.P. *Nature 265:* 42-45.

Gasse, F. 1977b. Les groupements de diatomées planctoniques: Base de la classification des lacs quaternaires de l'Afar Central. In: *Recherches francaises sur le Quaternaire hors de France.* Comité National Français de l'INQUA, Paris, pp.207-234.

Gasse, F. & G.Delibrias 1977. Les lacs de l'Afar Central (Ethiopie et TFAI) au Pléistocène supérieur. In: S.Horie (ed.), *Palaeolimnology of Lake Biwa and the Japanese Pleisto-cene* (4): 529-575.

Gasse, F., J.-Ch.Fontes & P.Rognon 1974. Variations hydrologiques et extension des lacs holocène du désert danakil. *Palaeogeogr., Palaeoclimatol., Palaeoecol. 15:* 109-148.

Gasse, F., P.Rognon & F.A.Street 1980. Quaternary history of the Afar and Ethiopian Rift lake basins. In: M.A.J.Williams & H.Faure (eds.), *The Sahara and the Nile.* A.A. Balkema, Rotterdam.

Gasse, F. & L.Stieltjes 1973. Les sédiments du Quaternaire récent du lac Asal (Afar Central, TFAI). *Bull. Bur. Rech. géol., min. (2e sér.) section IV* (4): 229-245.

Gasse, F. & F.A.Street 1978. Late Quaternary lake-level fluctuations and environments of the northern Rift Valley and Afar region (Ethiopia and Djibouti). *Palaeogeogr., Palaeoclim., Palaeoecol. 24:* 279-325.

Glymph, L.M. 1973. Summary: sedimentation of reservoirs. In: W.C.Ackermann, G.F. White, E.B. Worthington & J.L.Young (eds.), *Man-made lakes: Their problems and environmental effects.* Am. Geophys. Un., Geophys. Monogr. 17, pp.342-348.

Gregory, K.J. & D.E.Walling 1973. *Drainage basin form and process: A geomorphological approach.* Edward Arnold, London.

Grove, A.T., F.A.Street & A.S.Goudie 1975. Former lake levels and climatic change in the rift valley of southern Ethiopia. *Geogr. J. 141:* 177-202.

Halley, E. 1715. On the causes of the saltness of the ocean, and of the several lakes that emit no rivers; with a proposal, by help thereof, to discover the age of the world. *Phil. Trans. R. Soc. Lond. 29:* 296-300.

Hardie, L.A. & H.P.Eugster 1970. The evolution of closed-basin brines. *Mineral. Soc. Am. Spec. Paper 3:* 273-290.

Harvey, T.J. 1976. The paleolimnology of Lake Mobutu Sese Seko, Uganda-Zaire: the last 28 000 years. Unpubl. Ph.D. dissertation, Duke Univ., N.Carolina.

Hay, R.L. 1968. Chert and its sodium silicate precursors in sodium carbonate lakes of East Africa. *Contrib. Mineral. Petrol. 17:* 255-274.

Hecky, R.E. & P.Kilham 1973. Diatoms in alkaline saline lakes: Ecology and geochemical implications. *Limnol. Oceanogr. 18:* 53-91.

Holdship, S.A. 1976. The paleolimnology of Lake Manyara, Tanzania: A diatom analysis of a 56 meter sediment core. Unpubl. Ph.D. dissertation, Duke Univ., N.Carolina.

Hughes, G.H. 1974. Water-level fluctuations of lakes in Florida. *Florida Bureau of Geology, Tallahassee, Map Series 62.*

Jackson, J. 1833. *Observations on lakes being an attempt to explain the laws of nature regarding them; the cause of their formation and gradual diminution; the different phenomena they exhibit etc., with a view to the advancement of useful science.*Bossange, Barthés and Lovell, London.

Jones, B.F., H.P.Eugster & S.L.Rettig 1976. Hydrochemistry of the Lake Magadi Basin, Kenya. *Geochim. Cosmochim. Acta 41:*53-72.

Jones, B.F., S.L.Rettig & H.P.Eugster 1967. Silica in alkaline brines. *Science 158:*1310-1314.

Jones, B.F. & A.S.van Denburgh 1966. Geochemical influences on the chemical character of closed lakes. In: *Hydrology of lakes and reservoirs.* Int. Ass. Hydrol. Sci. Publ. 70.

Kamau, C.K. 1977. The Ol Njorowa Gorge, Lake Naivasha, Kenya. In: D.C.Greer (ed.), *Desertic terminal lakes.* Utah Water Research Laboratory, Logan, Utah, pp.297-307.

Kendall, R.L. 1969. An ecological history of the Lake Victoria Basin. *Ecol. Monogr. 39:* 121-176.

Krishnamurthy, K.V. & A.M.Ibrahim 1973. Hydrometeorological studies of Lakes Victoria, Kyoga and Albert. In: W.C.Ackermann, G.F.White & J.L.Young (eds.), *Manmade lakes: Their problems and environmental effects.* Am. Geophys. Un., Geophys. Monogr. 17, pp.272-277.

Kuznetsov, S.I. 1975. The role of micro-organisms in the formation of lake bottom deposits and their diagenesis. *Soil Science 119:*81-88.

Langbein, W.B. 1961. Salinity and hydrology of enclosed lakes. *US Geol. Surv., Prof. Paper 412.*

Livingstone, D.A. 1975. Late Quaternary climatic change in Africa. *Ann. Rev. Ecol. System. 6:*249-280.

McLachlan, A.J. & S.M.McLachlan 1969. The bottom fauna and sediments in a drying phase of a saline African lake (Lake Chilwa, Malawi). *Hydrobiologia 34:*401-413.

McLachlan, A.J. & S.M.McLachlan 1971. Benthic fauna and sediments in the newly created Lake Kariba (central Africa). *Ecology 52:*800-809.

Maglione, G. & S.Servant 1973. Signification des silicates de sodium et des cherts néoformés dans les variations hydrologiques et climatiques holocènes du bassin tchadien. *C.r. Acad. Sci., Paris, 277D:*1721-1724.

Makin, M.J., T.J.Kingham, A.E.Waddams, C.J.Birchall & B.W.Eavis 1976. Prospects for irrigation development around Lake Zwai, Ethiopia. *Land Resources Study 26,* Land. Res. Div. Ministry Overseas Devel., Tolworth.

Mörth, H.T. 1967. Investigation into the metorological aspects of the variations in the level of Lake Victoria. *Mem. East Afr. met. dept. 4.*

Mower, R.W. 1968. Ground-water discharge toward the Great Salt Lake through valley fill in the Jordan Valley, Utah. *US Geol. Surv., Prof. Paper 600D:*D71-D74.

Müller, G., G.Irion & U.Förstner 1972. Formation and diagenesis of inorganic Ca-Mg carbonates in the lacustrine environment. *Naturwissenschaften 59:*158-164.

Neal, J.T. 1972. Playa surface features as indicators of environment. In: C.C.Reeves, jr. (ed.), *Playa Lake symposium proceedings.* ICASALS publ. 4. Texas Tech. Univ., Lubbock, Texas, pp.107-132.

Peck, E.L. & E.A.Richardson 1966. Hydrology and climatology of Great Salt Lake. In: W.L.Stokes (ed.), *The Great Salt Lake.* Utah Geol. Soc., Guidebook to the Geology of Utah, 20, Salt Lake City, Utah, pp.121-134.

Peng, T.-H., J.G.Goddard & W.S.Broecker 1978. A direct comparison of ^{14}C and ^{230}Th

ages at Searles Lake, California. *Quatern. Res. 9:*319-329.

Reeves, C.C., jr. 1968. *Introduction to paleolimnology.* Elsevier, Amsterdam.

Richardson, J.L., T.J.Harvey & S.A.Holdship 1978. Diatoms in the history of shallow East African lakes. *Polish Arch. Hydrobiol. 25:*341-353.

Richardson, J.L. & A.E.Richardson 1972. History of an African Rift lake and its climatic implications. *Ecol. Monogr. 42:*499-534.

Roche, M.A. 1977. Lake Chad: A subdesertic terminal basin with fresh waters. In: D.C. Greer (ed.), *Desertic terminal lakes.* Utah Water Research Laboratory, Logan, Utah, pp.213-223.

Rooney, T.P., B.F.Jones & J.T.Neal 1969. Magadiite from Alkali Lake, Oregon. *Amer. Min. 54:*1034-1043.

Scholl, D.W. 1960. Pleistocene algal pinnacles at Searles Lake, California. *J. sedim. Petrol. 30:*414-431.

Scholl, D.W. & W.H.Taft 1964. Algal contributors to the formation of calcareous tufa, Mono Lake, California. *J. sedim. Petrol. 34:*309.

Schnitnikov, A.V. 1974. Water balance variability of Lakes Aral, Balkhash, Issyk-Kul and Chany. In: *Hydrology of lakes.* Int. Ass. Hydrol. Sci. Publ. 109, pp.130-169.

Servant, M. 1973. Séquences continentales et variations climatiques: Evolution du Bassin du Tchad au Cénozöique Supérieur. D.Sc. thesis, ORSTOM, Paris.

Snyder, C.T. 1962. A hydrologic classification of valleys in the Great Basin, western United States. *Int. Ass. Hydrol. Sci. Bull. 7:*53-59.

Street, F.A. 1979. Late Quaternary precipitation estimates for the Ziway-Shala Basin, southern Ethiopia. *Palaeoecology of Africa 11:*135-143.

Street, F.A. in press. Chronology of Late Pleistocene and Holocene lake-level fluctuations, Ziway-Shala Basin, Ethiopia. In: *Proc. 8th Panafr. Congr. Prehist. Quat. Stud., Nairobi.*

Szestay, K. 1974. Water balance and water level fluctuations of lakes. *Int. Ass. Hydrol. Sci. Bull. 19:*73-84.

Taieb, M. 1974. Evolution quaternaire du bassin de l'Awash (Rift éthiopien et Afar). D.Sc. thesis, Univ. Paris-VI.

Talling, J.F. & I.B.Talling 1965. The chemical composition of African lake waters. *Int. Rev. Ges. Hydrobiol. 50:*421-463.

Thurber, D.L. 1972. Problems of dating non-woody material from continental environments. In: W.W.Bishop & J.A.Miller (eds.), *Calibration of hominoid evolution.* Scottish Academic Press, Edinburgh, pp.11-17.

Thurber, D.L. & W.S.Broecker 1970. The behaviour of radiocarbon in the surface waters of the Great Basin. In: I.U.Olsson (ed.), *Radiocarbon variations and absolute chronology.* Nobel Symposium 12, Wiley Interscience, New York, pp.379-400.

Truckle, P.H. 1976. Geology and late Cainozoic lake sediments of the Suguta trough. *Nature 263:*380-383.

United Nations (FAO) 1965. *Report on survey of the Awash River Basin.* UN Spec. Fund. Proj. (FAO/SF:10/ETH), Imp. Ethiopian Government, Addis Ababa, 5 vols.

United Nations (UNDP) 1973. *Investigation of geothermal resources for power development: Geology, geochemistry and hydrology of hot springs of the East African Rift System within Ethiopia.* Tech. Report (DP/SF/UN/116). UNDP, New York.

Viner, A.B. & I.R.Smith 1973. Geographical, historical and physical aspects of Lake George. *Proc. R. Soc., Lond. B184:*235-270.

Ward, R.C. 1967, *Principles of Hydrology,* McGraw-Hill, London.

Winograd, I.J. 1962. Interbasin movement of groundwater at the Nevada test site, Nevada. *US Geol. Surv., Prof. Paper 450C:*C108-C111.

Young J.A.T. & R.W.Renaut 1979. A radiocarbon date from Lake Bogoria, Kenya Rift Valley, *Nature 278*, 243-245.

Yuretich, R.F. 1976. Sedimentology, geochemistry and geological significance of modern sediments in Lake Rudolf (Lake Turkana), Eastern Rift Valley, Kenya. Unpubl. Ph.D. dissertation, Princeton Univ.

Zektzer, I.S. 1974. Studying the role of groundwater flow in water and salt balances of lakes. In: *Hydrology of Lakes*, Int. Ass. Hydrol. Sci. Publ. 109, pp.197-201.

Zemljanitzyna, L.A. 1974. Inflow to lakes of the semi-arid zone of the USSR from groundwater. In: *Hydrology of Lakes*, Int. Ass. Hydrol. Sci. Publ. 109, pp.185-190.

Ziminova, N.A. 1974. Sediment balance of the Volga reservoirs. In: *Hydrology of Lakes*, Int. Ass. Hydrol. Sci. Publ. 109, pp.428-431.

ISOTOPIC IDENTIFICATION OF SAHARIAN GROUNDWATERS, GROUNDWATER FORMATION IN THE PAST*

C. SONNTAG / U. THORWEIHE / J. RUDOLPH / E. P. LÖHNERT
/ CHR. JUNGHANS / K. O. MÜNNICH / E. KLITZSCH /
E. M. EL SHAZLY / F. M. SWAILEM

INTRODUCTION

In the great Sahara Basins (Western Desert of Egypt, Cufra-, Syrte- and Mur-zuq-Basin) and the Grand Ergs, huge amounts of groundwater are found. Estimates of these Saharian groundwater reserves have increased from 15 x 10^6 Mill. m^3 in 1966 (Ambroggi 1966) to 60 x 10^6 Mill. m^3 in 1976 (Gischler 1976). If this amount of water is assumed to be homogeneously distributed over the whole Sahara area of 4.5 Mill. km^2, these figures correspond to a hypothetic water layer of 3.3 m and 13 m thickness respectively, which compares with the worldwide average thickness of 55 m (Sonntag 1978). Thirteen years ago, the groundwater of the inner Sahara was still believed to be recharged by 4 x 10^3 Mill. m^3 per year under the assumption of long-range groundwater movement from infiltration areas in rainy mountainous terrain at the periphery (Atlas Mountains, Tibesti, Ennedy, Darfur). For that case, the ratio between the groundwater inventory and annual recharge would yield a mean residence time or mean groundwater age of about 4 000 years and 15 000 years respectively. Moreover, a significant age increase with the direction of this long-range groundwater flow was to be expected. However, isotope dating (^{14}C, tritium, deuterium and ^{18}O measurements) did not yield this increase in groundwater age in the flow direction previously assumed. Moreover, serious geological arguments stand against a long-range ground-water motion through the various Sahara basins. More recently, the ground-water of the inner Sahara rather is believed not to be recharged under the present climate conditions, i.e. the water is to be looked upon as a fossil deposit, and its exploitation is equivalent to mining.

* This work was prepared by C.Sonntag, J.Rudolph, K.P.Münnich and Chr.Junghans, Institut für Umweltphysik, Universität Heidelberg, Federal Republic of Germany; U. Thorweihe and E.Klitzsch, Geologisches Institut der Technischen Universität Berlin; E.P.Löhnert, Geology Dept., University of Ife, Nigeria; E.M.El Shazly, Nuclear Materials Corporation, Cairo, Egypt; and F.M.Swailem, Middle Eastern Regional Radioisotope Centre, Cairo, Egypt.

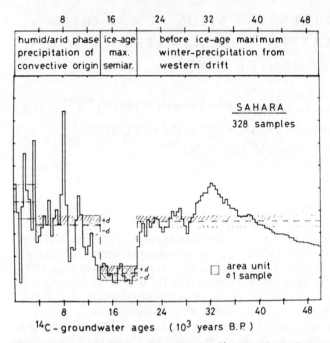

Figure 1. Frequency distribution of apparent ^{14}C ages of Saharian groundwaters based on an initial ^{14}C-content of 85 per cent modern, no carbonate or other age corrections made. The unit area representing one sample is always a rectangle; its width on the time axis is the ± 1 sigma dating uncertainty. Therefore, at low dating precision (high age) the area representing 1 sample is broad and flat, at high precision narrow and high. The ± σ range on the ordinate indicates the statistical error of the frequency distribution for the individual age periods.

Frequency histograms of the ^{14}C groundwater age occurrence in various regions of the Sahara reflect the alternating sequence of humid and arid periods in the late Pleistocene and Holocene. The deep groundwaters from continental deposits of Paleozoic till Upper Cretaceous Age have mainly been formed in a long humid period between more than 50 000 and 20 000 years B.P. (Great Pluvial), whereas shallow groundwaters, which are found here and there in late Tertiary and Quaternary deposits, show ^{14}C ages less than 12 000 years B.P. The time slice between 20 000 and 14 000 years B.P. is less populated, which frequency minimum is interpreted as representing a long dry period (Sonntag 1978).

Three years ago, a significant west-east decrease of the heavy stable isotope content (deuterium and oxygen-18) of fossil Saharian groundwaters has been found, which is similar to the one observed in European winter-precipitation and groundwater, known as 'Continental Effect' (Sonntag 1976, Sonntag

160

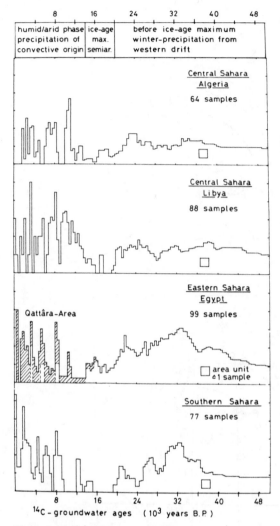

Figure 2. Regional frequency distributions of apparent ^{14}C ages of Saharian groundwaters. The Southern Sahara diagram includes the Sahel Zone. Algeria and Southern Sahara contain data taken from Gonfiantini (1974), Castany (1974) and Mabrook (1976).

1978). Interpretation of the west-east decrease in the case of the Sahara waters as the 'continental effect in groundwater' leads to the conclusion that North Africa must have been influenced by the western drift in the past. This is especially true in the case of the very old waters (age > 20 000 years B.P.). The westerly winds must thus be assumed to have brought rain by carrying

161

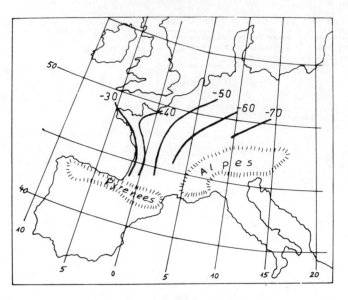

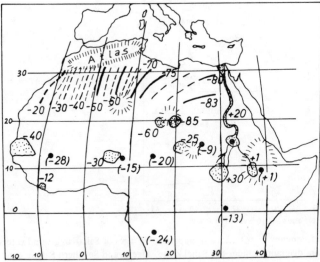

Figure 3. Preliminary δD isoline presentation of modern European and fossil Saharian groundwaters respectively, data of Gonfiantini (1974) included. For Central Africa, the mean δD of modern (and fossil) groundwater (in the dotted areas) and of mean weighted annual precipitation (heavy full dots, numbers in brackets) is shown. The European data points uniquely fall into distinct isozones, the resolution across the isolines (excluding the influence of altitude effects) may turn out to be not much more than $\sigma \approx \pm 50$ km or about ± 2 ‰ in δD.

162

wet Atlantic air masses across the Sahara. A model treatment of the continental effect based on successive Rayleigh condensation steps in a closed air-mass system (no vertical water vapor exchange, and no exchange between rain and water vapor assumed) gave an estimate of the mean paleowinter precipitation as well as its west to east variation across the Sahara. As an example, a paleowinter precipitation of 600 mm at Agadir would lead to about 250 mm in the Murzuq Basin which agrees surprisingly well with the precipitation estimate of Pachur (Pachur 1969) obtained from ^{14}C dated fluvial and limnic deposits and from the water demand of the paleo-fauna.

Since groundwater, once formed, does not change its heavy stable isotope content, the continental effect in Saharian groundwaters indicates that these water bodies were essentially formed by local rainfall. Long-range groundwater movement can only be assumed to exist along stable isotope isolines, for example along the isoline from Tibesti via Cufra and Farafra to Bahariya (Figure 3). A significant ^{14}C age increase along these lines, however, has not been found yet.

On the whole, the isotopes in groundwater contain considerable paleoclimatic information. This encouraged us to undertake a more detailed isotope study of Saharian groundwaters and of modern European groundwater for comparison. During the last two years, the amount of isotope data of Saharian groundwaters has approximately been doubled. In particular, there is now also data available from the Western and Southern Sahara and the adjacent Sahel Zone. All isotope data is stored in a computerized data collection, which also contains geographical, hydrogeological as well as hydrochemical data, relevant for reconstruction of the history of Saharian groundwater formation. In this paper, a short extract of this data collection is presented.

[14]C GROUNDWATER AGES

Figure 1 shows an updated frequency distribution of the ^{14}C ages of 328 Saharian groundwaters. These ages are based on an initial ^{14}C content of 85 per cent modern carbon, no carbonate exchange or other age corrections have been made. This statistical presentation is similar to a previous one based on 190 groundwaters only (Sonntag 1978). However, the significance of the frequency minimum in the time slice between 20 000 and 14 000 years B.P. is more obvious now. The average groundwater population of this time interval is significantly lower than that of the adjacent time periods, and there is no doubt that during maximum glaciation groundwater formation in the Sahara was low.

As can be seen from the regional groundwater age distributions (Figure 2), this frequency minimum seems to exist everywhere in the Sahara, even in the Southern Sahara and the Sahel Zone, although the statistical significance is relatively poor there. The long dry period (20 000-14 000 years B.P.) in the region south of the Sahara is also indicated by low levels of African

163

Lakes, Lake Tchad for example (Gasse 1979). The age spectrum of the Southern Sahara includes the ^{14}C groundwater data of the Ferlo Basin/Senegal (Castany 1974, Klussmann). Its deep groundwater in the Maastrichtian formation shows an age increase which indicates groundwater flow from the rivers Senegal and Gambie at the periphery towards the basin center, with a main flow from SE to NW. We believe that the deep groundwater of Ferlo Basin mainly originates from these rivers. This conclusion is supported on the one hand by the hydrochemical data which changes from bicarbonate-type groundwater in the south-eastern part of the basin to sulfate-type and finally chloride-type groundwater in the basin center under continuous increase of the total dissolved substance (TDS) from 250 ppm to 2 000 ppm (Castany 1974, Klussmann). On the other hand, the groundwaters show a uniform deuterium and oxygen-18 content, which is expected for rain in the catchment areas of the rivers, but being too low for rainfall in the Ferlo Basin. It is surprising that the ^{14}C age spectrum of the Ferlo groundwaters seems to show also the frequency minimum in the time slice of the ice-age maximum. If so, reduced flow rate of the rivers is to be concluded for that time period.

Altogether the ^{14}C groundwater data indicate that this long semi-arid or even arid period has affected the whole of North Africa down to about 20 degrees northern latitude.

CONTINENTAL EFFECT IN DEUTERIUM AND OXYGEN-18

New deuterium- and oxygen-18 data of fossil groundwaters from the Western and Southern Sahara and modern European groundwater (France, Great Britain, Germany, Poland), as well, confirm the similarity in the spatial variation of these heavy stable isotopes and thus the similarity in the present winter-rain pattern of Europe with that of the Northern Sahara in the past. Figure 3 shows isoline-presentations of the deuterium data; the spatial variation of ^{18}O is equivalent because of a linear correlation between δD^* and $\delta^{18}O^*$ (see below). These isoline-presentations show the western drift influence to groundwater formation (west-east decrease in the heavy stable isotope content = increasing negative δ-values towards east. In the case of the fossil Sahara waters the western drift seems to have reached down to about 20 degrees northern latitude only. At lower latitudes, the heavy stable isotope data of paleo-waters and of modern groundwaters (bomb tritium!) from the Sokoto-, Tchad- and Bara Basin shows meridional variations which indicate their tropical convective origin. These waters show also a slightly higher deuterium excess (see below) in comparison to the Northern Saharian paleo-waters, higher deuterium excess being typical for tropical summer rains. This means, the region south of the Sahara, in particular the Sahel Zone, has received tropical rain in all

---

* D- and ^{18}O-content presented as per mill deviation δ from that of Standard Mean Ocean Water (SMOW).

time periods. However, the paleo-waters of the Sahel Zone are considerably depleted isotopically (by about −15 to −20 per cent in deuterium) if compared with modern groundwaters at the same locality. This phenomenon has also been observed in paleo-waters and modern groundwaters from England, Western Germany and Hungary (Bath 1978, Geyh 1972). We do not believe that the lower heavy stable isotope content of the paleowaters has been caused by greater isotopic fractionation due to lower temperatures in the past. It might rather be due to a steeper inland decrease of the mean annual precipitation, which could have led to a steeper continental effect in D and ^{18}O in the past. Whether this is true or not, we believe the different heavy stable isotope content of local paleo-water and modern groundwater to be primarily due to a change of the atmospheric circulation.

DEUTERIUM/OXYGEN-18 -RELATIONSHIP

The heavy stable isotope data of the Saharian paleo-waters and of modern European groundwaters is presented in the δD-versus δ^{18}O diagrams of Figure 4. Figure 5 shows the diagram of the data of paleo- and modern groundwaters sampled at our last expedition from Algier to Tchad Basin. As can be seen from these figures, the European groundwaters fall on the classical 'meteoric water line' (MWL), δD = 8·δ^{18}O + 10, which at δ^{18}O = O shows a deuterium excess of d = +10 per mille. The variance of the data points from this regression line is very small. The Sahara waters, however, fall on a straight line parallel to this, cutting the δD-axis at d = +5‰ only. The higher variance of the data points around this line may indicate slight kinetic re-evaporation effects as to be expected under Savanna climate — evaporation loss of rain drops falling through a relatively dry atmosphere and/or evaporation loss at the surface before infiltration into the ground. In both cases, the residual rain water, which was originally on the MWL, will become isotopically enriched along evaporation lines of gradient 3-6 in the δD/δ^{18}O-diagram, i.e. the data points are shifted to the right from the meteoric water line (see Figure 5). We do, however, not believe that the lower apparent deuterium excess of the Saharian paleo-waters is exclusively due to kinetic re-evaporation since the eastern Sahara data (Cufra Basin, Western Desert of Egypt) shows a variance from the regression line as low as the European waters, but nevertheless a smaller deuterium excess.

It is commonly assumed that the deuterium excess of meteoric water is due to kinetic isotope fractionation in the evaporation of sea water. The data of a wind/water tunnel experiment in this laboratory is in agreement with the issue that the deuterium excess decreases with decreasing moisture deficit of the air over the ocean (Münnich 1978). The smaller d of the Sahara waters would correspond to a moisture deficit of only 15 per cent in the marine atmosphere to be compared with approximately 22 per cent today. Present paleoclimatic ideas are not in contradiction to this conclusion. The ground-

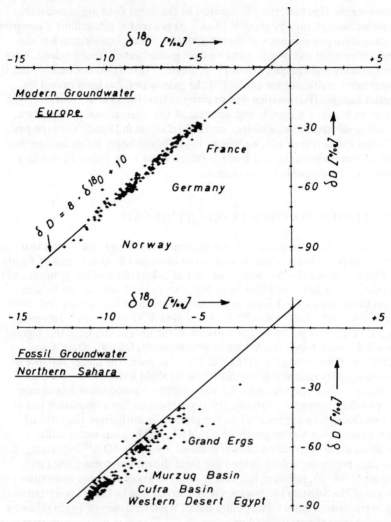

Figure 4. δD versus $\delta^{18}O$ diagram of modern European and fossil Saharian groundwaters (further data from England and Poland and from the Western and Southern Sahara not yet included). All European data points measured by Christel Junghans; the spread around the regression line is dominantly due to our present analytical precision of ± 1 ‰ for deuterium. The isotope data from the Grand Ergs was taken from Gonfiantini (1974).

waters south of the Sahara – modern groundwaters and paleo-waters below 20 degrees northern latitude – show slightly higher deuterium excess values than the waters previously discussed. As can be seen from Figure 5, modern groundwaters from the Sahel Zone and from Central Nigeria, as well as tropical rain at Ife/Nigeria (Löhnert), have d-values between +12 and +15‰, whereas the paleowaters of the Sahel are slightly below the Meteoric Water Line (MWL). However, their deuterium excess of about +8‰, nevertheless, is slightly higher than that of the North Saharian paleo-waters. The higher deuterium excess in tropical rain is due to water vapor originating from the subtropical ocean, where the relative humidity of the atmosphere is considerably lower (moisture deficit thus higher) than in other ocean areas. The higher kinetic fractionation at evaporation from the subtropical ocean can be seen from the meridional variation of the δ^{18}O-data of ocean surface water and marine vapor (Craig 1966). Its correlation with the moisture deficit of the marine atmosphere is in accordance with the wind/water tunnel experiment mentioned above.

PRESENT AND FUTURE PROGRAMME WITH RESPECT TO RECONSTRUCTION OF GROUNDWATER FORMATION HISTORY

Our isotope investigation of Saharian groundwaters goeś on. The isotope analyses will be supplemented by noble gas measurements which hopefully will supply information about the local temperature of the lower atmosphere and of the ground at time of groundwater formation, i.e. the spatial paleo-temperature pattern of the Sahara. The thermometer function of dissolved noble gases in groundwater is due to the fact that the physical solubility of the heavy noble gases argon, krypton and xenon in water varies considerably with temperature. For example, the solubility of argon decreases by near a factor of 2 from 10°C to 50°C, in the case of krypton and xenon the decrease is even steeper. The temperature dependence of the solubility of the light noble gases He and Ne, however, is negligible. Thus Ne and He can be used as indicators for possible outgassing loss. Experience has shown (Bath 1978, Mazor 1972) that even in hot springs the noble gas content is conserved, and no outgassing influences the noble gas temperature data. Noble gas data of Saharian groundwaters, which have been sampled last year, will soon be available.

At present, we try to model the various Saharian groundwater bodies. Estimates of the aquifer mass M and thus the groundwater mass $V = \Sigma \cdot M$ (porosity Σ = open model parameter) are based on spotwise available borehole data. This information is interpolated by suitable polynomial fits to give a general picture of the geological situation in the area. To this end, a three-dimensional computer presentation of the geological structure is used. A first presentation of the Western Desert of Egypt yields geological cross-

167

sections, which look, as expected, quite similar to those hand-drawn with geological intuition. The computer presentation has, however, the advantage that the masses of the various sediment layers (aquifers and aquicludes) can be obtained rapidly. Moreover, the geological picture can be continuously improved by incorporation of new geological information (borehole data) replacing interpolated data.

For reconstruction of the groundwater formation history the annual recharge R of the various groundwater bodies is needed. Then groundwater mass V and recharge R yield the mean residence time $T = V/R$ (years) which is to be compared with the ^{14}C model ages (exponential-, piston flow- and dispersion-model). The mean residence time is the link between the classical hydrogeological and the isotope data of groundwater. The annual recharge $R = I \cdot F$ is obtained from the component I of the local paleoprecipitation P, which deeply infiltrates and thus forms groundwater, and from the extension F of the potential infiltration areas of the various aquifers. Estimates of the infiltration areas F might be obtained from satellite pictures. For an improved model estimate of the local paleo-rain P, we try to simulate the continental effect in D and ^{18}O of groundwater by a general water vapor transport model which considers the local vertical distribution of the heavy stable isotopes in atmospheric vapor as well as the molecular exchange between the local rain and this vapor. This model is more realistic than our previous Rayleigh Condensation model which has yielded the above-mentioned estimate of the spatial variation of the paleo-winter rain across the Sahara. Special attention will be paid to the deuterium excess, which seems to be a tracer for the origin of the water vapor. In this new model treatment, the isotope data of the IAEA/WMO Global Precipitation Network will play an essential role.

The infiltration rate I results from the alternating play between local rain P and evapotranspiration E. This water balance at the interface atmosphere/ ground is governed by the local (paleo-) climate. On the one hand, the actual evapotranspiration E and thus the infiltration $I = P - E$ depends on the meteorological parameters, as there are mean annual rainfall and its distribution over the year, temperature, humidity, cloudiness. On the other hand, E and I depend on the local geomorphology, such as local relief, soil-type and its infiltration properties (porosity, suspended water content), vegetation. Regional water balances yield estimates for the average evapotranspiration and infiltration which, unfortunately, may differ considerably from the local data needed for our groundwater recharge estimate. Therefore, we go different ways. Based on daily weather records, we compute short-time model predictions for the actual evaporation and thus infiltration. This model data is then compared with field observation, which yield soil moisture change and downward displacement of a tritium tracer peak. The results obtained for various bare and plant covered soil types at the moderate climate of Heidelberg area (Thoma 1978) encouraged us to extend this combined study to arid and tropical climates, in the Negev Desert/Israel and tropical Nigeria/ Ife. These studies, hopefully, will improve the theoretical insight into the

168

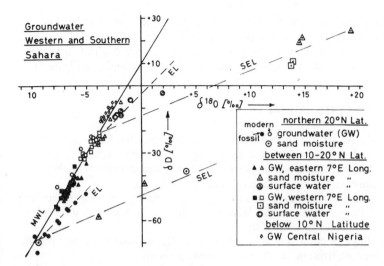

Figure 5. $\delta D/\delta^{18}O$ diagram of groundwaters sampled during our Algier/Tchad Basin expedition in 1978.
Key: EL = Evaporation Line due to isotope enrichment in evaporation from an open water surface (lake, dug well); SEL = Sand Evaporation Line due to isotope enrichment of the residual sand moisture in the course of drying up.

water balance of the uppermost soil layer, so that more realistic estimates of rain water infiltration and thus groundwater recharge under various (paleo-) climatic conditions can be expected. In our Dahna sand dune study of 1972 (Dincer 1974) and 1978 (Sonntag 1978), we have measured the vertical distribution of bomb tritium and the heavy stable isotopes D and ^{18}O in sand moisture down to 6 m depth. From the position of the bomb tritium maximum due to rain of 1963/64, a local groundwater recharge of 20 mm/y has been derived, which is in agreement with a model estimate based on the climatic data of Riyadh (about 70 mm annual rainfall, rain season December-April). Under the assumption, however, that this annual rainfall is homogeneously distributed over the whole year, the model estimate yields no recharge. This shows the sensitivity of groundwater recharge to rainfall distribution, i.e. rain intensity. Only soil water seeping down deeper than about 1 m below ground level will not be lost by evaporation. This can also be shown from the D- and ^{18}O vertical profiles which yield a steep increase from about 1 m depth towards sand surface. This isotope enrichment is due to different vapor pressures and to kinetic isotope fractionation in water vapor diffusion through the uppermost sand layer into the atmosphere, in which process also the light water molecule species $H_2^{16}O$ is favored.

As a by-product, the Dahna sand dune study has encouraged us to test a sand moisture sampling method for D- and ^{18}O-analyses in the residual mois-

169

ture which shall provide information about the variation of D and ^{18}O in groundwater across the Sahara under the present hydrometeorological pattern of sporadic rainfall. This isotope pattern is to be compared with that in the past (first isotope data of Saharian sand samples are presented in Figure 5).

ACKNOWLEDGEMENTS

Special thanks are due to Dr M.A.Geyh, Niedersächsisches Landesamt für Bodenforschung, Hannover, who has supplied all his unpublished ^{14}C and tritium data of Saharian groundwaters and has provided aliquots of these water samples for stable isotope analyses in Heidelberg.

This investigation was supported by the Deutsche Forschungsgemeinschaft, the Heidelberger Akademie der Wissenschaften and the Internationales Büro der Kernforschungsanlage Jülich.

Dipl.-Phys. M. Münnich, Miss A.Ebert and Miss V.Grossmann took care of the sample preparation for ^{14}C and ^{13}C measurements.

REFERENCES

Ambroggi, R.P. 1966. Water under the Sahara. *Scientific American 214:*21.

Bath, A.H. et al. 1978. Paleoclimatic trends deduced from the hydrochemistry of a Triassic sandstone aquifer, United Kingdom. In: *Isotope hydrology*. IAEA, Vienna, pp.545ff.

Castany, G. et al. 1974. Etude par les isotopes du milieu du régime des eaux souterraines dans les aquifères de grandes dimensions. In: *Isotope techniques in groundwater hydrology*. IAEA, Vienna, pp.243ff.

Craig, H. & L.I.Gordon 1965. Deuterium and oxygen-18 variations in the ocean and the marine atmosphere. In: *Stable isotopes in oceanographic studies and paleotemperatures*. Spoleto Pisa.

Dincer, T. et al. 1974. Study of the infiltration and recharge through the sand dunes in arid zones with special reference to the stable isotopes and thermonuclear tritium. *J. Hydrology 23:*79.

Gasse, F. 1979. Late Quaternary diatom record of reference sections from Ethiopia and Djibouti Territory, regional climatic interpretation. Symp. Sahara. Mainz, 1.-4.4.1979.

Geyh, M.A. 1972. Basic studies in hydrology and ^{14}C and ^{3}H measurements. *24th Int. Geol. Congress, Section 11, Montreal 1972*, pp.227ff.

Gischler, C.E. 1976. *Present and future trends in water resources development in Arab countries*. UNESCO Report.

Gonfiantini, R. et al. 1974. Etude isotopique de la nappe du Continental Intercalaire et de ses relations avec les autres nappes du Sahara. In: *Isotope techniques in groundwater hydrology*. IAEA, Vienna, pp.227ff.

Klussmann, D., Büro Dr. Pickel, Fuldatal, private communication.

Löhnert, E.P., Geol. Dept. Univ. of Ife, Nigeria, to be published.

Mabrook, B. & M.Sh.Abdel Shafi 1977. Hydrological and environmental isotope studies of Bara Basin, Central Sudan. *Symposium on trace elements in drinking water, agriculture and human life*. Middle Eastern Radioisotope Centre and Goethe Institut, Cairo.

Mazor, E. 1972. Paleotemperatures and other hydrological parameters deduced from noble gases in groundwaters; Jordan Rift Valley, Israel. *Geochim. Cosmochim. Acta* 36:1321.

Münnich, K.O. et al. 1978. Gas exchange and evaporation studies in a circular wind tunnel, . . . In: A.Favre & K.Hasselmann (eds.), NATO conference series V, Air-sea interactions, Vol. 1. Plenum Press, New York.

Pachur, H.J. 1979. The flat areas of Central Sahara in the early Holocene. Symp. Sahara. Mainz, 1.4.4.1979.

Sonntag, C. et al. 1976. Zur Paläoklimatik der Sahara: Kontinentaleffekt im D- und ^{18}O-Gehalt pluvialer Saharawässer. *Naturwissenschaften* 63(10):479.

Sonntag, C. et al. 1978. Paläoklimatische Information im Isotopengehalt ^{14}C datierter Saharawässer: Kontinentaleffekt in D und ^{18}O. *Geol. Rundschau* 67:413.

Sonntag, C. et al. 1978. Palaeoclimatic information from D and ^{18}O in ^{14}C dated North Saharian groundwaters; groundwater formation in the past. In: *Isotope hydrology.* IAEA, Vienna, pp.569ff.

Sonntag, C. et al. 1978. Environmental isotopes in North African groundwaters. The Dahna Sand Dune study (Saudi Arabia). *IAEA, Advisory group meeting on the application of isotope techniques to arid zones hydrology, Vienna* (in press).

Thoma, G. et al. 1978. New technique of in situ soil moisture sampling for environmental isotope analysis applied at Pilat Dune near Bordeaux. HETP modelling of bomb tritium propagation in the unsaturated and saturated zones. In: *Isotope hydrology.* IAEA, Vienna, pp.753ff.

171

INDICATORS OF CLIMATE
IN MARINE SEDIMENTS

CLIMATE INDICATORS IN MARINE SEDIMENTS OFF NORTHWEST AFRICA – A CRITICAL REVIEW

EUGEN SEIBOLD

Geologisch-Paläontologisches Institut, Universität Kiel

Changes of air temperatures, wind systems, precipitation and evaporation on land, of water temperatures, salinities, oxygen contents and currents in the ocean due to climatic variations can be partly recorded in sediments. Principally the continents are areas of erosion by water and wind as dramatically illustrated by the widespread concentration of paleolithic to recent artefacts on Sahara surfaces. The ocean bottom, however, is the final dump area of terrestrial and marine input.

Climatic factors more directly affect erosion and sedimentation on land. In the ocean they are buffered by sea water. On the other hand, on land local conditions may cause important interference of the climatic changes in space and time. In the ocean generally more uniform conditions and widespread distribution processes tend to characterize regional and sometimes even global climatic changes. The following is a short introduction to the marine environment and sedimentation off NW Africa. Special attention will be given to difficulties in using sediment layers as pages of a book in deciphering the history of climates for this area.

INTRODUCTION

The margin of the Northwest African continent illustrates a climatic situation typical for various subtropical areas bordering the Eastern side of oceans (Figure 1). The Canary Current, similar to other Eastern Boundary Currents off California, Chile/Peru, Southwest Africa, and to some degree, off West Australia, transports cooler water from higher to lower latitudes. The water masses in the Southern Hemisphere are directly connected with Antarctic waters. The connections with Arctic waters, however, are restricted, and were even more so during glacial periods with lowered sea levels. Trade winds generate·coastal upwelling off NW Africa producing an additional local lowering of surface water temperatures and therefore a decrease in precipitation. This *hydroclimate* and its changes are indicated in marine sediments by siliceous and calcareous algae as diatoms and coccoliths, by planktonic and benthic foraminifera and other marine organisms such as radiolarians and

Figure 1. Oceanic surface currents and arid continental regions.
1 = desert; 2 = steppe; 3 = cold; 4 = warm surface currents (after Schwarzbach 1974)

pteropods, particularly by oxygen isotopes in carbonate shells. Details are discussed in several special symposium papers, for example Diester-Haass, Duplessy, Lutze & Pflaumann, Sarnthein et al., Shackleton & Thiede.

The adjoining continents are typically arid zones with transitions to more humid conditions to the North and South (Figure 1). Naturally land relief causes regional differences, resulting in semi-arid mountains in the Sahara and in other platform deserts.

The modern zonation of *continental climates* in NW Africa is schematically illustrated in Figure 2. Temperatures, rainfall and their variations are partly reflected on land by products of weathering and soil formation, both long lasting processes (stained quartz, clay minerals), and by plant particles (pollens, phytholiths). However, land climate strongly influences transport of these terrigenous particles, too. As is well known, increased aridity, for example, not only means deficiency but also unreliability and more drastic variations in precipitation and therefore transport. River input north and south of the Sahara might reflect the geological and climatological source-area conditions more directly than does eolian input. Additionally, both agents are changing the material itself by partial destruction and sorting. Therefore a quantitative approach to eolian vs. river input studies is difficult, as demonstrated by the contributions of Diester-Haass and Sarnthein (this volume).

Surface sediments and sediment cores off Northwest Africa have been investigated by the Kiel Marine Geological Group since 1967 based on several cruises (Figure 2): Meteor-8/1967 off Morocco (Closs et al. 1969); Meteor-25/1971 off Sahara and Central Senegal (Seibold 1972); Meteor-39/1975 and Valdivia-10/1975 from Morocco to South Senegal (Seibold & Hinz 1976). Summaries of the results are given in Seibold 1978 (1975), Seibold et al. (1976) and Seibold & Fütterer (1980).

In the marine sediments off Northwest Africa nearly all sediment particles were originally contributed by wind, rivers and organisms. Generally, surface sediments in water depths down to about 1 000 m (Figure 3) contain about two-thirds and more terrigenous material off the Senegal river and less than one-third off the Sahara. In water depths of more than 3 000 m roughly half of the sediments are of terrigenous origin. The rest, i.e. carbonates are contributed by marine organisms, mostly planktonic foraminifera and coccoliths in deeper water, and benthic molluscs on the shelf. Additionally, proximal dust from the northern Sahara (N of $21°N$) contains up to some 20 per cent carbonates (Koopmann et al. 1979), a fraction which is difficult to distinguish in hemipelagic sediments. With the exception of sandy shelf sediments 40-60 weight per cent consists of silt (2-60 μm) fractions. Higher clay contents are reported from some shelf areas as off the Senegal mouth and from the Continental Rise more distant from land (Fütterer 1977 and Sarnthein et al., this volume). These are broad generalisations with local exceptions, especially on the shelf and upper slope due to Pleistocene relict particles both of biogenous and terrigenous origin.

177

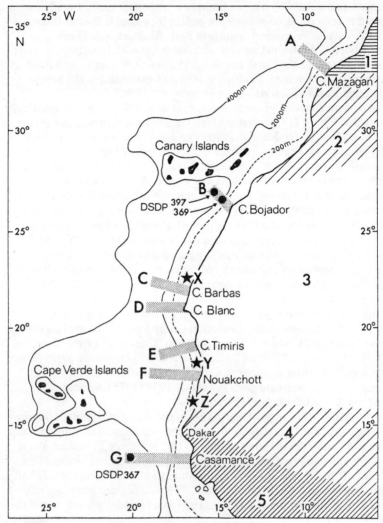

Figure 2.
1. Standard profiles, Northwest African continental margin, Geological Institute, Kiel University.

A = 'Meteor' cruise 8-1967; B, C = 'Meteor' cruise 25-1971; D = 'Meteor' cruise 39-1975; E, F, G = 'Valdivia' cruise 10-1975; X, Y, Z = detailed shelf studies.

2. Zonation of vegetation and climate:

1 = Mediterranean scrub, warm-temperate, winter rains;	3 = Desert, hot, dry;
	4 = Steppe, hot, winter dry;
2 = Steppe, hot, summer dry;	5 = Savanna, tropical, winter dry.

Senegal mouth near Z (15°47'N). (After Seibold & Fütterer 1979).

178

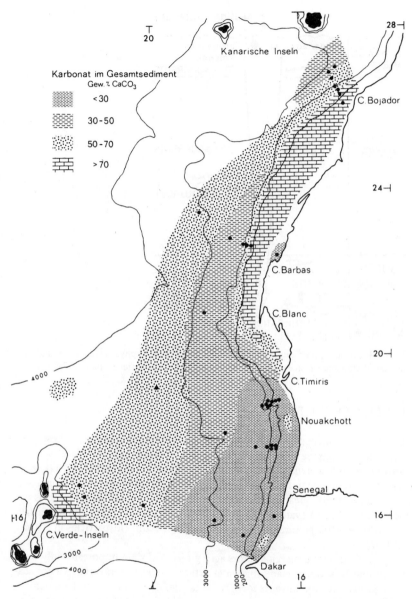

Figure 3. Carbonate content of surface sediments (weight % CaCO₃) (after Seibold &
Fütterer 1979).

179

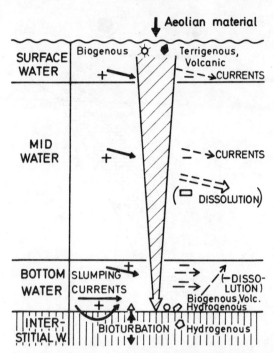

Figure 4. Sediment particles input (+) and losses (−) due to physical, chemical and biological filters from surface waters to the marine sediments (after Seibold 1979b).

FILTER EFFECTS

Before finally incorporated into the oceanic bottom sediments all these particles have to pass several filters. During this pathway they are transported downward and sidewise. They are altered or completely destroyed. These filters are acting in surface, mid, bottom and interstitial waters. Mechanical, chemical and biological processes are involved (Figure 4). Therefore, it is necessary to try to evaluate the importance of these filters for the reconstruction of a realistic picture of climates from these sediment particles in space and time.

Near continental margins filters acting *physically* are important only for bottom waters. Surface and midwater effects such as lateral transport of suspended particles by currents seem to be negligible because of fast particle settling rates. Empty pteropods and planktonic foraminifera may reach more than 2 000 m day^{-1}, diatom frustrules and coccolithophore plates around 10 m day^{-1}. Often these smaller particles become incorporated in faecal pellets with settling rates of some 100 m day^{-1}. Therefore, a lateral transport of only some kilometers to tens of kilometers may be the result of these currents. Even very fine grained eolian dust clay minerals, pollen and spores may partly settle down as aggregates bound together by coatings of organic material. Clay mineral distribution on the ocean floor for example roughly follows long-term climatic weathering zonations on the continents (Chamley et al. 1977).

Bottom water movements by currents and waves cause erosion or non-deposition on wide shelf areas off Northwest Africa. Some examples are given by Einsele et al. (1977), Newton et al. (1973), Siedler & Seibold (1974). Ancient sediments are reworked and therefore climatic indicators of different areas and periods get mixed even in the patchy areas of finer shelf sediments. Absolute ages from these sediments are difficult to determine. Organic particles are damaged within the sands. Normally, fine particles are winnowed out and are transported over the shelf edge to the continental slope and rise.

During recent years, poleward-flowing contour currents were observed on the Northwest African slope from Senegal up to C.Bojador in water depths between 100 and 600 m (Johnson et al. 1975, Mittelstaedt 1976, 1978). In 150-200 m water depths daily mean speeds of 7-20 cm sec^{-1} were measured. Normally higher velocities are required to stir up and erode finer particles. They may be reached by current pulses and by internal waves as described by Fahrbach & Meincke (1974) off Sierra Leone. Even weak currents can transport sediment particles if they are brought into suspension, e.g. by organisms touching the sea bottom or living on and in it. Continuous television profiles and extremely high bioturbation effects in the slope and upper rise sediments indicate the importance of this mechanism off Northwest Africa.

The most important effect is a 'grain by grain' downslope transport with a general enrichment of finer grains with increasing water depths as demonstrated in Figure 5.

It is interesting to note a sharp downslope decrease of the sand fraction and an increase of the silt fraction around 600 to 800 m as a result of winnowing by the poleward undercurrent.

The Kiel group concentrated their investigations of climatic indicators in sediments from water depths below 1 000 m because in finer sediments normally core lengths are highest and 'noises' by physical filters are reduced.

Certainly some of these indicators are transported northward, too. Lange (1975) and Diester-Haass & Müller (1979) give examples for river muds.

Quantitative analysis of coarser grains originally deposited in shallow water as glauconite and other shelf relict materials, thick shelled benthic shallow water pelycopods (Diester-Haass 1975); red algae, ascidiae particles and cliona boring chips (Fütterer 1977) and benthic foraminifera (Lutze et al. 1979) confirm the importance of the 'grain by grain' downslope transport. At the steep slope off C. Bojador up to about one-third of the sediment in 2 800 m water depth may have been transported downslope by this mechanism (Lutze et al. 1979). Therefore again mixing of different climatic indicators is evident. On the other hand, local conditions are averaged.

181

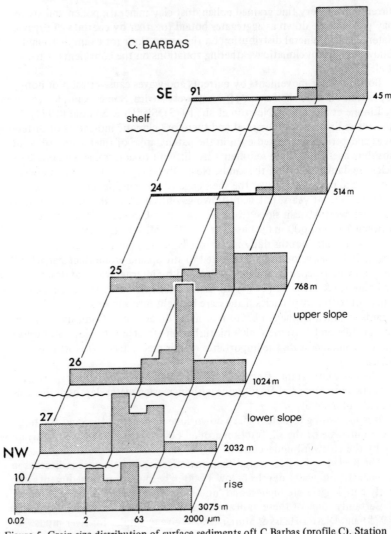

Figure 5. Grain size distribution of surface sediments off C.Barbas (profile C). Station numbers: Geol. Inst. Univ. Kiel (91 = 12391 etc.). Silt fractions are 2-6, 6-20 and 20-63 μm (after Seibold & Fütterer 1979).

Chemical filters cause weakening of tests by partial dissolution or even complete losses of siliceous and carbonatic organic remains. Higher temperatures and lower silica concentrations within the uppermost few hundred meters of the ocean ('silica corrosion zone', Berger 1976) are responsible for dissolution of empty thin-walled diatom tests as observed by Richert (1975)

182

and others. Some few of them may bypass this filter if contained in faecal pellets (Schrader 1971, 1972). Further dissolution in interstitial waters may partly be prevented by saturation as a result of very high supply of siliceous tests. In water depths of 1 000-3 000 m where most of the investigated cores have been taken, midwater and possibly even bottom water losses by dissolution of both aragonitic and calcitic tests seem to be small. Planktonic foraminifera tests, however, demonstrate aggressive interstitial waters depending on organic matter content together with bulk sedimentation rates, which are also responsible for the absence of aragonitic shells below 500 m (Diester-Haass & Müller 1979). Decreasing resistance against dissolution seems to be illustrated by the succession: benthic foraminifera – planktonic foraminifera – coccoliths (calcite) – pteropods (aragonite) and: radiolarians – diatoms (opaline silica) (see more detailed discussion in Seibold 1979a).

Biological filters are acting from surface waters to the sediments. Metabolism of foraminifers can fractionate the stable isotopic balance of carbonate shells. Predators in the water column damage tests but preserve them in faecal pellets, too. However, most effects are to be expected by bioturbation within the sediments. Bottom dwellers partly destroy or damage particles and transport them up and down. Because of high food supply in upwelling waters very high densities of these benthic animals can be expected on the upper continental slope off Northwest Africa, decreasing oceanward. Therefore, most of the primary sediment structures are destroyed. Andreas Wetzel (Ph.D.-Thesis, Kiel University, personal communication) distinguishes 25 different morphological types of borrows. They are arranged in storeys (Figure 6) reaching some 0.9 m sediment depth in 250 m water depths and some 0.3 m in 2 000 m water depths. This storey-like arrangement seems to be confined to sediments with 0.5-2.0 % C_{org} and sedimentation rates of about 20-200mm 1 000 yr^{-1}. Bioturbation off Northwest Africa homogenises the uppermost 3 cm of sediment completely. In general, upward particle transport works over larger distances than downward transport with a less effective transport of coarser grains. Therefore sediment ages determined by methods evaluating biogenous particles generally are too high by up to 1 500 years, especially the very surface layer as demonstrated by ^{14}C ages (Berger & Heath 1968, Sarnthein 1972, Peng et al. 1977, Berger & Johnson 1978). Much more detailed investigations of sediment samples selected after radiographs from cores and taking into account sedimentation rates and bioturbation are necessary to diminish this stratigraphic noise and to improve time resolution.

Even in an area off Northwest Africa with Holocene sedimentation rates between 30 and 100 mm yrs^{-1} in 800-2 000 m water depths compared with some 10 mm 1 000 yrs^{-1} in distal deep sea carbonate oozes, time resultion up to now may have *optimal* values of some 300-500 to 1 000-2 000 years, depending on sedimentation rates and bioturbation types. Therefore, some short climatic fluctuations may escape from the sedimentary record.

As a result of these complicating 'filter' effects and of the fact that only a few indicators for climates on land and in the ocean can be used in marine

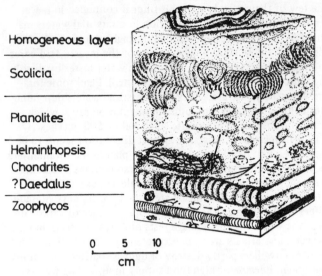

Homogeneous layer

Scolicia

Planolites

Helminthopsis
Chondrites
?Daedalus

Zoophycos

0   5   10
cm

Figure 6. Biogenic sedimentary structures off Northwest Africa (after Wetzel, Geol. Inst. Univ. Kiel, Ph.D. Thesis, unpublished).

sediments it is not easy to reconstruct some climate parameters even from modern sediments. The reconstruction of fossil climates from continental margin sediments is even more complicated.

GLACIAL CONDITIONS

During glacial stages, the repeated sea level fall of about 100 m exposed the Northwest African shelf enlarging the Sahara region to the West. Dunes reached the present shelf edge causing eolian sand turbidites off Senegal (Sarnthein & Diester-Haass 1977). Formation of soils and cementation by calci-crusts characterized the shelf, now covered with water. Certainly land climate was influenced locally or even regionally by this extension of the Sahara to the West.

Near the glacial shelf edge water dynamics were intensified. Upwelling was stronger (Diester-Haass, this volume). Possibly, the poleward undercurrent and 'grain by grain' transport were intensified, too. Activation of even more massive downslope sediment transport is indicated by turbidites originating from submarine canyons or from the open slope, by mass movements as slides and slumps occurring up to about 11 000 years B.P. (Pflaumann, Haake in Seibold & Fütterer 1979), i.e. during the phase of maximal sea level rise.

The wind system changed (Sarnthein & Walger 1974, Sarnthein 1978) and certainly eolian sediment supply was intensified, at least south of C. Blanc.

The Canary current (Thiede 1977) and upwelling were stronger. Both changes certainly influenced the climate in coastal zones. Until now the effects of possibly more vigorous bottom currents have not been well investigated off Northwest Africa. Most of these factors increased sedimentation rates by a factor of about 2-3 for terrigenous and biogenous particles together with organic carbon.

The Holocene sea level rise was not continuous as indicated in Northwest Africa by the 'Nouakchottian' transgression with a sea level about 2,5 m higher than at present (Delibrias 1974, Einsele et al. 1973) again influencing local coastal climates and sedimentation.

Up to the present, information is not sufficient to detect any possible relations between the volcanism of the Islands off Northwest Africa and the climate of this region.

CONCLUSIONS

Further research is required to

1. Improve the definition, quality of record and quantitative significance of climatic indicators found in sediments, as for example wind vs. river input. What kind of particles and how many are transported to the sea under different conditions of precipitation, vegetation, infiltration, run off? Are short pulses quantitatively more important than longer periods of less extreme conditions? A better correlation between research results on land and offshore is needed.

2. Distinguish climatic differences in time and space between the various parts of the Northern and Southern Sahara, and between coastal and inland areas. Was coastal upwelling in the northern part really intensified in spite of hypothetically stronger westerlies during glacial periods?

3. Better understand the mentioned filter effects, especially the alteration of primary features by transport, sorting and differential dissolution.

4. Define the stratigraphy by further using quantitative evaluation of marine faunas and the distribution of isotopes in marine organic remains and by considering new results of bioturbation in sediment cores with continuous sedimentation and high but varying sedimentation rates.

ACKNOWLEDGEMENTS

As can be seen from the bibliography, most results and remarks are based on the co-operation with the other members of the Kiel Marine Geological Group.

The help given by colleagues from the Bundesanstalt für Geowissenschaften und Rohstoffe in Hanover and financial assistance by the Deutsche Forschungsgemeinschaft and the Bundesministerium für Forschung und Technologie is gratefully acknowledged.

Finally, many aspects were added to the manuscript after the lectures and discussions during the Mainz Symposium. Dr Terry Healy corrected the English.

REFERENCES

Berger, W.H. 1976. Biogenous deep sea sediments: Production, preservation and interpretation. In: J.P.Riley, R.Chester (eds.), *Chemical oceanography 5*, 2nd ed. Academic Press, London, pp.265-387.

Berger, W.H. & G.R.Heath 1968. Vertical mixing in pelagic sediments. *J. Mar. Res. 26:* 134-143.

Berger, W.H. & R.F.Johnson 1978. On the thickness and the ^{14}C age of the mixed layer in deep-sea carbonates. *Earth Planetary Sci. Letters 41:* 223-227.

Chamley, H., L.Diester-Haass & H.Lange 1977. Terrigenous material in East Atlantic sediment cores as an indicator of NW African climates. *'Meteor' Forsch.-Ergebn. C 26:* 44-59.

Closs, H., G.Dietrich, G.Hempel, W.Schott & E.Seibold 1969. Atlantische Kuppenfahrten 1967 mit dem Forschungsschiff 'Meteor' – Reisebericht. *'Meteor' Forsch-Ergebn. A 5:* 1-71.

Delibrias, G. 1974. Variation du niveau de la mer sur la côte ouest-africaine depuis 26 000 ans. *Colloque int. CNRS, 219:* 127-134.

Diester-Haass, L. 1975. Sedimentation and climate in the Late Quaternary between Senegal and the Cape Verde Isles. *'Meteor' Forsch-Ergebn. C 20:* 1-32.

Diester-Haas, L. & P.J.Müller 1979. Processes influencing sand fraction composition and organic matter content in surface sediments off Africa (12-19°N). *'Meteor' Forsch-Ergebn. C* (in press).

Einsele, G., P.Elouard, D.Herm, F.C.Kögler & H.U.Schwarz 1977. Source and biofacies of Late Quaternary sediments in relation to sea level on the shelf off Mauritania, West Africa. *'Meteor' Forsch.-Ergebn. C 26:* 1-43.

Einsele, G., D.Herm & H.U.Schwarz 1973. Holocene eustatic (?) sealevel fluctuation at the coast of Mauretania. *'Meteor' Forsch.-Ergebn. C 18:* 43-62.

Fahrbach, E. & J.Meincke 1974. High frequency velocities fluctuations near the bottom over the continental slope. *'Meteor' Forsch.-Ergebn. A 20.*

Fütterer, D. 1977. Die Feinfraktion (Silt) in marinen Sedimenten des ariden Klimabereichs: Quantitative Analysenmethoden, Herkunft und Verbreitung. Habil. Schrift Univ. Kiel.

Johnson, D.R., E.D.Barton, P.Hughes & C.N.K.Mooers 1975. Circulation in the Canary Current upwelling region off Cabo Bojador in August 1972. *Deep Sea Res. 22:* 547-558.

Koopmann, B., A.Lees, P.Piessens & M.Sarnthein 1979. Skeletal carbonate sands and wind-derived silty marls off the Saharan coast: Baie du Levrier, Mauretania. *'Meteor' Forsch.-Ergebn. C.* 15-57.

Lange, H. 1975. Herkunft und Verteilung von Oberflächensedimenten des westafrikanischen Schelfs und Kontinentalhanges. *'Meteor' Forsch.-Ergebn. C22:* 61-84.

Lutze, G.F., M.Sarnthein, B.Koopmann, U.Pflaumann, H.Erlenkeuser & J.Thiede 1979. Meteor cores 12309: Late Pleistocene reference section for interpretation of the Neogene of Site 397. *Initial Reports DSDP, 47A.* Washington D.C. (in press).

Mittelstaedt, E. 1976. On the currents along the northwest African coast south of 22° north. *Dt. hydrogr. Z. 29*(3):97-117.

Mittelstaedt, E. 1978. Physical oceanography of coastal upwelling regions with special reference to Northwest Africa. *Canary current symposium, Las Palmas, April 1978,* Abstr. 40.

Newton, R.S., E.Seibold & F.Werner 1973. Facies distribution patterns on the Spanish Sahara continental shelf mapped with sidescan sonar. *'Meteor' Forsch.-Ergebn. C 15:* 55-77.

Peng, T.H., W.S.Broecker, G.Kipphut & N.Shackleton 1977. Benthic mixing in deep sea cores as determined by ^{14}C dating and its implications regarding climate stratigraphy and the fate of fossil fuel CO_2. In: N.R.Andersen & A.Malahoff (eds.), *The fate of fossil fuel CO_2 in the Oceans*. Plenum Press, New York, pp.355-373.

Richert, P. 1975. Die räumliche Verteilung und zeitliche Entwicklung des Phytoplanktons, mit besonderer Berücksichtigung der Diatomeen, im nordwestafrikanischen Auftriebswassergebiet. Diss. Univ. Kiel.

Sarnthein, M. 1972. Stratigraphic contamination by vertical bioturbation in Holocene shelf sediments. *24th Int. Geol. Congr.* Sect. 6, pp.432-436.

Sarnthein, M. 1978. Eolomarine sediments. *Abstr. 10th Int. Congr. Sedim., Jerusalem.* pp.575-576.

Sarnthein, M. & L.Diester-Haass 1977. Eolian-sand turbidites. *J. Sed. Petrol. 47:*868-890.

Sarnthein, M. & E.Walger 1974. Der äolische Sandstrom aus der W-Sahara zur Atlantikküste. *Geol. Rundschau 63:*1065-1087.

Schrader, H.J. 1971. Fecal pellets: Role in sedimentation of pelagic diatoms. *Science 174:* 55-57.

Schrader, H.J. 1977. Kieselsäure-Skelette in Sedimenten des ibero-marokkanischen kontinentalrandes und angrenzender Tiefsee-Ebenen. *'Meteor' Forsch.-Ergebn. C8:* 10-36.

Schwarzbach, M. 1974. *Das Klima der Vorzeit,* 3rd ed. Stuttgart.

Seibold, E. 1972. Cruise 25/1971 of R.V. 'Meteor': Continental Margin of West Africa. General report and preliminary results. *'Meteor' Forsch.-Ergebn. C 10:*17-38.

Seibold, E. 1978. Tiefseesedimente als Klimazeugen vor Nordwestafrika. *Leopoldina, Mitt. Dt. Akad. Naturf. Leopoldina,* Series 3, Yr 21 (1975), Halle.

Seibold, E. 1979a. Sediments in upwelling areas, particularly off Northwest Africa. *Rapports et procès Verbaux Series.* ICES, Kopenhagen (in press).

Seibold, E. 1979b. Mechanical processes influencing the distribution of pelagic sediments. *Micropaleontology* 24: 407-421.

Seibold, E., L.Diester-Haass, D.Fütterer, M.Hartmann, F.C.Kögler, H.Lange, P.J.Müller, U.Pflaumann, H.-H.Schrader & E.Suess 1976. Late Quaternary sedimentation off the Western Sahara. *Acad. brasil. Cienc., 48*(Supl.): 287-296.

Seibold, E. & D.Fütterer 1980. Sediment dynamics on the NW African continental margin. In: *Heezen Memorial Volume.* Wiley & Chichester (in press).

Seibold, E. & K.Hinz 1976. German cruises to the continental margin of North West Africa in 1975: General reports and preliminary results from 'Valdivia' 10 and 'Meteor' 39. *'Meteor' Forsch.-Ergebn. C 25:*47-80.

Siedler, G. & E.Seibold 1974. Currents related to sediment transport on the Ibero-Moroccan continental shelf. *'Meteor' Forsch.-Ergebn. A 14:*1-12.

Thiede, J. 1977. Aspects of the variability of the glacial and interglacial North-Atlantic eastern boundary current (last 150 000 years). *'Meteor' Forsch.-Ergebn. C 28:*1-36.

MARINE CLAY SEDIMENTATION AND CLIMATE IN THE LATE QUATERNARY OFF NORTHWEST AFRICA (ABSTRACT)

HERVE CHAMLEY
Sedimentology and geochemistry, Lille, France

Quaternary clay materials in the Atlantic realm are known to be mainly of detrital origin, but their dependence on sources, transport agents and climate is not very clear everywhere. With this end in view, the clay mineralogy of Riss to Holocene sediments off Northwest Africa has been studied from seven cores taken by R/V Meteor on the Atlantic margin between 19° and 27°N, and between 1 060 and 3 320 m water depth. Methods used are x-ray diffraction on less than 2 μm non-calcareous particles, and transmission electromicroscopy on less than 8 μm particles.

Minerals identified are numerous: smectite (40% of clay fraction), illite (20%), kaolinite (15%), chlorite (0-10%), various irregular mixed-layers (chiefly illite-smectite, chlorite-smectite), palygorskite (attapulgite), quartz, feldspars, goethite, pyrite, subamorphous iron oxides (Chamley & Diester-Haass 1976, Chamley et al. 1977). Most of them are detrital, including smectite and palygorskite (see Griffin et al. 1968, Lange 1975, and discussion in op. cit.).

The distribution of clay assemblages changes in three ways according to space and time. When considering sediments of similar age in increased distance from coasts, smectite and palygorskite amounts increase while illite and iron oxides amounts decrease. The cause is a differential settling of clay minerals, small-sized and low-flocculable smectite and fibrous clays being favoured in deep sediments, far away from continental sources.

Another change concerns the latitudinal distribution. Kaolinite and smectite progressively increase towards the South, while illite and chlorite decreases. This change reflects the main pedogenic trend on the nearby landmasses (Pédro 1968), due to increasing climatic alteration towards the humid tropical zone. As a consequence, the influence of northsouth marine currents or dust supplies seems to be of minor importance compared to the direct latitudinal influence through eolian and fluvial contribution from the African continent. Palygorskite abundance locally increases near 21°N, perhaps due to local continental outcrops rich in this mineral.

The third type of mineralogical variations affects the sedimentary column, independently of the depth of burial. As a rule in 19° to 25°N cores, chlorite and palygorskite abundance increases in sediments of interglacial age, and irregular mixed-layers amounts increase in sediments of glacial age. A very good correlation arises from comparison of clay mineral data and coarse fraction data (Chamley et al. 1977, Diester-Haass, this volume), from isotopic stages 6 to 1. Considering the sensibility of clay minerals to continental alteration (Millot 1964) and the present latitudinal zonation of marine and continental surficial clays, these variations are interpreted as consequences of late Quaternary climatic changes in the northern Sahara. In this area, periods of interglacial age such as the present one (stages 1, 5) seem to be marked by stronger desert conditions than periods of glacial age (stages 2, 3, 4, 6). Data presently available from the investigation of marine sediments point to the dominant influence of humidity/aridity alternations on the nearby land-masses, whereas the possible influence of distant marine or/and eolian supplies seems to be of minor importance.

Grants CNEXO (France) n° 77/5489, 78/5708.

REFERENCES

Chamley, H. & L.Diester-Haass 1976. Argiles marines détritiques et climats quaternaires au large de l'Afrique nord-occidentale. *4ème réun. ann. Sci. Terre* (Soc. géol. Fr.), Paris, 104.

Chamley, H., L.Diester-Haass & H.Lange 1977. Terrigenous material in East Atlantic sediment cores as an indicator of NW African climates. *'Meteor' Forsch.-Ergebn. C* 26:44-59.

Diester-Haass, L. 1979. Upwelling and climate off NW Africa during the late Glacial. *Palaeoecology of Africa 12.*

Griffin, J.J., H.Windom & E.D.Goldberg 1968. The distribution of clay minerals in the world ocean. *Deep-Sea Res., 15:*433-459.

Lange, H. 1975. Herkunft und Verteilung von Oberflächensedimenten des westafrikanischen Shelfs und Kontinentalhanges. *'Meteor' Forsch.-Ergebn. C 26:*44-59.

Millot, G. 1964. *Géologie des Argiles.* Masson, Paris.

Pédro, G. 1968. Distribution des principaux types d'altération chimique à la surface du globe. Présentation d'une esquisse géographique. *Rev. Géogr. phys. Géol. dyn. 10:* 457-470.

VARIATIONS OF THE SURFACE WATER TEMPERATURES ALONG THE EASTERN NORTH ATLANTIC CONTINENTAL MARGIN (SEDIMENT SURFACE SAMPLES, HOLOCENE CLIMATIC OPTIMUM, AND LAST GLACIAL MAXIMUM)

UWE PFLAUMANN

Geologisch-Paläontologisches Institut der Universität,. Kiel

INTRODUCTION

Planktonic foraminifera have served as indicators of marine paleoenvironmental parameters since the pioneer work of Schott (1935). Since then qualitative and quantitative investigations of many authors have increased our knowledge of climatic variations and of the changes in ocean circulation patterns during the Quaternary. The statistical method introduced by Imbrie & Kipp (1971) and refined by Kipp (1976) was used by the members of the CLIMAP project (McIntyre et al. 1976) for a reconstruction of surface temperature and salinity maps of the North Atlantic ocean during the last glacial maximum at 18 000 years ago.

This paper is concerned with the paleotemperatures of two time slices, viz. the Holocene climatic optimum and the last glacial maximum. These were investigated along the West African continental margin based on planktonic foraminiferal taphocoenoses in deep sea-cores as compared to the dispersal pattern in (modern) sediment surface samples.

Most samples were obtained from the collections of the German research vessels 'Meteor' and 'Valdivia' (Seibold 1972, Seibold & Hinz 1976). Additional material came from 'Vema' and 'Discoverer' expeditions.

METHODS

Sediment-surface samples and core samples have been treated as described by Pflaumann (1975). But the lower limit of grain sizes was shifted from 160 to 150 μm in diameter.

Counting procedures follow the species concept described by Pflaumann & Krashenninikov (1978). To adapt this concept to a form suitable for the transfer functions (Kipp 1976), only simplications were necessary. The basic data will be given in the appendix of a next paper (Pflaumann, in preparation).

The basis for the reconstruction of paleoecological maps is an adequate chronostratigraphy. A great many cores have been dated absolutely by radiocarbon ages (Geyh 1976, Erlenkeuser et al. in preparation). Another method utilising the oxygen isotope ratios of benthic foraminifera, has evolved into a useful tool because it is almost independent of the geographic location of the studied cores, and is controlled predominantly by fluctuations in the total amount of the ice deposits on land (Shackleton & Opdyke 1973). The $\delta^{18}O$ record of planktonic foraminifera quite closely parallels that of the benthics (Emiliani 1955, 1958, 1966, 1972, Emiliani et al. 1961, Erlenkeuser et al. in preparation). However, the tool is sensitive to stratigraphic gaps and/or other sedimentological discrepancies, and has to be controlled by traditional strati-graphical methods such as biostratigraphy (Pflaumann & Krashenninikov 1978) or by locally restricted lithostratigraphy (Hays & Perruzza 1972) or ecostratigraphy (Ericson et al. 1961, Ruddiman 1971, Wollin et al. 1971, Pflaumann 1975, Zobel & Ranke 1978). In recent years, it has been con-firmed that climatic changes occur within very short time intervals of less than 1 000-2 000 years (Broecker et al. 1960, Lutze et al. 1979). So one has to be very careful in estimating ages, where absolute age datings are not available.

The quantitative evaluation of the planktonic foraminiferal taphocoenoses is based on the transfer function technique of Imbrie & Kipp (1971) which was successfully used in various pelagic parts of the world oceans (McIntyre et al. 1976, Imbrie et al. 1973, Thunnell 1979). However, difficulties arise, when the methods are applied to regions outside the pelagic realm (Thiede 1978, Molina Cruz & Thiede 1977), such as the hemipelagic conditions within the area under investigation near the continental margin. For the estimation of past conditions, I used the regression formulae derived by Thiede (1977), which are based on the evaluation of selected surface samples from the East-ern North Atlantic off Europe and West Africa. Thiede has reduced the species number to 19 categories, a number small enough to allow the use of a Compu-corp 325 Scientist calculator with a 392 magnetic tape unit for the calculations.

Basic data are derived from quantitative counts of planktonic-foraminifer species from the sea bottom sediment. These taphocoenoses are handled as projections of the environmental conditions of the sea surface waters, where the planktonic foraminifera are living. The CABFAC-analysis (Klovan & Imbrie 1971) performs the raw counts into varimax assemblages, which are indica-tive of a set of environmental parameters of an ecologically defined water mass. Surface water temperature is regarded as the most important factor.

To check the validity of the regression formulae used 39 'Meteor' and 'Valdivia', and 9 'Vema' sediment surface samples have been examined. Only in a few samples was the generally required minimum number of 300 speci-mens per count not reached. The results are plotted in Figures 2-11, together with some samples extracted from Thiede (1977, Appendix).

The communality (Figure 2) may be considered as a scale of the degree to which the samples are explained by the transfer functions. In our region, the communality is high in the northern half, while in the southern part some samples are off limits of the applicability of the equations.

The factor loadings (relatively large absolute values) of the five associa-

192

tions show that in our region essentially only four of them contribute to the model. The Subpolar association (Figure 7) may be neglected, and considered as statistical noise.

MODERN DISTRIBUTION PATTERNS

The maps of the factor loadings (Figures 3-7) show considerable noise, but a number of general trends are obvious: High loadings of the Tropical and Subtropical assemblages to the west and south, high loadings of the Polar to Temperate assemblage around Cape Blanc. A synopsis is given in Figure 8, where the dominant assemblages are shown. These patterns are obviously related to the cold Canary current, which turns the normally longitudinally directed gradients toward a latitudinal direction (Cifelli & Stern Benier 1976). Upwelling effects are indicated by additional isolated high loadings of the Polar to Temperate factor, that are outside their normal northern distribution.

According to Thiede (1977), the Transitional planktonic foraminiferal assemblage today is confined to the central North Atlantic water mass and is mainly formed by species that prefer greater water depths. The analysis of Thiede indicates that distributions of this factor are similar to those of areas of modern coastal upwelling. Our estimates, however, do not fully support this. The zone of upwelling is more or less near the shelf edge, and there the subtropical factor plays the dominant role (Figure 8). The Transitional assemblage usually is distributed further offshore and marginal to the upwelling area rather than within it. As the dominant assemblage the Transitional is mainly found between the dominant Tropical and the dominant Subtropical assemblages in the region off Cape Blanc and off Cape Vert, while Factor 1, the Polar to Temperate assemblage, covers the region of the main upwelling features.

Figures 9 and 10 compare the calculated with the 'measured' temperatures the latter being derived from the U.S. Naval Office Atlas, which summarizes measurements over about the last 100 years. In order to give an indication of the short-term regional gradients that are possible, measurements from ten days sections of the 1972/73 portion of the 'A.v.Humboldt' expedition (Schemainda et al. 1975) are inserted in the region between Cape Blanc and Cape Vert.

Figure 11 shows the differences between the estimated and 'measured' temperatures for summer and winter. Around Cape Blanc, the estimates in general appear to be too low, while further offshore they are in quite good agreement with the measured data, the differences being randomly distributed. In addition, the surface water temperature estimates for winter fit better than those for summer. However, throughout the year a regional pattern in the deviation between calculated and observed temperatures is indicated with a difference of more than $5°C$ around Cape Blanc and Cape Vert. There are

193

three possible explanations for all of these deviations: 1) the applied formulae are not sufficient, 2) the 'measured' values are too crude, 3) the taphocoenoses are not representative of actual surface water temperatures.

To 1): the regional distribution of the error suggests the existence of an additional factor, possibly related to upwelling features. However, further results will be presented more extensively in Pflaumann, in preparation.

To 2): the region under discussion is characterized by the neighbourhood of very different water masses: the cool Canary current, a warm northward directed current off Cape Vert, and the more or less intense but short-lived eddies of the upwelling region. The boundaries between these water masses are never fixed (Shaffer 1974). So the measured temperatures averaged over long time may include considerable noise.

To 3): because this is a region where redeposition at the continental slope is a quite common feature, one cannot exclude the possibility that the samples, though carefully checked, contain older fauna (Seibold 1976, Bein & Fütterer 1977, Lutze et al. 1979). Also bioturbation mixes the 'surface' sediment with older horizons. And finally, the sample interval and the estimated sedimentation rates indicate that the sediment surface samples comprise a time slice of at least 200 years.

These discrepancies require further investigations and restrict the following results to preliminary ones.

PALEOTEMPERATURE ESTIMATES

For the reconstruction of past sea-surface conditions two, climatically very different levels, were selected: the Holocene climatic optimum and the last glacial maximum. The positions of these time slices in the cores have been determined by radiocarbon datings, and/or oxygen-isotope curves, and/or by micropaleontological analysis. Some core data from Thiede (1977) have been added in order to make a more complete picture.

HOLOCENE DISTRIBUTION PATTERNS

The Holocene climatic optimum is included in 15 cores. If we take the age of the minimum $\delta^{18}O$ values as the age of the Holocene maximum of deglaciation (Shackleton 1967, 1969, Shackleton & Opdyke 1973, Thiede 1978, Pastouret et al. 1978, Lutze et al. 1979, Erlenkeuser et al. in preparation) some problems arise in the exact positioning of a minimum value. Most of the curves are scattered, due to sampling position and sizes, and the dates of minimum $\delta^{18}O$ value lie within a time interval of 500 to 1 500 years, which is near the frequency of the rapid fluctuations found by Lutze et al. (1979). However, the extremes in our region are centered more closely around 5 500 than 6 000 years B.P. (Lutze et al. 1979, Erlenkeuser et al. in preparation.

194

The list of samples selected for the Holocene climatic optimum is given in Table 1. Core 12345-5 could not be included as exceptionally high abundances of the Globorotalia cultrata complex do not allow use of the transfer fluctions (Kipp 1976).

The communalities from the Holocene climatic optimum samples used as a scale for the reliability of the model (Figure 12), show a trend similar to the sediment surface samples, viz. a southward directed decrease. The estimates for the winter temperatures (Figure 14b) have a similar pattern as the modern ones in the north, but with generally 1-2°C higher temperatures. To the south, a northward directed lobe of warm water is indicated near the slope north of Cape Vert similar to modern summer conditions (Figure 9a). The summer temperatures of the climatic optimum (Figure 14a) were generally about 2-3°C higher than at present, and even higher in the southern region. Core 12310-4 forms an exception showing cooler conditions compared to modern ones, which may indicate the variability of water types influencing the fossil assemblages.

Among the dominant assemblages (Figure 13) the Polar to Temperate factor is shifted northward and replaced by the Subtropical or by the Transitional assemblages at the Holocene climatic optimum. If Factor 1, the Polar to Temperate planktonic foraminiferal assemblage of Thiede (1977), is related to the intensity of the Canary current (Thiede 1977), its reduction in the dominant assemblages points to a less developed current intensity during the Holocene climatic optimum in comparison to the modern one. The Transitional assemblage alternates with the Subtropical one in dominance along the continental margin.

Table 1. List of samples used as representatives for the Holocene climatic optimum and their main transfer function technique results

Core	Refer. sample	Commu-nality	Factor loadings 1	2	3	4	5	Temp. estim. T_S	T_W
12309-2	37.5	.9784	.6843	.3074	−.5241	.3754	.0026	18.27	14.42
12310-4	20.5	.9656	.7916	.1961	−.4865	.2064	−.1457	17.67	12.51
12328-5	50	.9571	.4073	.1394	−.6674	.5649	.0850	20.72	13.54
12329-6	12	.9284	.3907	.3430	−.4929	.6272	.1475	24.49	16.88
12331-4	8	.7456	.1185	.7714	−.2309	.1723	.2313	24.23	20.68
12337-5	20	.7484	.2166	.4661	−.4705	.4986	.1193	23.72	17.22
12347-2	40	.8492	.1170	.6373	−.5249	.3219	.2242	24.47	17.83
12379-1	37.5	.9528	.5293	.1654	−.6756	.4309	−.0562	18.56	13.45
12392-1	25	.9664	.4556	.2509	−.7202	.4200	−.0279	19.67	14.63
13289-3	20	.8815	.4536	.2131	−.4809	.6053	.1808	21.81	14.67
13291-1	20	.9151	.3655	.2262	−.6450	.5589	.0435	21.49	14.51
13292-3	40	.8864	.1839	.5718	−.2110	.6896	−.0739	28.53	21.47
13255-3	38	.8298	.1924	.4476	−.4673	.6081	.0656	26.24	18.14
TAG 72/1	3	.8853	.3398	.3451	−.4082	.6718	.1811	26.59	18.04

Table 2. List of samples used as representatives for highest estimated surface water temperatures shortly after the Holocene climatic optimum and their main transfer function technique results. Dominant assemblages are in italics.

Core	Refer. sample	Delay (Ka)	Commu-nality	Factor loadings 1	2	3	4	5	Temp. estim. T_S	T_W
12309-2	17.5	1.7	.9534	.1663	*.7155*	−.5445	.3386	.0519	23.54	18.12
12310-4	6.5	2.1	.9022	.2778	.4402	−.4532	*.6349*	−.1507	23.30	16.78
12328-5	40	1.2	.9228	.1630	.1376	−.6795	.6006	.2342	24.13	14.69
12329-6	12		.9284	.3907	.3430	−.4929	*.6272*	.1475	24.49	16.88
12331-4	8		.7456	.1185	*.7714*	−.2309	.1723	.2313	24.23	20.68
12337-5	20		.7484	.2166	.4661	−.4705	.4986	.1193	23.72	17.22
12347-2	30		.8712	.1528	.4812	−.4858	*.5757*	.2213	28.13	19.07
12379-1	27.5	0.5	.9050	.3812	.2154	−.5998	*.6262*	−.0878	21.53	14.03
12392-1	20	1.7	.9541	.2811	.3076	−.7190	.5116	−.0425	22.28	15.34
13289-3	10	2.4	.9313	.4265	.1883	−.4835	*.6901*	.0628	.23.40	15.22
13291-1	10		.8935	.4533	.1698	−.4819	*.6521*	.0421	21.73	14.42
13292-3	40		.8864	.1839	.5718	−.2110	*.6896*	−.0739	28.53	21.47
13255-3	38		.8298	.1924	.4476	−.4673	*.6081*	.0656	26.24	18.14
TAG 72/1	3		.8853	.3398	.3451	−.4082	*.6718*	.1811	26.59	18.04

Table 3. List of samples used as representatives for the last glacial maximum and their main transfer function technique results.

Core	Refer. sample	Commu-nality	Factor loadings 1	2	3	4	5	Temp. estim. T_S	T_W
12309-2	202.5	.9451	.6192	.0971	−.6332	.2663	.2836	13.97	9.93
12310-4	96	.9846	.7341	.0968	−.6164	.2275	.0676	16.11	11.35
12328-5	220	.8661	.4948	.1347	−.4486	.5795	.2570	19.87	13,34
12329-6	50	.8560	.6209	.1571	−.4487	.4463	.2127	16.76	11.84
12331-4	57	.8851	.4387	.2853	−.5645	.4178	.3437	18.94	13.37
12347-2	360	.9604	.4203	.2257	−.6873	.5021	.0908	20.56	14.31
12345-5	470	.9105	.4036	.1891	−.6145	.5616	.1375	20.98	13.96
12379-1	232	.9575	.7335	.0452	−.6059	.2133	.0697	15.29	10.59
12392-1	182.5	.9718	.6657	.0638	−.6996	.1611	.0961	15.79	11.27
13289-3	120	.9547	.5506	.1370	−.7196	.3258	.0939	17.15	12.45
13291-1	110	.9284	.6181	.0524	−.6506	.3220	.1289	15.24	10.75

The distribution patterns described and plotted on Figures 13 and 14 deals with the estimates around the climatic optimum, as derived from oxygen isotopes and absolute datings (Geyh 1976, Geyh et al. 1974, Lutze et al. 1979, Erlenkeuser et al. in preparation). However, using the record of planktonic foraminifera to predict the surface water temperatures, a considerable delay is indicated. There are about 500 to more than 2 000 year delays in reaching maximum estimated temperatures. The highest surface water values are reached later in the north than in the south. As shown in Table 2, the number of the dominant factors is reduced, and the modern upwelling indicating Factor 1 is replaced by the Transitional assemblage as the dominant. The calculated surface water temperatures at this maximum exceed those of the climatic optimum by more than 5°C for summer conditions or up to more than 4°C for winter conditions. These differences are actually near the outer range of the confidence limits as given by Thiede (1977) and cannot be neglected. The length of the time interval of the 'optimal' surface water temperatures is difficult to estimate because adequate absolute age determinations are yet rare and insufficient due to present day bioturbation and coring technique.

Additional investigations are necessary to determine the reason for the delay. However, it seems, that warm water faunas invade the northerly regions more slowly than temperatures as indicated by oxygen isotope values of planktonic foraminifera (Erlenkeuser et al. in preparation) would allow. It cannot be excluded that unfavourable life conditions, like too many eddies per life cycle, may suppress the dispersion rate over longer time intervals (comp. Bé et al. 1977). This could be seen as an indication that tropical or even warm water adapted species are more sensitive to relatively cool conditions than species which tolerate cooler waters.

GLACIAL DISTRIBUTION PATTERNS

The 18 000 years level is accepted by many investigators as the time of the last glacial maximum (McIntyre et al. 1976). In our region, this level is reached only by a limited number of 11 suitable cores (Table 3). The communalities of the samples from this time period (Figure 15) are slightly elevated as compared to the modern patterns and show that the model works better with faunal assemblages of higher latitudes or of lower surface water temperatures.

In glacial times, the dominant assemblages are the Polar to Temperate and the Subtropical (Figure 16). The Transitional assemblage is dominant at one site only (12328-4). The Tropical assemblage never is dominant in the cores studied, as in the modern situation.

On the average, the estimated summer and winter temperatures (Figure 17a, b) for 18 000 years B.P. are about 3.5°C lower than today. The isotherms do not point to a near coastal warm water lobe as today, though the number of observations is limited. As compared to the Holocene climatic optimum, the temperatures are about 4°C lower in the mean for winter and about 5°C lower in the mean for summer (Figure 18a, b). Site 12328 off Cape Blanc shows nearly no differences between the mentioned time slices.

The widespread dominance of the Polar to Temperate assemblage confirms the assumption of a more intense Canary current during glacial periods (Thiede 1977, McIntyre et al. 1976).

Like in the Holocene, there is a delay of minimum surface water temperatures relative to the oxygen isotope values (Erlenkeuser et al. in preparation).

197

Table 4. List of samples used as representatives for lowest estimated surface water temperatures shortly after the glacial maximum and their main transfer function technique results. Dominant assemblages are in italics.

Core	Refer. sample	Delay (Ka)	Commu- nality	Factor loadings 1	2	3	4	5	Temp. estim. T_S	T_W
12309-2	202.5	.1	.9451	.6192	.0971	−.6332	.2663	.2836	13.97	9.93
12310-4	96		.9846	.7341	.0968	−.6164	.3275	.0676	16.11	11.35
12328-5	210	.5	.8842	.6471	.0634	−.5338	.4463	.1627	15.02	10.47
12329-6	50		.8560	.6209	.1571	−.4487	.4463	.2127	16.76	11.84
12331-4	57		.8851	.4387	.2853	−.5645	.4178	.3437	18,94	13,37
12347-2	350		.9694	.6039	.1673	−.5553	.4984	.1416	18.64	12.93
12345-5	470		.9105	.4036	.1891	−.6145	.5616	.1375	20.98	13,96
12379-1	212.5	1.0	.9220	.7891	.0424	−.4731	.2290	.0731	14.49	9.78
12392-1	172.5	.4	.9844	.8168	.0603	−.5344	.1653	.0284	15.95	10.59
13289-3	110	.5	.9223	.6030	.0630	−.6452	.3371	.1571	15.23	10.73
13291-1	110		.9284	.6181	.0524	−.6506	.3220	.1289	15.24	10.75

But the time interval is very short, and may fall within the confidence limits. However, arrangement of dominant assemblages (Table 4) is much simpler and there are only two: the Polar to Temperate and the Subtropical. The cold-water region reaches further to the south. If we compare the lowest estimated temperatures of the glacial time with that of the warmest ones of the Holocene there are mean differences of about 8°C for the summer and 5°C for the winter along the continental margin (Figure 19a, b).

Though we have to consider the limitations on the estimates as mentioned above and the reliability of the transfer functions used which is better during cooler times, the estimates of the surface water temperatures during the last glacial maximum are in line with the results of McIntyre et al. (1976). The additional samples strengthen the hypothesis of the existence of a cold water lobe off West Africa. But its source cannot be differentiated between an intensified mass transport of cold glacial surface water or a marked increase of upwelling (Gardner & Hays 1976) as responsible for an extension of the isotherms towards the glacial equator (McIntyre et al. 1976). However, the close proximity of the Subtropical and the Polar to Temperate assemblages during the time of minimal surface water temperatures in the glacial time suggests the existence of a narrow but intensive longitudinal arranged upwelling region as also inferred by other methods (Diester-Haass 1975, Pflaumann 1976, Sarnthein & Walger 1974, Gardner & Hays 1976).

ACKNOWLEDGEMENTS

I am indebted to the members of the Kiel working group on isotope stratigraphy for contributing unanimously their working results and for co-operation, especially H.Erlenkeuser, B.Koopmann and M.Sarnthein. Discussions with the members of the Mainz Symposium stimulated me to publish these preliminary results, though many questions are open for further investigations. I would like to thank the core curator of the Lamont-Doherty Geological Observatory, New York for generously furnishing deep-sea core samples. T.Healy and S.McLean procured the English version. This study was supported by the Deutsche Forschungsgemeinschaft.

REFERENCES

Bandy, O.L. 1956. Paleotemperatures of Pacific bottom waters and multiple hypotheses. *Science 123*(3104 –:459-460.
Bé, A.W.H. & D.S.Tolderlund 1971. Distribution and ecology of living planktonic foraminifera in surface waters off the Atlantic and Indian Oceans. In: B.M.Funnel & W.R. Riedel (eds.), *The micropaleontology of oceans.* Cambridge Univ. Press, 105-149.
Bé, A.W.H., J.Damuth, L.Lott & R.Free 1976. Late Quaternary climatic record in western equatorial Atlantic sediments. *Geol. Soc. Amer. Mem. 145:*165-200.
Bé, A.W.H., C.Hemleben, O.R.Anderson, M.Spindler, J.Hacunda & S.Tuntivate-Choy 1977. Laboratory and field observations of living planktonic foraminifera. *Micropaleont. 23:*155-179.
Bein, A. & D.Fütterer 1977. Texture and composition of continental shelf to rise sediments off the northwestern coast of Africa: An indication for downslope transportation. *'Meteor' Forsch.-Ergebn. C 27:*46-74.
Berger, W.H. 1969. Ecological patterns of living planktonic foraminifera. *Deep-sea Res. 16:*1-24.
Berger, W.H. 1971. Sedimentation of planktonic foraminifera. *Marine Geol. 11:*325-358.
Boltovskoy, E. 1969. Distribution of planktonic foraminifera as indicators of water masses in the western part of the Tropical Atlantic. In: *Proc. Sympos. Oceanography and Fisheries Res. of the Tropical Atlantic.* Unesco, Paris, 45-55.
Boltovskoy, E. 1971, Distribution patterns of living planktonic foraminifera in the uppermost layer of the Drake Passage and their relation to the surface hydrology. In: *Proc. Joint Oceanogr. Assembly:*430-431.
Boltovskoy, E. 1973. Reconstruction of Postpliocene climatic changes by means of planktonic foraminifera. *Boreas 2:*55-68.
Broecker, W.S., M.Ewing & B.C.Heezen 1960. Evidence for an abrupt change in climate close to 11 000 years ago. *Amer. J. Sci. 258:*429-448.
Caralp, M., J.Duprat, J.Moyes & C.Pujol 1974. La stratigraphie du Pleistocène superieur et de l'Holocène dans le Golf de Gascogne: Essai de synthèses des critères actuellement utilisables. *Boreas 3:*35-40.
Cifelli, R. & C.Stern Benier 1976. Planktonic foraminifera from near the West African coast and a consideration of faunal parcelling in the North Atlantic. *J. Foram. Res. 6:* 258-273.
Cita, M.B. & M.A.Chierici 1962. Crociera Talassografica Adriatica 1955. V. Ricerche sui Foraminiferi contenuti in 18 carote prelevate sul fondo del mare Adriatico. *Arch. Oceanogr. Limnol., Venezia 12:*297-359.
Diester-Haass, L. 1975. Sedimentation and climate in the Late Quaternary between Senegal and Cape Verde Islands. *'Meteor' Forsch.-Ergebn. C 20:*1-32.
Emiliani, C. 1955. Pleistocene temperatures. *J. Geol. 63:*538-578.

199

Emiliani, C. 1958. Paleotemperature analysis of core 280 and Pleistocene correlations. *J. Geol* *66:*264-275.

Emiliani, C. 1966. Paleotemperature analysis of the Caribbean cores P 6304-9 and P 6304-9, and a generalized temperature curve for the past 425 000 years. *J. Geol. 74:*109-124.

Emiliani, C. 1972. Quaternary paleotemperatures and the duration of the high-temperature intervals. *Science 178:*398-401.

Emiliani, C., T.Mayeda & R.Selli 1961. Paleotemperature analysis of the Plio-Pleistocene section at La Castella, Calabria, southern Italy. *Geol. Soc. Amer. Bull. 72:*679-688.

Ericson, D.B., M.Ewing, G.Wollin & B.C.Heezen 1961. Atlantic deep-sea sediment cores. *Geol. Soc. Amer. Bull. 72:*193-286.

Ericson, D.B. & G.Wollin 1956. Micropaleontological and isotopic determinations of Pleistocene climates. *Micropaleont. 2:*257-270.

Ericson, D.B. & G.Wollin 1968. Correlation of six cores from the equatorial Atlantic and the Caribbean. *Deep-Sea Res. 3:*104-125.

Erlenkeuser, H., B.Koopmann & M.Sarnthein (in preparation). Oxygen isotope stratigraphy of 'Meteor' cores at the Eastern Atlantic continental margin.

Gardner, J.K. & J.D.Hays 1976. The eastern equatorial Atlantic: Sea-surface temperature and circulation responses to global climatic change during the past 200 000 years. *Geol. Soc. Amer. Mem. 145:*221-246.

Geyh, M.A. 1976. [14]C routine dating of marine sediments. *Nineth Radiocarbon Conf., Proc., Los Angeles.*

Geyh, M.A., W.E.Krumbein & H.-R.Kudrass 1974. Unreliable ^{14}C dating of long-stored deep-sea sediments due to bacterial activity. *Marine Geol. 17:*45-50.

Hays, J.D. & A.Perruzza 1972. The significance of calcium carbonate oscillations in Eastern Equatorial Atlantic deep-sea sediments for the end of the Holocene warm interval. *Quatern. Res. 2:*355-362.

Imbrie, J. & N.G.Kipp 1971. A new micropaleontological method for quantitative paleoclimatology: application to a Late Pleistocene Caribbean core. In: K.K.Turekian (ed.), *The Late Cenozoic glacial ages.* Yale Univ. Press, New Haven and London, 71-181.

Kipp, N.G. 1976. A new transfer function for estimating past sea-surface conditions from sea-bed distribution of planktonic foraminiferal assemblages in the North Atlantic. *Geol. Soc. Amer. Mem. 145:*3-41.

Klovan, J.E. & J.Imbrie 1971. An algorithm and Fortran-IV program for large-scale Q-mode factor analysis and calculation of factor scores. *J. Int. Ass. Math. Geol. 3:*61-77.

Lutze, G.F., M.Sarnthein, B.Koopmann, U.Pflaumann, H.Erlenkeuser & J.Thiede 1979. Meteor cores 12309: Late Pleistocene reference section for interpretation of the Neogene of Site 397. In: U.v.Rad & W.B.F.Ryan et al., *Initial Reports DSDP, 47/1:727-739.*

McIntyre, A., N.G.Kipp, A.W.H.Bé, T.Crowley, T.Kellogg, J.V.Gardner, W.Prell & W.F. Ruddiman 1976. Glacial North Atlantic 18 000 Years ago: A CLIMAP Reconstruction. *Geol. Soc. Amer. Mem. 145:*43-76.

McIntyre, A., W.F.Ruddiman & R.Jantzen 1972. Southward penetrations of the North Atlantic Polar Front: Faunal and floral evidence of large scale surface water mass movements over the last 225 000 years. *Deep-Sea Res. 19:*61-77.

Molina Cruz, A. & J.Thiede 1978. The glacial eastern boundary current along the Atlantic Eurafrican continental margin. *Deeo-Sea Res. 25:*337-358.

Olaussen, E. 1965. Evidence of climatic changes in North Atlantic deep-sea cores, with remarks on isotopic temperature analysis. *Progr. Oceanogr. 3:*221-251.

Parker, F.L. 1958. Eastern Mediterranean foraminifera. *Swed. Deep-Sea Exped., Report 8:* 219-283.

200

Parker, F.L. 1962, Planktonic foraminiferal species in Pacific sediments. *Micropaleont. 8:* 219-254.

Pastouret, L., H.Chamley, G.Delibrias, J.C.Duplessy & J.Thiede 1978. Late Quaternary climatic changes in Western Tropical Africa deduced from deep-sea sedimentation off the Niger delta. *Oceanologica Acta 1:*217-232.

Pflaumann, U. 1975. Late Quaternary stratigraphy based on planktonic foraminifera off Senegal. *'Meteor' Forsch.-Ergebn. C 23:*1-46.

Pflaumann, U. (in preparation). Results of the 'Meteor' and 'Valdivia' expeditions off West Africa: Planktonic foraminifera, ecostratigraphical and paleoenvironmental evaluations.

Pflaumann, U. & V.A.Krashenninikov 1978. Quaternary stratigraphy and planktonic foraminifers of the Eastern Atlantic, Deep Sea Drilling Project, Leg 41. In: Y.Lancelot & E.Seibold, et al., *Initial Report, DSDP, Vol. 41.* Suppl.:883-911.

Phleger, F.B., F.L.Parker & J.F.Peirson 1953. North Atlantic Foraminifera. *Swed. Deep-Sea Exped., Report 7:*1-122.

Ruddiman, W.F. 1971. Pleistocene sedimentation in the Equatorial Atlantic: Stratigraphy and faunal paleoclimatology. *Geol. Soc. Amer. Bull. 82:*283-302.

Ruddiman, W.F. & A.McIntyre 1973. Time transgressive deglacial retreat of polar waters from the North Atlantic. *Quatern. Res. 3:*117-130.

Sarnthein, M. & E.Walger 1974. Der äolische Sandstrom aus der Westsahara zur Atlantik-küste. *Geol. Rundschau 63:*1065-1087.

Schemainda, R., D.Nehring & S.Schulz 1975. Ozeanologische Untersuchungen zum Produktionspotential der nordwestafrikanischen Wasserauftriebsregion 1970-1973. *Akad. Wiss. DDR, Geodät. Geophys. Veröff., R. 4, H. 16:*1-85.

Schott, W. 1935. Die Foraminiferen in dem äquatorialen Teil des Atlantischen Ozeans. *Wiss. Ergebn. Dt. Atlant. Exped. 'Meteor' 1925-1927, 3*(3):43-134.

Seibold, E. 1972. Cruise 25, 1971 of R.V. 'Meteor': Continental margin of West Africa. General Report and preliminary results. *'Meteor' Forsch.-Ergebn. C 10:*17-38.

Seibold, E., L.Diester-Haass, D.Fütterer, M.Hartmann, F.C.Kögler, H.Lange, P.J.Müller, U.Pflaumann, H.-J.Schrader & E.Suess 1976. Late Quaternary sedimentation off the Western Sahara. *Acad. brasil. Cienc. 48*(Supl.):287-296.

Seibold, E. & K.Hinz 1976. German Cruises to the Continental Margin of NW Africa in 1975: General reports and preliminary results from 'Valdivia' 10 and 'Meteor' 39. *'Meteor' Forsch.-Ergebn. C 25:*47-80.

Shackleton, N.J. 1967. Oxygen isotope analyses and Pleistocene temperatures reassessed. *Nature 25:*15-17.

Shackleton, N.J. 1969. The last Interglacial in the marine and terrestrial records. *Proc. R. Soc. London, B 174:*135-154.

Shackleton, N.J. & N.D.Opdyke 1973. Oxygen isotope and palaeomagnetic stratigraphy of Equatorial Pacific core V 28-238. Oxygen isotope temperatures and ice volumes on a 10^5 year and 10^6 year scale. *J. Quatern. Res. 3:*39-55.

Shaffer, G. 1974. On the Northwest African Upwelling System. Diss. Univ. Kiel.

Thiede, J. 1977. Aspects of the variability of the Glacial and Interglacial North Atlantic eastern boundary current (last 150 000 years). *'Meteor' Forsch.-Ergebn. C 28:*1-36.

Thiede, J. 1978. A glacial Mediterranean. *Nature 276:*680-683.

Thunnell, R.C. 1979. Pliocene-Pleistocene paleotemperature and paleosalinity history of the Mediterranean Sea: Results from DSDP Sites 125 and 132. *Marine Micropaleont. 4:*173-187.

Todd, R. 1958. Foraminifera from western Mediterranean deep-sea cores. *Swed. Deep-Sea Exped. Report 8:*167-215.

US Naval Oceanographic Office 1967. *Oceanographic Atlas of the North Atlantic Ocean.* Section II: Physical properties.

Wollin, G., D.B.Ericson & M.Ewing 1971. Late Pleistocene climates recorded in Atlantic and Pacific Deep-Sea sediments. In: K.K.Turekian (ed.), *The Late Cenozoic glacial ages.* Yale Univ. Press, New Haven and London, 199-214.

Zobel, B. 1973. Biostratigraphische Untersuchungen an Sedimenten des indisch-pakistanischen Kontinentalrandes (Arabisches Meer). *'Meteor' Forsch.-Ergebn. C 12:*9-73.

Zobel, B. & U.Ranke 1978. Zusammensetzung, Stratigraphie und Bildungsbedingungen der Sedimente am Kontinentalhang vor Sierra Leone (Westafrika). *'Meteor' Forsch.-Ergebn. C 29:*21-74.

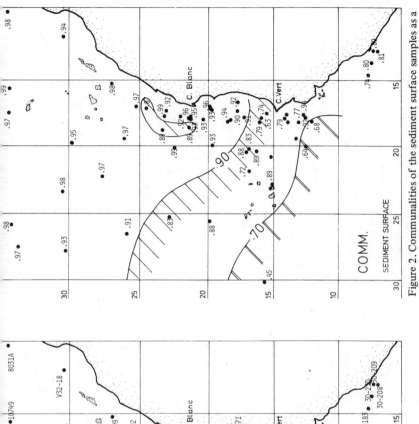

Figure 2. Communalities of the sediment surface samples as a scale of the applicability of the model.

Figure 1. Station locations.

203

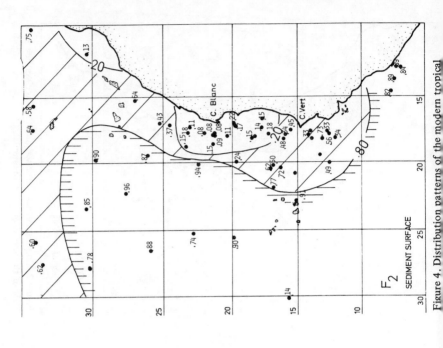

Figure 4. Distribution patterns of the modern tropical

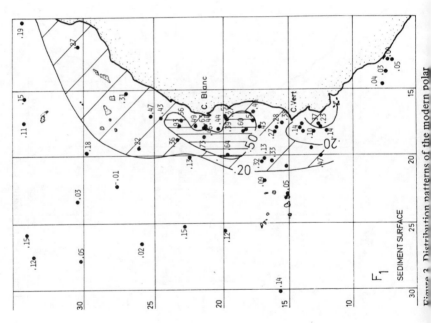

Figure 3. Distribution patterns of the modern polar

204

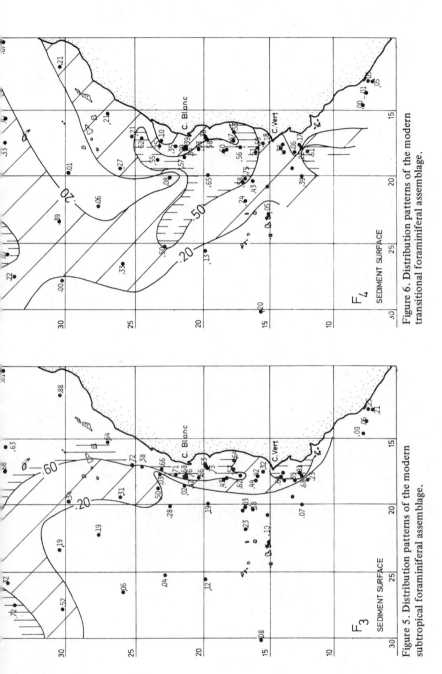

Figure 6. Distribution patterns of the modern transitional foraminiferal assemblage.

Figure 5. Distribution patterns of the modern subtropical foraminiferal assemblage.

205

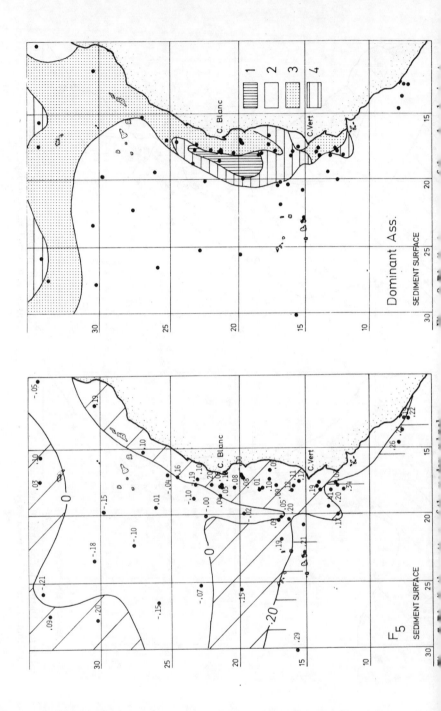

Dominant Ass.

SEDIMENT SURFACE

F_5

SEDIMENT SURFACE

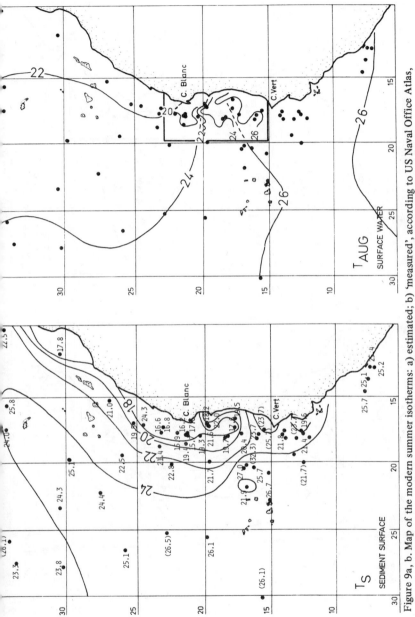

Figure 9a, b. Map of the modern summer isotherms: a) estimated; b) 'measured', according to US Naval Office Atlas, within the frame between Cape Blanc and Cape Vert 10 days measurements of Schemainda et al. (1975).

207

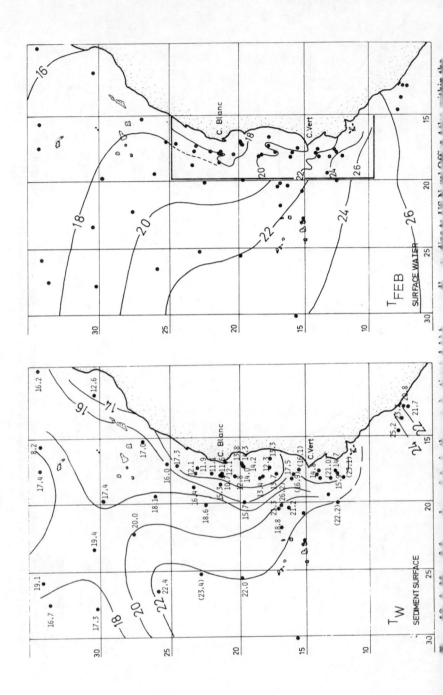

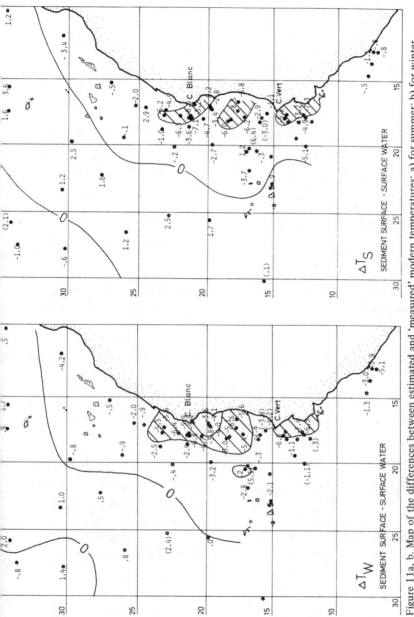

Figure 11a, b. Map of the differences between estimated and 'measured' modern temperatures: a) for summer; b) for winter.

209

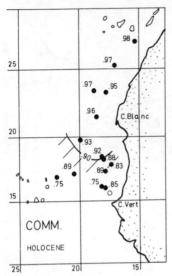

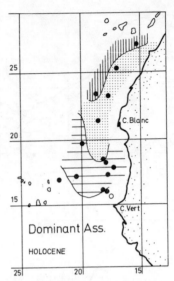

Figure 12. Communalities of the samples of the Holocene climatic optimum (compare Figure 2).

Figure 13. Distribution patterns of the dominant foraminiferal assemblages during the Holocene climatic optimum (for legend, see Figure 8).

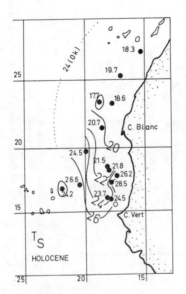

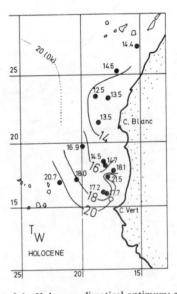

Figure 14a, b. Map of the estimated isotherms of the Holocene climatical optimum: a) for summer; b) for winter.

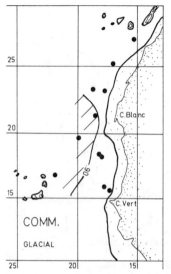

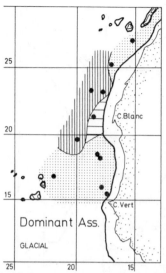

Figure 15. Communalities of the samples of the last glacial maximum (compare Figure 2). The shoreline is near the modern shelf edge due to eustatical sea level drop.

Figure 16. Distribution patterns of the dominant foraminiferal assemblages during the last glacial maximum (for legend, see Figure 8).

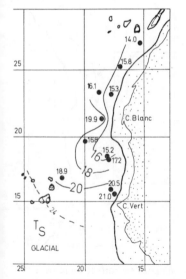

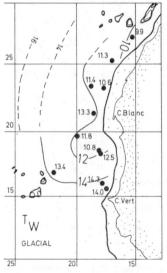

Figure 17a, b. Map of the estimated isotherms of the last glacial maximum: a) for summer; b) for winter. Stippled lines are estimates of McIntyre et al. (1976).

211

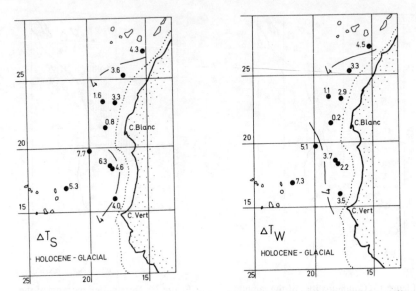

Figure 18a, b. Differences in estimated sea surface temperatures between Holocene climatic optimum and the last glacial maximum: a) for the summer; b) for the winter.

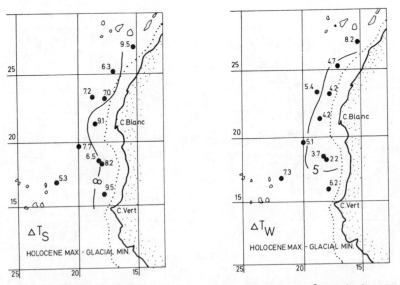

Figure 19a, b. Map of maximum differences of the estimated sea surface water temperatures between the Holocene and the last glacial climaxes as deduced from planktonic foraminiferal taphocoenoses: a) for summer; b) for winter.

THE LATE QUATERNARY MARINE PALEO-ENVIRONMENTS BETWEEN EUROPE AND AFRICA

JÖRN THIEDE
Department of Geology, University of Oslo

1. INTRODUCTION

The late Quaternary environments in southern Europe and northern Africa have gone through important changes which have affected the faunas and floras of this region, last but not least also the human civilizations which have flourished around the Mediterranean since early times (Butzer 1958). In this paper I will attempt to 1) compare the modern oceanography of the Mediterranean Sea and the adjacent part of the northeast Atlantic Ocean to the Glacial one (approximately 18 000 years ago) close to the peak of the last ice age, and 2) compare the results of the paleo-oceanographic reconstructions with evidence from the land record from southern Europe, northern Africa and the western part of Arabian-Asian land masses. Three sets of data will be used for this purpose: 1) modern hydrographic (Figure 1) and climatic (Figure 2) data, 2) estimates of past sea surface temperatures which have been reconstructed by means of planktonic foraminifers (Figures 3 and 4) and the transfer function technique (Figure 5), and 3) some pertinent data which have been published about Quaternary sections from the land areas adjacent to the northeast Atlantic and Mediterranean Sea.

2. MODERN HYDROGRAPHY AND CLIMATE OF THE NORTHEAST ATLANTIC-MEDITERRANEAN SEA REGION

The Mediterranean Sea and the part of the northeast Atlantic Ocean adjacent to it (Figure 1) are situated close to a very sensitive climatic boundary between the dry, arid and warm North Africa and the humid to semiarid, subtropical to temperate portion of southern Europe (Figure 2). The Mediterranean is only one of a string of marine basins which stretch in a zonal belt from the Atlantic towards the Indian Ocean (Black, Caspian and Red Seas, Gulf of Aden). Some of these basins are much deeper than marginal seas usually are because they have been generated due to platetectonic processes along the plate margins between Africa and Eurasia or between Africa and

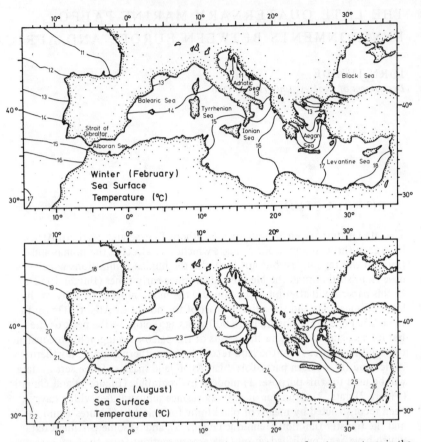

Figure 1. Modern winter (February) and summer (August) sea surface temperatures in the northeast Atlantic Ocean (after Deutsches Hydrographisches Institut 1967) and in the Mediterranean (after Bruns 1958).

Arabia. The Mediterranean can indeed be considered to represent a true small ocean basin as can be seen from the water depths of its western and eastern subbasins which are separated from each other by the shallow strait between Sicily and North Africa (Strait of Tunis). The Mediterranean has exchanged its water masses with the Black Sea during Quaternary intervals of high sea levels (such as today) through the Bosporus or through the fossil Sakaria depression (Pfannenstiel 1944), but it became a brackish or fresh water lake during intervals of low sea levels because the threshold depths in both regions are estimated to lie about 45 m below the present sea level (Butzer 1958). The marine water exchange of the Mediterranean occurred throughout the Quaternary with the Atlantic through the more than 300 m deep Strait of

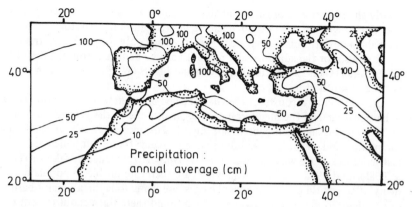

Figure 2. Mean annual rainfall over the land areas adjacent to the northeast Atlantic and Mediterranean Sea region (after Lamb 1972).

Gibraltar. Because of the deficit of the water balance in the Mediterranean and because of its peculiar mode of deep-water formation (Gascard 1978), its water masses are more salty and denser than those of the Atlantic Ocean and of the Black Sea.

The Mediterranean balances the excess surface water inflow from both regions by a subsurface outflow as one would expect from a marginal marine basin in an arid (semiarid) climatic region (Seibold 1970). A water mass of Mediterranean origin has been observed to occur widespread over vast regions of the Atlantic Ocean (Madelain 1976, Zenk 1970) effectively sealing its deep-water from its surface water masses. If we today find a marine environment in the Mediterranean Sea close to a normal oceanic one despite the narrow and shallow Straits of Gibraltar and Tunis, then this must be the result of a delicate balance which might be disturbed by only slight climatic fluctuations. In this study, I will therefore address the response of the Mediterranean and northeast Atlantic hydrographies to the late Quaternary climatic fluctuations which have affected both these sea regions and the adjacent land masses.

This sea region is filled today by temperate to tropical water masses, but the temperature gradients in the Atlantic and in the Mediterranean are quite different (Figure 1). While this part of the northeast Atlantic is characterized by the sluggish and broad eastern boundary current system along the Eurafrican continental margins with its meridional temperature gradient, the zonal Mediterranean Sea reveals a temperature structure with warm surface water in the eastern Levantine Basin, but with water masses which become increasingly more temperate in a westerly direction. In general the sea surface temperatures are only slightly warmer than in the adjacent parts of the North Atlantic Ocean, but the seasonal differences are higher over the eastern Mediterranean than over the part of the northeast Atlantic which is considered here.

The distribution of the mean annual precipitation over this region reveals

215

clearly the importance of the zonal Mediterranean Sea which is situated just north of a region of a very steep decrease in precipitation separating it from the Sahara desert. This has wide consequences not only for the Mediterranean Sea, but also for the adjacent land region as can be deduced from the distribution of soils (Gannsser & Hädrich 1965) and plants (Wright 1976) to the north and south of it.

The method to reconstruct the Glacial paleo-environments of the northeast Atlantic Ocean and the Mediterranean Sea, makes use of the pelagic sediments of both regions and their contents of microfossils, more precisely of planktonic foraminifers. Planktonic foraminifers which are eurytherm but which are restricted to surface water masses of a relatively narrow range of salinities (Bé & Tolderlund 1971, Bé 1977) and which are therefore unable to populate many of the world oceans' marginal basins, are living today in all major basins of the Mediterranean. They have also been found there throughout the entire Quaternary even though the passages from the western to the eastern Mediterranean Sea and to the Atlantic Ocean were much narrower and shallower during the glacial stages. These facts have directed me to look rather for temperature than for salinity changes. Because of the presence of these microfossils the Mediterranean Sea has already been the locality for early attempts of quantitative Pleistocene temperature estimates (Emiliani 1955). I will, therefore, first consider planktonic foraminiferal shell assemblages from the modern sea bed to find out how the sediment record reflects the modern oceanography, then I will discuss some Glacial species distributions and finally try to reconstruct the Glacial paleo-oceanography of this sea region.

If planktonic foraminifers merely drift from the North Atlantic into the Mediterranean or if they are able to establish true autochthonous faunas is one of the most important questions. The distribution of the tropical species *Globigerinoides ruber* (Figure 3a) is probably a good argument that autochthonous faunas are present in the Mediterranean and that planktonic foraminifers do not merely drift into the Mediterranean as inhabitants of the northeast Atlantic surface water masses. This had already been deduced from plankton tow studies (Cifelli 1974). It is interesting to note that this species supports the gradient of sea surface temperatures between the eastern and the western Mediterranean basin (Figure 1), while it is rare in the Atlantic just outside Gibraltar and in the Bay of Biscay. *Globoquadrina pachyderma* on the other hand which is typical for relatively cool or cold surface water masses (Bé 1977) reveals a particularly interesting distribution pattern. Although on Figure 5a left and right coiling specimens have been lumped, the dominant phenotype present in this area is the right coiling one which is typical for cool but not cold surface water masses. It is interesting to note that it is frequent in the entire Bay of Biscay, that it can be traced as a common species along the Eurafrican continental margin and that it increases in frequency again under the influence of the coastal upwelling off West Africa (Thiede 1977). In the Mediterranean the only area of frequent occurrence is in the northern part of the western Mediterranean.

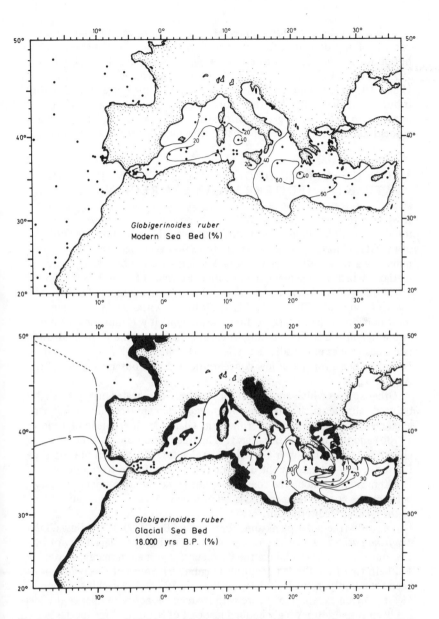

Figure 3. Distribution of *Globigerinoides ruber* in the northeast Atlantic Ocean and in the Mediterranean Sea: a) in Recent sediments (from Thiede 1977; and Thunell 1978); b) in Glacial sediments (deposited during the last Glacial maximum, 18 000 years ago, from Thiede, 1980 and Molina-Cruz & Thiede 1978).

217

The total planktonic foraminiferal fauna of the sea region under study consists of approximately 20 different species whose surface sediment distribution reflect the present day oceanography in surprising detail (Loubere 1977, Thunell 1978, Thiede 1977, Molina Cruz & Thiede 1978, Thiede 1980). The patterns shown illustrate very well that the biogeographical distributions of the single planktonic foraminiferal species reveal a coherent relationship to the hydrography of the surface water masses, and that truly autochthonous faunas are established in this marginal small ocean basin.

3. THE GLACIAL HYDROGRAPHY OF THE NORTHEAST ATLANTIC-MEDITERRANEAN SEA REGION

If we now turn to the fossil record we will observe the response of the planktonic foraminiferal faunas to the different Glacial conditions. The core levels of the Glacial maximum, approximately 18 000 years ago, have been determined by a variety of oxygen isotopic, radiometric and biostratigraphic methods which will be discussed in detail elsewhere (Thiede 1980). The distribution of *Globigerinoides ruber* is shown as an example of a tropical species (Figure 3b). During the Glacial, it had to withdraw into the innermost portion of the eastern Mediterranean, while it was much less common in the Ionian Sea and rare to almost entirely lacking in the western Mediterranean as well as in the adjacent Atlantic. Despite its high frequency in the Levantine Sea a tongue of low concentrations can be observed south of the Aegean Sea.

Other species which are typical for relatively cool water masses of slightly reduced salinities such as *Globigerina quinqueloba* are frequent in the marginal region along the northern rim of the Glacial Mediterranean (Thiede 1978b) and just outside Gibraltar. It is interesting to note that the Strait of Gibraltar despite the lower Glacial sea level whose effect upon the coastline I have tried to indicate by the dark rim (corresponding to the modern − 100 m isobath) along the coasts does not mark a sharp faunal boundary between the cool to temperate Glacial northeast Atlantic faunas and those of the Glacial Mediterranean.

The only true polar planktonic foraminifer (Bé 1977), left coiling *Globoquadrina pachyderma,* is almost entirely confined to the Glacial Bay of Biscay, but it can be traced along the Portuguese continental margin to the southern tip of the Iberian peninsula (Figure 4b). South of this region, it has been found only in small abundances (Thiede 1977). It is lacking in the Mediterranean except in one core with frequencies above ten per cent which has been found interestingly enough just east of Sardinia. This species is usually only found in areas where the sediments receive a considerable contribution of ice rafted material such as in the Glacial Bay of Biscay. However, ice rafted material is also found along the entire portion of the Eurafrican continental margin (Kudrass 1973) shown in the figures of this paper, and we

218

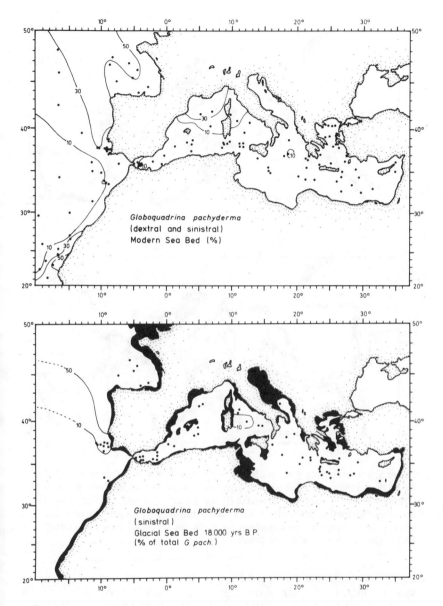

Figure 4. Distribution of *Globoquadrina pachyderma* in the northeast Atlantic and in the Mediterranean Sea: a) sinistral and dextral specimen in Recent sediments (compiled from Thiede 1977, and Thunell 1978); b) proportion of sinistral specimen in the total population of *G.pachyderma*.

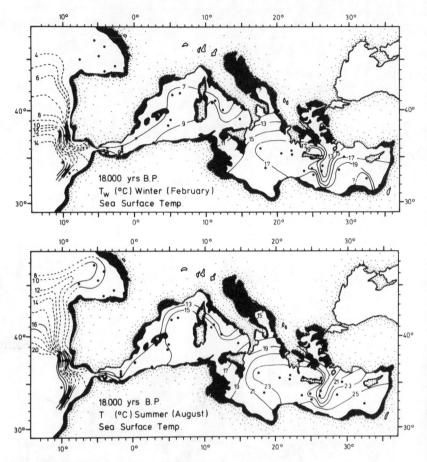

Figure 5. Sea surface temperatures in the Mediterranean and in the northeast Atlantic Ocean during the peak of the last Glacial approximately 18 000 years ago: a) during the Glacial winter (February); b) during the Glacial summer (August). Dots mark the positions from where Glacial core levels have been obtained. The blackened areas mark schematically changes of the coast lines due to the lowered Glacial sea levels (after Molina-Cruz & Thiede 1978, and Thiede 1978a).

have to conclude that sea ice is drifting southwards at least during certain seasons following the path of the Glacial eastern boundary current.

The planktonic foraminiferal faunas from the Bay of Biscay and Mediterranean Sea can also be used for a quantitative reconstruction of certain aspects of the Glacial sea surface hydrography. The technique of using transfer functions is well-known (Imbrie & Kipp 1971). It has to be mentioned that I have applied a redesigned northeast Atlantic transfer function (Kipp

220

1976) which was heavily biased to the specific conditions close to continental margins by loading it with samples from the Bay of Biscay and from the Atlantic continental margin off the Iberian peninsula and off Northwest Africa (Molina Cruz & Thiede 1978). The detailed counts of the Glacial Mediterranean foraminiferal samples will be discussed somewhere else (Thiede 1980), but in general it can be said that the Glacial planktonic foraminiferal faunas of the Mediterranean are very similar to certain modern northeast Atlantic faunas and that they are therefore well-suited for this statistical technique (Imbrie & Kipp 1971). Admittedly, there are also difficulties with this approach. Especially, it has to be mentioned that there seems to be a no analog problem in the Adria, in the Aegean Sea and along the northern shore of the western Mediterranean. The transfer function developed here allowed us to estimate paleotemperatures of the ocean surface (Figure 5).

Glacial summer sea surface temperatures ranged from 8°C in the Bay of Biscay to more than 25°C in the eastern Mediterranean. The extension of the contours into the open ocean which are stippled in Figure 5, have been drawn using the Glacial reconstruction of Climap (cf. Climap project members 1976, McIntyre et al. 1976). The steep temperature gradients close to the Ibero-Moroccan continental margin and the surface waters along the European coast of the western Mediterranean are in contrast to each other because the Mediterranean Sea appears to have been a few degrees warmer than the adjacent Bay of Biscay. The eastern Mediterranean Sea was filled by a pool of warm water with a relatively steep T-gradient across the shallow area between Sicily and North Africa. This pool of warm surface water reveals considerable structure because a tongue of cool and probably relatively fresh surface water (Thunell et al. 1977, Williams et al. 1978) is obviously entering the eastern Mediterranean from the Aegean Sea. During the Glacial winter (Figure 5), the Bay of Biscay is filled with cold polar surface water masses and the entire northeast Atlantic cools considerably to temperatures less than 4°C to more than 14°C. The western Mediterranean sea surface temperatures dropped to between 7 and 11°C while the eastern Mediterranean was filled by relatively warm water masses with an influx of cold water from or through the Aegean Sea as has been discussed above. We have no evidence that the water exchange with the Atlantic changed in any drastic way as argued for by Stanley et al. (1975) and the results of detailed studies of Alboran Sea cores support the previously published ideas of a Glacial surface inflow into and bottom water outflow from the Mediterranean (Diester-Haass 1974). On the other hand, we know that the intensity of the eastern boundary current regime in the Atlantic fluctuated rapidly during the late Quaternary (Thiede 1977) but it seems impossible to trace these fluctuations in the Alboran Sea.

To assess the seasonal differences, let us consider both Glacial winter and summer sea surface temperatures. The most drastic changes occurred in the Atlantic just outside Gibraltar and in the Bay of Biscay. In summer time, this region is filled by relatively temperate water masses with steep temperature gradient parallel to the polar front which was running at that time in an east-westerly direction across the Atlantic. The boundary current system was circulating in a clockwise direction around the central subtropical water mass of the Atlantic and was probably flowing from west to east in the open Atlantic Ocean. It was probably breaking up into two branches under the influence of the continental margin with a relatively small counterclockwise gyre in the Bay of Biscay and a narrow southward flowing one which was developing into the eastern boundary current (Molina-Cruz & Thiede 1978). During winter time, we cannot see anything of the temperate water masses in the Bay of Biscay which was then filled by true polar waters which drifted along the Iberian peninsula to the south. The seasonal differences in the Mediterranean were between 5 and 7°C, and we can trace how rapidly the Mediterranean surface became warmer towards the east. I earlier indicated that the western Mediterranean was between 3 and 5°C warmer than the Bay of Biscay, even close to the areas of continental glaciation in the Alps. There is also a jump of sea surface temperatures between the eastern and western Mediterranean of some 5-7°C, but the most exciting information came probably from the data south of the Aegean Sea which seem to indicate the presence of a tongue of cool and fresh water which was present during both seasons. This feature is probably related to an important source of cool and probably relatively fresh water in the hinterland of the Aegean Sea, probably also of the Black Sea (Thunell 1979).

The differences between the modern and Glacial hydrography can be evaluated by comparing Figure 1 and 5, respectively the foraminiferal distributions shown in Figures 3 and 4. The range of sea surface temperatures in the northeast Atlantic has probably not changed so much, but the temperature distribution has changed drastically probably indicating a reorganization of the sluggish eastern boundary current system into a relatively narrow but more rapid flowing current regime along the continental margin. The western Mediterranean has been considerably cooler during the Glacial than it is today while the subtropical temperatures which make the modern Mediterranean Sea such a pleasant place, were restricted to the eastern basin.

4. THE GLACIAL PALEO-ENVIRONMENT OF NORTH AFRICA, SOUTHERN EUROPE AND THE ASIAN-ARABIAN LANDMASSES ADJACENT TO THE MEDITERRANEAN

It is impossible here to review the vast body of data collected by Quaternary geologists from land sections along the coast of the northeast Atlantic Ocean and Mediterranean Sea. They have been summarized in recent papers (Wright 1976, Farrand 1971, Lauer & Frankenberg 1979). However, there seems to be a major discrepancy between the results presented in Figure 5 and

222

the overall temperature increase of 10-12°C since the Glacial which has been estimated from land records (Brunnacker 1979).

It seems even more difficult to assess the amount of precipitation during the last Glacial along the coastlines of the northeast Atlantic and of the Mediterranean Sea. Circumstantial evidence from North African pollen studies (Rossignol-Strick & Duzer 1979), lake investigations (Street 1979), isotopic ratios of northern Saharan groundwaters (Sonntag et al. 1979), and reconstruction of Glacial dune distributions (Sarnthein 1978) point to arid conditions during the last Glacial maximum. Although southern Europe and southeastern Asia have been classical areas to define pluvials to correspond to the Glacial intervals (Butzer 1958), increasingly more evidence suggests aridity during the Glacial maximum in these regions too (Bonatti 1966, van der Hammen et al. 1971 and many others). The fact that indications for reduced salinities of the surface water masses have been found along the northern rim of the Mediterranean, will therefore have to be evaluated very carefullly. The freshwater influx into the Aegean Sea can be explained with an outflow from the Black Sea which, during the Glacial, was the recipient of the drainage from the vast periglacial areas in eastern Europe. The Adriatic Sea and the northern part of the western Mediterranean were then situated south of a region of important alpine glaciation. The steep gradient of decreasing precipitation which is found today along the southern Mediterranean shore over North Africa (Figure 2) was during the Glacial maximum displaced to the north which allowed the Mediterranean to maintain a hydrography favourably for the existence of autochthonous planktonic foraminiferal faunas. The Glacial hydrography of this sea region can now be modelled (Bethoux 1977) and will have to be evaluated carefully in future studies.

6. ACKNOWLEDGEMENTS

Part of this study has been supported by the Climap program of the International Decade of Ocean Exploration (IDOE) through the U.S. National Science Foundation (grant no. IDO-75-22133). R.Thunell kindly gave me access to his counts of planktonic foraminifers from Mediterranean surface sediments. I gratefully acknowledge the invitation to participate in the symposium 'Sahara and adjacent seas: climatic change and sediments' which has been held 1-4 April 1979 at the Academy of Sciences and Literature in Mainz/Germany.

REFERENCES

Bé, A.W.H. 1977. An ecological, zoogeographic and taxonomic review of recent planktonic foraminifera. In: A.T.S.Ramsay (ed.), *Oceanic micropaleontology*. Academic Press, London, Vol. 1, pp.1-100.

Bé, A.W.H. & D.S.Tolderlund 1971. Distribution and ecology of living planktonic foraminifera in surface waters of the Atlantic and Indian Oceans. In: B.M.Funnell & W.R.
Riedel (eds.), *Micropaleontology of oceans*. Cambridge Univ. Press, London, pp.105-
149.

Bethoux, J.P. 1978. Les actions du climat et de l'Ocean Atlantique sur le régime de la Mer
Méditerranée, notamment au cours des périodes glaciaires. In: *Evolution des atmosphéres
planetaires et climatologie de la terre*. Centre Nat. Êtud. Spat., Toulouse, pp.213-219.

Bonatti, E. 1966. North Mediterranean climate during the last Würm glaciation. *Nature
209:*984-985.

Brunnacker, K. 1979. Les Loess du Pleistocène récent comme indicateur climatique dans
les régions méditerranéennes. In: *Symp. Sahara and surrounding seas – sediments and
climatic changes*. Acad. Wiss. Lit. Mainz. April 1-4, 1979, Abstract vol.

Bruns, E. 1958. *Ozeanologie*, Vol. 1, VEB Dt. Verl. Wiss., Berlin.

Butzer, K.H. 1958. Quaternary stratigraphy and climate in the North East. *Bonner Geogr.
Abh. 24:*1-157.

Cifelli, R. 1974. Planktonic foraminifera from the Mediterranean and adjacent Atlantic
waters (Cruise 49 of the Atlantis II, 1969). *J. foram. Res. 4:*171-183.

CLIMAP Project Members 1976. The surface of the Ice-Age Earth. *Science 191:*1131-1144.

Deutsches Hydrographisches Institut 1967. *Monatskarten für den Nordatlantischen Ozean.*
Hamburg, 4th ed. Nr. 2420.

Diester-Haass, L. 1974. No current reversal at 10 000 years B.P. at the Strait of Gibraltar:
A reply. *Marine Geol. 17:*174-177.

Emiliani, C. 1955. Pleistocene temperature variations in the Mediterranean. *Quaternaria 2:*
87-98.

Farrand, L.R. 1971. Late Quaternary paleoclimates of the Eastern Mediterranean area. In:
K.K.Turekian (ed.), *Late Cenozoic glacial ages*. Yale Univ. Press. New Haven and London, pp.529-564.

Gannsser, R. & F.Hädrich 1965. *Atlas zur Bodenkunde*. Bibliogr. Inst., Mannheim.

Gascard, J.C. 1978. Mediterranean deep water formation, baroclinic instability and
oceanic eddies. *Oceanol. Acta 1:*315-330.

Imbrie, J. & N.G.Kipp 1971. A new micropaleontological method for quantitative paleoclimatology: application to a Late Pleistocene Caribbean core. In: K.K.Turekian (ed.),
Late Cenozoic glacial ages. Yale Univ. Press. New Haven and London, pp.71-181.

Kipp, N.G. 1976. New transfer function for estimating past sea-surface conditions from
sea-bed distribution of planktonic foraminiferal assemblages in the North Atlantic.
*Geol. Soc. Amer. Mem. 145:*3-41.

Kudrass, H.-R. 1973. Sedimentation am Kontinentalhang vor Portugal und Marokko im
Spätpleistozän und Holozän. *'Meteor' Forsch.-Ergebn. C 13:*1-63.

Lamb, H.H. 1972. *Climate: Present, past and future*. Vol. 1: *Fundamentals and climate
now*. Methuen & Co. Ltd., London.

Lauer, W. & P.Frankenberg 1979. Zur Klima und Vegetationsgeschichte der westlichen
Sahara. *Akad. Wiss. Lit.Mainz, Abh. Math.-Naturw. Kl. 1:*1-61.

Loubere, P. 1977. The sediment surface distribution of foraminifera as indication of
western Mediterranean oceanography for the present day and for the last glacial
maximum, 18 000 years B.P. *Geol. Soc. Amer. Abstr. Progr. 9:*1074.

Madelain, F. 1976. Circulation des eaux d'origine méditerranée au niveau du Cap St.
Vincent: Hydrologie et courrents de densité. CNEXO Publ. Centr. Nat. Explor. Oc.,
Rapp. Scient. Techn., 24:1-73.

McIntyre, A., G.Kipp, A.W.H.Bé, T.Crowley, T.Kellogg, J.V.Gardner, W.Prell & W.F.

224

Ruddiman 1976. Glacial North Atlantic 18 000 years ago: A CLIMAP reconstruction. *Geol. Soc. Amer. Mem. 145:*43-76.

Molina-Cruz, A. & J.Thiede 1978. The glacial eastern boundary current along the Atlantic Eurafrican continental margin. *Deep-Sea Res. 25:*337-356.

Pfannenstiel, M. 1944. Diluviale Geologie des Mittelmeergebietes. *Geol. Rundschau 34:* 342-434.

Rossignol-Strick, M. & D.Duzer 1979. Late Quaternary West African climate inferred from palynology of Atlantic deep-sea cores. *Palaeoecology of Africa 12.*

Sarnthein, M. 1978. Sand deserts during glacial maximum and climatic optimum. *Nature 272:*43-46.

Seibold, E. 1970. Nebenmeere im humiden und ariden Klimabereich. *Geol. Rundschau 60:*73-105.

Sonntag, C. et al. 1979. Isotopic identification of Saharian groundwaters, groundwater formation in the past. *Palaeoecology of Africa 12.*

Stanley, D.J., A.Maldonado & R.Stuckenrath 1975. Strait of Sicily depositional rates and patterns, and possible reversal of currents in the late Quaternary. *Palaeogeogr., Palaeoclim., Palaeoecol. 18:*279-291.

Street, F.A. 1979. The relative importance of climate and hydrological factors in influencing lake-level fluctuations. *Palaeoecology of Africa 12.*

Thiede, J. 1977. Aspects of the variability of the Glacial and interglacial North Atlantic eastern boundary current (last 150 000 years). *'Meteor' Forsch.-Ergebn. C 28:*1-36.

Thiede, J. 1978a. A Glacial Mediterranean. *Nature 276:*680-683.

Thiede, J. 1978b. The Glacial Mediterranean and Bay of Biscay. In: *Evolution des atmosphères planetaires et climatologie de la terre.* Centre Nat. Étud. Spat., Toulouse, pp.121-132.

Thiede, J. 1980. Late Quaternary planktonic foraminifera and paleooceanography of the Mediterranean Sea. *Oceanol. Acta.* In preparation.

Thunell, R.C. 1978. Distribution of recent planktonic foraminifera in surface sediments of the Mediterranean Sea. *Marine Micropal. 3:*147-173.

Thunell, R.C. 1979. Eastern Mediterranean Sea during the last Glacial Maximum; an 18 000 years B.P. reconstruction. *Quatern. Res. 11:*353-372.

Thunell, R.C., D.F.Williams & J.P.Kennett 1977. Late Quaternary paleoclimatology, stratigraphy and sapropel history in eastern Mediterranean deep-sea sediments. *Marine Micropal. 2:*371-388.

Van der Hammen, T., T.A.Wijmstra & W.H.Zagwijn 1971. The floral record of the late Cenozoic of Europe. In: K.K.Turekian (ed.), *The Late Cenozoic glacial ages.* Yale Univ. Press, Hew Haven and London, pp.391-424.

Williams, D.R., R.C.Thunell & J.P.Kennett 1978. Periodic fresh-water flooding and stagnation of the Eastern Mediterranean Sea during the late Quaternary. *Science 201:* 252-254.

Wright, H.E. 1976. The environmental setting for plant domestication in the Near East. *Science 194:*385-389.

Zenk, W. 1970. On the temperature and salinity structure of the Mediterranean water in the Northeast Atlantic. *Deep-Sea Res. 17:*627-631.

LATE QUATERNARY WEST AFRICAN CLIMATE INFERRED FROM PALYNOLOGY OF ATLANTIC DEEP-SEA CORES

M. ROSSIGNOL-STRICK
Laboratoire de Palynologie, Museum National d'Histoire Naturelle, Paris / Lamont Doherty Geological Observatory, Palisades, N.Y.

D. DUZER
Laboratoire de Palynologie, Faculté des Sciences, Montpellier

West African vegetational and climatic history since 22 500 B.P. is suggested by pollen extracted from three Meteor deep-sea cores off Senegal (1, 6), whose stratigraphy was previously described (2, 3). Pollen from north of the Sahara (Central and Northern Saharan desert and steppe, Pre-Saharan steppe and Mediterranean forest) points to a Mediterranean climate of summer dryness and winter/spring rainfall with transport by the surface NE trade wind system including the continental dry hot Harmattan and maritime-cool wind parallel to the coast. Comparison with eolian pollen transport through the Sahara (4, 5), and the present day wind-pattern over West Africa, suggests that the variation of this pollen group reflects more eolian transport efficiency than phytogeographical zones migration. Pollen from south of the Sahara (Sahelian steppe, 200 to 600 mm rain/year, Soudanese savanna, 600 to 1 000 mm, Soudano-Guinean open forest, 1 000 to 1 750 mm) originates in the tropical area of monsoonal summer rains. Transport to the sea is predominantly by the Senegal River and also by easterly winds.

From 22 500 to 19 000 B.P., more general aridity than today is indicated by very small amount of tropical pollen and moderate amount of Mediterranean pollen. From 19 000 to 12 500 B.P., a very arid phase brought more pollen from north of the Sahara, and eliminated the tropical pollen input. With the general atmospheric circulation, more increased in the Northern than in the Southern Hemisphere, and a more active northern trade-wind system, the Intertropical Convergence Zone is pushed southward in summer, possibly as far as today in winter ($8°N$). Consequently, the desert expands far south of its present $17°N$ limit, the monsoonal rain belt regresses southward. From 12 500 to 5 500 B.P., the pollen from north of the Sahara disappears very suddenly, and the tropical pollen increases. The Soudano-Guinean pollen reaches a peak at 12 000 B.P., Soudanese around 7 000 B.P., Sahelian at 5 500 B.P. with each group decreasing immediately after having reached its peak. With the global atmospheric warming, and a decrease of the temperature gradient stronger in the Northern than the Southern Hemisphere, the northern atmospheric circulation weakens more markedly than the southern one. The decrease of the northern trade winds, in summer, allows the ITCZ to migrate poleward. The resulting expansion of the monsoonal rainbelt appears combined with a pattern of decreasing pluviosity after the initial peak is reached.

Until 5 500 B.P., the Sahara southern limit still lies further south than today; only after 5 500 B.P. does the northward Sahelian steppe expansion bring it to its present position.

On the coast, a Chenopodiaceae steppe develops in sebkhas in the arid phase when glacio-eustatic sea-level lowering is maximum, while during the post-glacial optimum, mangrove thrives on the coast and along the lower course of the Senegal river, swollen by the monsoonal rainfall peak.

REFERENCES

1. Rossignol-Strick, M. & D.Duzer 1979. Late Quaternary pollen and dinoflagellate analysis of marine cores off West Africa. *'Meteor' Forsch.-Ergebn. C 30:*1-14.
2. Diester-Haass, L. 1975. Sedimentation and climate in the Late Quaternary between Senegal and the Cape Verde Islands. *'Meteor' Forsch.-Ergebn. C 20:*1-32.
3. Pflaumann, U. 1975. Late Quaternary stratigraphy based on planktonic foraminifera off Senegal. *'Meteor' Forsch.-Ergebn. C 23:*1-46.
4. Cour, P. & D.Duzer 1976. Persistance d'un climat hyperaride au Sahara central et meridional au cours de l'Holocene. *Rev. Geogr. phys. Geol. dyn. 18*(2-3):175-198.
5. Van Campo, M. 1975. Pollen analyses in the Sahara. In: F.Wendorf & A.E.Marks (eds.), *Problems in prehistory, North Africa and the Levant.* Southern Methodist Univ. Press, pp.45-64.
6. Rossignol-Strick, M. & D.Duzer 1979. West African vegetation and climate since 22 500 B.P. from deep-sea cores palynology. *Pollen et Spores,* XXI, 1-2:105-134.

UPWELLING AND CLIMATE OFF NORTHWEST AFRICA DURING THE LATE QUATERNARY

LISELOTTE DIESTER-HAASS

Geologisches Institut der Universität, Würzburg

1. INTRODUCTION

The investigation of marine sediments taken off Northwest Africa can give results on the oceanography of the area and on climatic changes on the continent. Shells from organisms living in the ocean may reflect oceanographic changes by means of species variations and changes in their accumulation rates. Besides temperature variations, indications as to currents and upwelling may be obtained. Terrigenous particles supplied from the continent may contain some information on the continental climate, which can be detected by means of mineralogical and grain size studies. Hints as to rainfall and river formation or to aridity and wind supply may be obtained.

These parameters are studied here in the Late Quaternary. Information on the upwelling and climatic history of older periods can be obtained from Diester-Haass (1979), Chamley & Diester-Haass (1979), Chamley & Giroud d'Argoud (1979) (Middle and Early Quaternary); Diester-Haass (1978), Diester-Haass & Chamley (1978), Sarnthein & Diester-Haass (1977) (Neogene); Diester-Haass (1980), Chamley & Diester-Haass (1980) (Oligocene).

2. UPWELLING

2.1 *Oceanography*

Today, upwelling off Northwest Africa is produced by the offshore blowing trade winds and the Canary current. The combined effect of both is an offshore movement of the near coastal surface waters, which is compensated by upwelling of deep water from about 200-300 m. This water comes from a depth where remineralization of organic matter takes place. Thus upwelling water brings nutrients to the photic zone and enhances a high organic production.

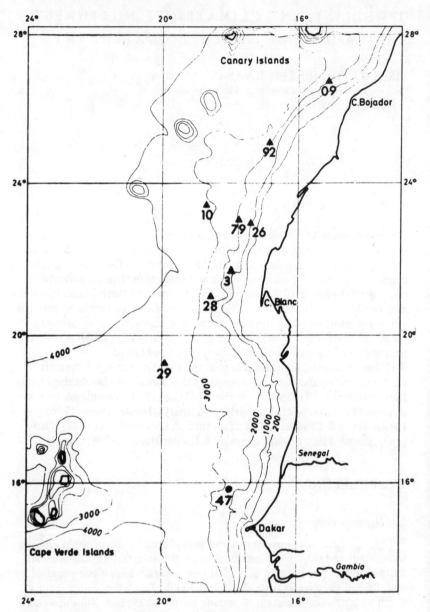

Figure 1. Location of cores, studied for climate and upwelling history off the northern Sahara (triangles) (only last two numbers of cores are given). Dot: core 12347-2 off Senegal (see Figure 4).

2.2 Upwelling influence in the sediments

The effect of upwelling off Northwest Africa can be detected in the underlying sediments by means of several factors (for details, see Diester-Haass 1978a). The fertility increase leads to increased organic carbon contents (Müller 1975). Diatoms, primary producers with siliceous tests, are highly increased in upwelling areas. They are preserved in the sediments when supplied in high amounts, whereas in normally fertile areas they are completely dissolved due to undersaturation of pore and sea water with respect to opal. The increased opal supply by diatoms leads to a better preservation also of radiolarian opal (Diester-Haass 1977). Planktonic foraminifers show increased accumulation rates in upwelling areas. The increased food supply to the sea-floor allows a higher production of benthonic organisms. The lower temperatures of the upwelled water may lead to cool water assemblages of planktonic foraminifers (Diester-Haass et al. 1974, Pflaumann 1975). However, phosphorite formation, increased accumulation rates of fish debris and enrichment of metals has not been found off Northwest Africa. Another method for the detection of upwelling influence in the sediments is the determination of the carbon isotopes. The light isotope ^{12}C is preferentially fixed during photosynthesis and subsequently released during remineralization. So, in shells from upwelling areas ^{12}C values are higher than in normal fertile areas (Berger et al. 1978).

The area of most intense upwelling today is off Cape Blanc ($20°N$), whereas north and south of this area upwelling influence decreases due to the seasonal shifting of the upwelling belt (Schemainda et al. 1975). Maps showing organic carbon and opal content in surface sediments (Diester-Haass 1978a) correlate with this observation. A maximum is found off Cape Blanc, whereas north of this area organic carbon and opal decrease considerably. Opal even disappears except for a small belt in 1 000 m water depth (op. cit.). No observations are given concerning upwelling influence south of $17°N$, because here rivers bring large amounts of not only solid but also dissolved load from intensely weathered tropical areas. Their effect can be similar to that of upwelling (Diester-Haass & Müller 1979).

2.3 Late Quaternary upwelling history

The Late Quaternary upwelling history has been studied in eight cores situated off Northwest Africa in $20-27°N$ (Figure 1) (cp. Diester-Haass 1976b, 1977, 1978a). One core (Meteor Core 12392-1) is shown here as an example, the other ones having similar trends, only absolute values are different (Figure 2).

The oxygen isotope stratigraphy (established by Shackleton 1977) shows that the core has a continuous sedimentation down to stage 6 (Figure 2a). The carbon isotope curve (also from Shackleton 1977) reveals low ^{12}C values in the interglacial (1,5) and high ones in the glacial stages (2-4,6), indicating a higher fertility in the latter periods.

231

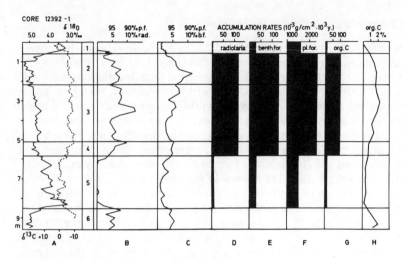

Figure 2. Meteor core 12392-1, upwelling parameters.

A. Heavy line: oxygen isotope curve and stages 1-6. Dotted line: carbon isotope curve. (Both curves from Shackleton 1977).

B. Opal content of the sand fraction, plotted as a ratio with planktonic foraminifers: (radiolarians/radiol. + planktonic foraminifers) x 100.

C. Benthos/plankton ratios of foraminifers, calculated as (benthos/benthos + plankton) x 100.

D-G. Accumulation rates of radiolarians, benthonic foraminifers, planktonic foraminifers and organic carbon in g/cm^2. 10^3 years in oxygen stage 1,2-4 and 5.

H. Percentage of organic carbon in total sediment (values concerning organic carbon from Müller 1975).

The opal content is plotted as a ratio of radiolarians versus planktonic foraminifers (see Diester-Haass 1978b) (Figure 2b). Interglacial periods 1 and 5 contain only very little opal, whereas in glacial periods, opal content is high.

The benthos foraminifer amounts are plotted as a ratio versus planktonic foraminifers (Diester-Haass 1978a) (Figure 2c). The ratio increases in all glacial periods compared to interglacial ones. This is due to the fact, that glacial accumulation rates of benthonic foraminifers increase by a factor of 4, compared to interglacial ones, whereas planktonic foraminifers increase less (factor 1.5) (Figure 2e, f).

Accumulation rates of sand-sized opal increase by a factor of 100 in glacial periods (Figure 2d). Carbon contents (Figure 2h) are less than one per cent in interglacial and one to two per cent in glacial periods. Accumulation rates of organic carbon (Figure 2g) increase by a factor of 10 (data from Müller 1975).

Benthic organisms like sponges, echinoderms, ostracods all show an

232

increase in accumulation rates in the Late Glacial compared to the interglacial periods (Diester-Haass 1976b).

From these results one can conclude that upwelling has been more intense off Northwest Africa during all the Late Glacial periods (oxygen stages 2, 3 and 4; 11-75 000 years B.P.) than in the Holocene (stage1) and Eem (stage 5).

The depth of maximum upwelling influence in the sediments, today situated in 1 000 m, was at about 2 000-3 000 m in the Late Glacial period (Diester-Haass 1978a).

3. CLIMATE

3.1 *Late Quaternary climatic history of the northern Sahara*

What is the reason for the increased upwelling influence during Late Glacial periods? It can be either an increase in trade wind velocity or an increase in the velocity of the Canary Current or both factors together. Trade winds most probably are not responsible, as proposed in Diester-Haass (1976a). Results obtained from an analysis of coarse terrigenous material (op. cit.) as well as from a clay mineral analysis (Chamley et al. 1977, Chamley, this volume), suggest that the Late Glacial climate (oxygen stage 2, 3 and 4) of the northern Sahara has been more humid than the present and Holocene one. A Late Glacial climate similar to the present Moroccan one, with rainfall in winter and trade winds in summer is suggested. Isotopic studies on ground-water also favour a humid Late Glacial period with west winds bringing rainfall (Sonntag et al. this volume).

There might have been small scale climatic variations in the order of magnitude of 1 000-2 000 years, as suggested by Sarnthein (1978), who belives that the 18 000 B.P. level was very arid and that the climatic optimum at about 6 000 B.P. was very humid (see also Jäkel 1978). These short period climatic variations can hardly be found in the cores, where accumulation rates are in general 2-6 mm/1 000 years (interglacial) and 4-9(20) cm/1 000 years (glacial) (Diester-Haass 1976b) where bioturbation is important and where samples have been taken in about 10 cm intervals.

Furthermore, the level 18 000 years B.P. seems to be not typical for the whole Glacial period. It represents a rather exceptional climatic phase (Rhodenburg & Sabelberg, Sonntag et al. this volume, Jäkel 1978).

The problem of the Early Holocene humid phase can perhaps be solved when considering regional climatic variations: after Gabriel (1978) and Petit-Maire (1979) the western coastal area of the Sahara (21-26°N, Petit-Maire 1979) was not humid during the Early Holocene, in contrast to the coastal area in 21-20°N and to the eastern continental area of the Sahara, where a considerable increase in rainfall has been described.

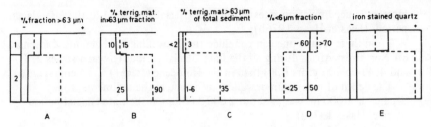

Figure 3. Schematic graph showing comparison of terrigenous fractions from cores off the northern Sahara (heavy line) and off Senegal (dotted line) during oxygen stage 1 and 2. (Data from Diester-Haass 1975, 1976a, Lange 1975 and unpublished results.)
A. Percentage of sand (>63 μm) fraction of total sediment.
B. Percentage of terrigenous matter in >63 μm fraction.
C. Percentage of sand-sized terrigenous matter of total sediment.
D. Percentage of <6 μm fraction of total sediment.
E. Amount of iron stained quartz.

3.2 Comparison of Late Quaternary climates of Senegal and of the northern coastal Sahara

Further evidence for a more humid Late Glacial period in the coastal area of the northern Sahara can be given when comparing results on terrigenous components in sediment cores taken off Senegal and off the northern Sahara (data from Diester-Haass 1975, 1976a and b, Diester-Haass et al. 1973, Chamley et al. 1977). In the Senegal area, a climatic trend opposite to that in the northern Sahara is unanimously accepted by all students of the area (Diester-Haass 1975, 1976a. Further literature see here and in Michel 1973, Sarnthein 1978, Rossignol-Strick 1979). In the Senegal area, a humid Holocene with intense river supply by the Senegal river is preceded by a very arid Late Glacial (oxygen stage 2).

Figure 3 shows a comparison of both areas during oxygen stage 1 and 2. Coming from stage 1 to stage 2, the sand fraction content, essentially controlled by terrigenous fine matter, decreases in the north due to dilution by river mud, and strongly increases in the south, due to the disappearance of river mud (Figure 3a). The amount of sand-sized terrigenous matter increases slightly in the north and very strongly in the south, where wind supplies eolian fine sand (Figure 3b). The sand-sized terrigenous particles form up to one-third of the total sediment in oxygen stage 2 off Senegal (Figure 3c), whereas off the northern Sahara the percentage of sand-sized particles remains about constant. The <6 μm fraction (data from Lange 1975, Chamley et al. 1977, Lange, unpublished results) strongly decreases during the arid oxygen stage 2 off Senegal compared to stage 1, because of the disappearance of the Senegal river, whereas in the north there is a slight or no decrease (Figure 3d). The amount of red stained quartz strongly increases off Senegal and strongly decreases off the northern Sahara, pointing to non-for-

234

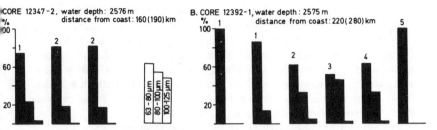

Figure 4. Comparison of grain size distribution of sand-sized terrigenous material off Senegal (A) and off the northern Sahara (B, for position of cores, see Figure 1), in Late Quaternary interglacial (oxygen stages 1, 5) and glacial (stages 2, 3, 4) periods.

mation of red stained quartz or a destruction of older red coatings in the latter area during glacial periods.

We have to mention a basic difference between the arid stage 1 of the northern Sahara and the arid stage 2 of the Senegal area. In stage 1, sea-level has been high and thus the distance from the continent relatively greater, whereas in stage 2 sea-level lowering led to a decrease in distance from the continent.

3.3 *Problem of fluviatile versus eolian supply*

Another criterion for the evaluation of continental climates might be the grain-size distribution of sand-sized terrigenous material. This fraction (763 μm) is the coarsest one supplied from the continent. Extremes in general are very sensitive to changes. Data published by Bagnold (1954) suggest that no quartz $>$ 80 μm can be transported by wind in suspension. An examination of eolian dusts collected over the Atlantic by Jaenicke and Schütz (1971) revealed that the few particles $>$ 50 μm were all aggregates of minerals and/or NaCl. So one might suppose that marine sediments containing eolian terrigenous material do not contain terrigenous fractions $>$ 80 μm.

A test was made to see whether sandy terrigenous material (without considering the light and easily dispersed mica) shows grain-size variations which can be related to climatic changes. Such a test can only be made in cores taken below about 2 000 m water depth, because bottom transport of biogenous and terrigenous particles 'grain by grain' from shelf and upper slope environments takes place down to these depths (own unpublished results and Lutze, 1979). Figure 4 shows an example of grain size data of sand-sized terrigenous material in two cores. One is situated off the northern Sahara, one off Senegal, in similar water depths; the distance from the continent is slightly smaller off Senegal (Figure 1).

The arid oxygen stage 2 identified off Senegal shows 80-83 per cent of 63-80 μm fraction, 20-17 per cent of 80-100 μm fraction and no material

> 100 μm. On the contrary, the oxygen stage 1 and 5 identified off the northern Sahara, interpreted also as arid periods, contain no material > 80 μm, except for one sample, taken 3 cm above the oxygen stage 2, which has much coarser sediments (probable supply through bioturbation). The difference between both grain size curves from arid periods can be explained by a sea-level lowering and increased wind velocities off Senegal (Sarnthein & Walger 1974) during stage 2, which both may be responsible for the presence of the 80-100 μm sized terrigenous material off Senegal.

The humid Holocene (stage 1) off Senegal shows an increase in 80-100 μm sized terrigenous fraction and the presence of 100-125 μm sized quartz. The glacial oxygen stages 2-4 off the northern Sahara contain considerable amounts of 80-100 and 100-125 μm sized quartz which could hardly be supplied by wind, when compared with oxygen stage 2 data off Senegal and stage 1 data off the northern Sahara. If wind could supply 100-125 μm sized terrigenous matter, we do not see why it is not found in the cores off Senegal, where a very arid climate and high wind velocities existed during stage 2. This interpretation of grain size data nicely correlates with clay mineralogy results (Chamley et al 1977, Chamley, this volume).

We suggest that rivers from areas without a dense vegetation cover have a broader grain size spectrum than material transported by the wind alone, rivers supplying finest fractions (< 6 μm) as well as coarsest fractions (> 80 μm). Grain size data of sandy terrigenous particles presented by Kudrass (1973) also show that terrigenous material supplied during humid periods can be very coarse. The transport mechanism of the sand sized terrigenous material in the sea, however, remains a problem.

4. REASON FOR THE INCREASED LATE GLACIAL UPWELLING

All results presented here show that the observed increase in upwelling influence in Late Glacial sediments from the continental slope off Northwest Africa is correlated to a more humid climate than the present one. A climate similar to the present Moroccan one is suggested, with winter rainfall by west winds and summer drought under the effect of trade winds. These conditions lead today to seasonal upwelling off Morocco, measurably by oceanographic methods. But the upwelling is not reflected in the underlying sediments.

The reason for the strongly increased Late Glacial upwelling influence is probably the increase in velocity of the Canary Current (Thiede 1977). The polar front shifted southwards to northern Spain during the maximum of the last glaciation (McIntyre et al. 1976). Thus the general circulation in the North Atlantic occurred in a much narrower belt and velocities of all currents were increased (op. cit.). The increased velocity of the near-coastal Canary Current leads to a stronger offshore movement of surface waters, which is compensated by stronger upwelling from deeper layers.

REFERENCES

Bagnold, R.A. 1954. *The physics of blown sand and desert dunes*. Methuen & Co., London.

Berger, W.H., L.Diester-Haass & J.S.Killingley 1978. Upwelling off North-west Africa: the Holocene decrease as seen in carbon isotopes and sedimentological indicators. *Oceanol. Acta 1*(1):3-7.

Chamley, H. 1979. Marine clay sedimentation and climate in the Late Quaternary off North West Africa. *Palaeoecology of Africa 12*.

Chamley, H., L.Diester-Haass & H.Lange 1977. Terrigenous material in East Atlantic sediment cores as an indicator of NW African climates. *'Meteor' Forsch.-Ergebn. C 26:*44-59.

Chamley, H. & L.Diester-Haass 1979. Upper Miocene to Pleistocene climates in NW Africa deduced from terrigenous components of Site 397 sediments (DSDP Leg 47A). In: B.Ryan, U.von Rad, et al. *Initial Reports DSDP 47*. Washington D.C. (in press).

Chamley, H. & G.Giroud d'Argoud 1979. Clay mineralogy of Site 397, south of the Canary Islands (DSDP Leg 47 A). In: B.Ryan, U.von Rad, et al., *Initial Reports DSDP 47*. Washington D.C. (in press).

Chamley, H. & L.Diester-Haass 1980. Oligocene climatic, tectonic and eustatic history off NW Africa (DSDP Leg 41, Site 369A). *Oceanol. Acta* (in press).

Diester-Haass, L. 1975. Sedimentation and climate in the Late Quaternary between Senegal and the Cape Verde Islands. *'Meteor' Forsch.-Ergebn. C 20:*1-32.

Diester-Haass, L. 1976a. Late Quaternary climatic variations in North-West Africa deduced from East Atlantic sediment cores. *Quatern. Res. 6:*299-314.

Diester-Haass, L. 1976b. Quaternary accumulation rates of biogenous and terrigenous components on the East Atlantic continental slope off NW Africa. *Marine Geol. 21:* 1-24.

Diester-Haass, L. 1977. Radiolarian/planktonic foraminiferal ratios in a coastal upwelling region. *J.Foram. Res. 7*(1):26-33.

Diester-Haass, L. 1978a. Sediments as indicators of upwelling. In: R.Boje & M.Tomczak (eds.), *Upwelling ecosystems*. Springer Verlag, pp.261-281.

Diester-Haass, L. 1978b. Influence of carbonate dissolution, climate, sea-level changes and volcanism on Neogene sediments off Northwest Africa (Leg 41). In: Y.Lancelot, E. Seibold, et al., *Initial Reports DSDP, 41:*1033-1045.

Diester-Haass, L. 1979. DSDP Site 397: climatological, sedimentological and oceanographic changes in the Neogene autochthonous sequence. In: B.Ryan, U.von Rad, et al., *Initial Reports DSDP, 47*. Washington D.C. (in press).

Diester-Haass, L. 1980. Oligocene at Site 369A (DSDP Leg 41): reduced carbonate accumulation rates lead to high opal content. *Marine Geol.* (in press).

Diester-Haass, L., H.-J.Schrader & J.Thiede 1973. Sedimentological and paleoclimatological investigations of two pelagic ooze cores off Cape Barbas, North-west Africa. *'Meteor' Forsch.-Ergebn. C 16:*19-66.

Diester-Haass, L. & H.Chamley 1978. Neogene paleoenvironment off NW Africa based on sediments from DSDP Leg 14. *J. Sed. Petrol. 48*(3):39-53.

Diester-Haass, L. & P.J.Müller 1979. Processes influencing sand fraction composition and organic matter content in surface sediments off W.Africa. *'Meteor' Forsch.-Ergebn. C* (in press).

Gabriel, B. 1978. Klima- und Landschaftswandel der Sahara. In: *Sahara. 10 000 Jahre zwischen Weide und Wüste*. Ausstellungskatalog, Museen der Stadt Köln, Köln, pp.22-34.

Jaenicke, R., C.Junge & H.J.Kanter 1971. Messungen der Aerosolgrössenverteilung über dem Atlantik. *'Meteor' Forsch.-Ergebn. B 7:*1-54.

Jäkel, D. 1978. Eine Klimakurve für die Zentralsahara. In: *Sahara. 10 000 Jahre zwischen Weide und Wüste.* Ausstellungskatalog. Köln, pp.382-396.

Kudrass, H.R. 1973. Sedimentation am Kontinentalhang vor Portugal und Marokko im Spätpleistozän und Holozän. *'Meteor' Forsch.-Ergebn. C 13:*1-63.

Lange, H. 1975. Herkunft und Verteilung von Oberflächensedimenten des westafrikanischen Schelfs und Kontinentalhanges. *'Meteor' Forsch.-Ergebn. C 22:*61-85.

Lutze, G. 1979. Foraminiferal record on stratigraphy and paleooceanography. In: *Symp. Sahara and surrounding seas – sediments and climatic changes.* Akad. Wiss. Lit. Mainz, April 1-4, 1979, Abstract vol.

McIntyre, A., N.G.Kipp et al. 1976. Glacial North Atlantic 18 000 years ago: A CLIMAP reconstruction. *Geol. Soc. Amer. Mem. 145:*43-76.

Michel, P. 1973 *Des bassins des fleuves Sénégal et Gambie. Etude géomorphologique.* Mém. 63. ORSTOM, Bondy, France.

Müller, P.J. 1975. Diagenese stickstoffhaltiger organischer Substanzen in oxischen und anoxischen marinen Sedimenten. *'Meteor' Forsch.-Ergebn. C 22:*1-60.

Petit-Maire, N. 1979. Prehistoric paleoecology of the Sahara Atlantic coast in the last 10 000 years: A synthesis. *J. Arid. Envir. 2:*85-88.

Pflaumann, U. 1975. Late Quaternary stratigraphy based on planktonic foraminifera off Senegal. *'Meteor' Forsch.-Ergebn. C 23:*1-46.

Rohdenburg, H. & U.Sabelberg 1979. Northern Sahara margin. Terrestrial stratigraphy of the Upper Quaternary and some paleoclimatic implications. *Paleoecology of Africa 12.*

Rossignol-Strick, M. 1979. Evolution of the atmospheric circulation in West Africa during full glacial/postglacial times and its influence on climate and vegetation. *Quat. Res.* (in press).

Sarnthein, M. 1978. Sand deserts during glacial maximum and climatic optimum. *Nature 272*(5648):43-46.

Sarnthein, M. & E.Walger 1974. Der äolische Sandstrom aus der West-Sahara zur Atlantikküste. *Geol. Rundschau 63*(3):1065-1087.

Sarnthein, M. & L.Diester-Haass 1977. Eolian sand turbidites. *J. Sed. Petrol. 47*(2):868-890.

Schemainda, R., D.Nehring & S.Schulz 1975. Ozeanologische Untersuchungen zum Produktionspotential der nordwestafrikanischen Wasserauftriebsregion 1970-1973. *Geod. Geoph. Veröff. IV 16:*1-88.

Shackleton, N.J. 1977. Carbon-13 in Uvigerina: Tropical rainforest history and the equatorial Pacific carbonate dissolution cycles. In: N.R.Andersen & A.Malahoff (eds.), *The fate of fossil fuel CO_2.* Plenum Press, New York, 401-427.

Sonntag, C., U.Thorweihe, J.Rudolph, E.P.Löhnert, C.Junghans, K.O.Münnich, E.Klitzsch, E.M.El Shazly & F.M.Swailem 1979, Isotopic identification of Saharian groundwaters, groundwater formation in the past. *Palaeoecology of Africa 12.*

Thiede, J. 1977. Aspects of the variability of the glacial and interglacial north Atlantic eastern boundary current (last 150 000 years). *'Meteor' Forsch.-Ergebn. C 28:*1-36.

238

LATE QUATERNARY DEEP-SEA RECORD ON NORTHWEST AFRICAN DUST SUPPLY AND WIND CIRCULATION

M. SARNTHEIN / B. KOOPMANN
Geologisch-Paläontologisches Institut, Universität Kiel

INTRODUCTION

Aeolian dust from North Africa contributes a large fraction of the Atlantic deep-sea sediments in northern subtropical latitudes. Thus long-term trends of the atmospheric circulation that supplies the dust and ultimately, the history of Saharan climates can possibly be deciphered from the sediments on the ocean floor, where modern methods of stratigraphic dating have achieved a high standard of precision. Nevertheless, various ways of interpreting the sources and transport of terrigenous sediments are still open for discussion.

During the past decades, both the dispersal pattern and composition of Saharan aerosols have been studied by many workers, more recently by Delany et al. (1967), Folger (1970), Prospero et al. (1970), Chester et al. (1971, 1972), Beltagy et al. (1972), Carlson & Prospero (1972), Prospero & Carlson (1972), Parmenter & Folger (1974), Lepple (1975), Szekielda (1977), Jaenicke & Schütz (1978) and by the Gothenburg workshop (Morales 1979). Likewise, several aspects of oceanic dust deposition have been described, for example, by Biscaye (1965), Parkin & Shackleton (1973), Parkin (1974), Bowles (1975), Lange (1975), Diester-Haass (1975, 1976), Diester-Haass et al. (1973), Windom (1975), Chamley et al. (1977), Johnson (1979) and Kolla et al. (1979).

A most comprehensive view of the meteorological setting and the dynamics of Saharan dust supply was published by Carlson & Prospero (1977). At the African continental margin, they show large-scale dust outbreaks with Saharan air plumes being centered at the 700 mb ('Harmattan') level and originating from the southern Sahara and northern Sahel zone during summer (Figure 1A, and Rognon 1976). Over the eastern Atlantic, in many cases the dust trajectory extends into an anticyclonic vortex. This is also substantiated by a number of Landsat imageries (e.g. cover piece of this book), and synoptic studies of meteorology by Tetzlaff & Wolter (1979).

This interesting pattern has been a basic incentive for the present study which is based on the terrigenous fraction of surface sediments and core samples between Northwest Africa and northern Brazil, and tries to attack

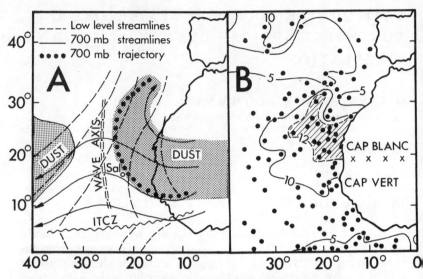

Figure 1. A. Schematic representation of the flow patterns associated with an easterly wave and a Saharan air outbreak (Carlson & Prospero 1977, slightly modified). B. Distribution of weight percent quartz (carbonate-free) in surface sediments of the eastern Atlantic (Kolla et al. 1979). x x x x = Sahel-Saharan boundary

the problems of aeolian-dust dispersal by a new set of grain-size data (Koopmann 1979). Utilizing the modern patterns of aeolian sediment distribution and of marginal river-sediment supply as baseline information, we also examined the distributions at 18 000 and 6 000 yrs B.P., and tried to find critical changes in the subtropical wind system during the last glacial maximum and during the climatic optimum.

METHODS OF STUDY

The grain-size distribution of the > 6 μm silt fraction in carbonate-free samples has been determined in detail by Coulter Counter (15 size fractions). Additionally, the proportion of biogenic opal has been measured by grain counts from SMEAR-slides (> 500 counted grains per size fraction) which have been prepared from six silt sub-fractions being separated by a settling-tube device (Koopmann 1979). Principal and secondary modes of the terrigenous grain sizes have been determined visually from frequency curves where opal has been removed. Results from processing the data by Q-mode multivariate techniques did not yield significant additional aspects and thus have not been included in this paper.

The maps of surface (modern) grain sizes are based on 80 sediment samples which were carefully selected from the uppermost centimeters of trigger weight

240

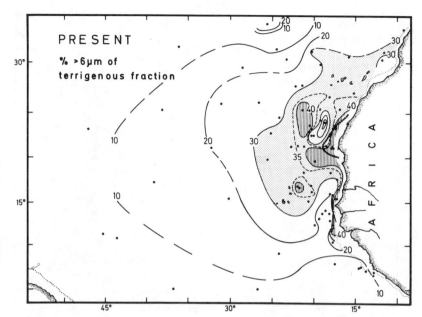

Figure 2. Distribution of percent terrigenous silt (>6 μm, carbonate- and opal-free) in surface sediments. Area with supply of unstained aeolian sediment (<15 per cent stained quartz) is surrounded by a dotted line.

tops and other cores. Some of the modern samples possibly are as much as 2 000 years old when taken from up to 8 cm depth, and when assuming a depth of the mixing layer of 8 cm, but most are considerably younger. The maps of the 6 000 yrs B.P. and 18 000 yrs B.P. levels have been compiled from 32 cores (avoiding turbidites). In 20 of them, an oxygen-isotope stratigraphy has been established (Emiliani 1955, Erlenkeuser et al. in preparation, Shackleton 1977). In the remaining cores, the stratigraphy is based on carbonate and foraminiferal curves, where in most cases the critical age level of the last Glacial maximum at 18 000 yrs B.P. has been determined by CLIMAP project members (Cline & Hays 1976). The 18 000 years B.P. level has tolerances of less than ±1 000 years in well-dated cores with high sedimentation rates, and ±3 000 years in poorly dated cores with low sedimentation rates. The 6 000 yrs B.P. level ranges from ± 500 to ± 2 500 years respectively.

RESULTS AND DISCUSSION

Modern (surface) dispersal patterns of terrigenous silt

High concentrations of the silt fraction (> 6 μm) are clearly confined to the

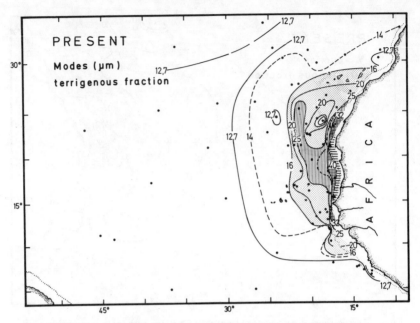

Figure 3. Distribution of modal grain sizes of terrigenous silt (>6 μm, carbonate- and opal-free) in surface sediments. Dotted area: >16 μm; hatched area: >25 μm; cross-shaded area: >32 μm; area hatched by solid lines: >40 μm.

eastern Atlantic margin (Figure 2). They particularly occur up to 500-1 000 km offshore of the arid belt of Mauritania. The silt concentration attains its maximum (more than 35-45 per cent by weight) in a hook-shaped zone which emanates westward south of Cape Blanc, subsequently bends towards north and northeast, and is finally directed to the Canary Isles and Morocco. Another restricted maximum lies northeast of the Cape Verde Isles. The silt percentage is typically low all over the distal regions of the ocean except for a few locations proximal to seamounts, and adjacent to lowland-river mouths the sediment output of which is dominated by fine rather than by coarse grained material. This occurs off the African tropical rivers, up to the Senegal River, and also off Moroccan wadis. Furthermore, silt is almost absent (down to <10 per cent) in an intriguing eye shaped area which fills the center of the silt maximum belt at about 23°N, and is approximately 450 km long and 250 km wide.

Coarse modal grain sizes directly coincide with high concentrations of silt (Figure 3). The principal modes range between 30-64 μm in a zone paralleling the lower continental slope, and at a low seamount (300-500 m above the surroundings) immediately adjacent (18°N, 18°W). Further off-shore the hook-shaped and other silt maxima are characterized by principal modal sizes of

20-30 μm. Outside a transitional belt which is about 600 km wide, terrigenous modal grain sizes are uniformly about 12 μm in the distal oceanic regions. The silt minimum 'eye' at 23°N is reflected by fine modes of 11-16 μm; likewise the sediments directly off Guinea, south Senegal and south Morocco are fine grained and show no difference to distal oceanic sediments. The resolution of this distribution pattern is far more detailed than that of Johnson (1979), because it is based on a denser sample grid.

We attribute the varying abundances and grain sizes of terrigenous silt in the deep sea to various forms of supply from fluvial and aeolian sources, to downslope transport at the continental margin, and to redistribution by bottom and other marine currents. Most fine grained sediments (with < 20-30 % > 6 μm; modes < 14 μm and rarely up to 25 μm, and particularly lacking coarse silt > 40 μm) which occur at the continental margin and upper rise, are confined to the neighbourhood of large rivers and thus can be related confidently to fluvial sources south and north of the Saharan desert. Similar fine grain sizes have been observed in muds at the Senegal River mouth (Kiper 1977). Another distinct feature is the narrow belt of abundant and coarse grained silt (> 40 % > 6 μm; modes coarser than 32 μm) which occurs all along the lower continental slope. Except for two more distant samples at 18°N, this zone must be understood as being strongly influenced by continuous downslope bottom transport winnowed from above such as supported by grain-size distribution (Bein & Fütterer 1977, Koopmann 1979) and by shallow-water benthic foraminifers serving as tracer particles (Lutze et al. 1979, Seibold, this volume).

The bulk of silt-rich sediments, particularly the hook-shaped silt maximum with modal grain sizes of 25-30 μm, occurs off the hyper-arid regions in North Africa: no river mouths and wadis are active in the northern Sahel zone and the Sahara desert from 17 to 28°N. But the marked distribution pattern of (coarse grained) silt abundance is very much similar to the outlined trajectories of Saharan dust plumes (Figure 1, Carlson & Prospero 1977, Tetzlaff & Wolters 1979). It is interesting to note that they also largely coincide with the distribution pattern of quartz in deep-sea sediments (Kolla et al. 1979, Figure 1B). These similarities and the absence of other sources suggest that the silt grains originate from Saharan dust outbursts. Apparently the silt maximum in the deep sea can approximately trace the long-term average streamlines of Saharan dust plumes, which are concentrated in the Harmattan winds of the summer season. They follow the sub-Saharan lowlevel Easterly Jet, pass around an upper-air heat high over the western Sahara, and finally point to the Canary Isles. Indeed, most dust falls which have been observed for about 30 years at the Izaña Observatory, Teneriffa (2 370 m above sea level), have arrived above the trade wind inversion from the western sectors during summer (Abel et al. 1969). From this we conclude that aeolomarine dust deposits do not so much indicate aridity on the nearby mainland, but rather the prevailing direction of the 700 mb wind regime carrying the dust. Kaolinite, montmorillonite, and other late-

243

13289-1/2

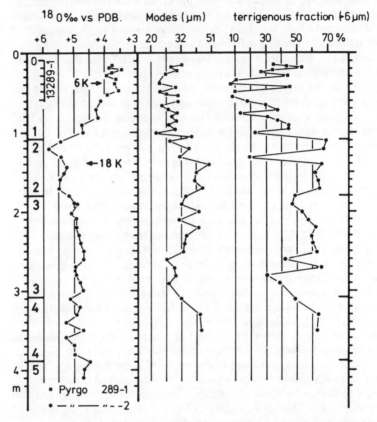

Figure 4. Vertical distribution of terrigenous sand plus silt ($>6\ \mu m$) and principal modal grain sizes (carbonate- and opal-free) in sediments of Valdivia core 13289-1 and 2, 18°04'N, 18°00'W. The oxygen-isotope stratigraphy is based on measurements in shells of benthic foraminifers *(Pyrgo murrhina)* by Dr H.Erlenkeuser (Erlenkeuser et al. in preparation).

ritic weathering products may serve as tracers for a dust flux originating from the Sahel zone (as shown on a large-scale distribution pattern by Biscaye 1965), whereas carbonate particles, palygorskite, and mixed layer clay minerals may indicate more central Saharan sources with chlorite coming from other sources even further north (Lange 1975, Sarnthein 1979).

The effect of this turbulent upper air system on deep-sea silt sedimentation appears to be considerably stronger than that of the northeasterly trade winds at the surface. And only minor amounts of clay sized particles fall out from the Harmattan level over the first 1 000 km (Jaenicke &

Schütz 1978). Settling silt grains can pass through the trade winds within a few hours. Thus the outfall pattern of silt is only little shifted, or elongated towards the south(-west) before they reach the sea surface. Also the superposition of marine currents over the dustfall pattern can only slightly alter the dispersal of aeolian silt grains which settle through the water column too quickly to be transported further off. Finally the resulting grain-size spectrum of silt from deep-sea sediments still basically correlates with the primary grain sizes modelled and observed within aerosols (Jaenicke & Schütz 1978), although aggregates of fine grains have been included in those curves.

In contrast, transport of clay by marine currents can be clearly distinguished locally by a few marked differences of the abundance pattern of silt coarser than 6 μm (Figure 2) as compared with the distribution of silt modal grain sizes (Figure 3): off Senegal, a lobe of enriched fine fraction finer than 6 μm coincides with comparatively coarse silt modes. This feature can indicate a current related long-distance transport of river-borne (tropical) clays from the south up to 19°N which are superimposed on the local high silt concentrations of the Harmattan-supplied dust load (first observed by Lange 1975).

Yet to be discussed is the striking eye-shaped feature of fine grained sediment at 23°N. Because there are no rivers within 500 km, it does not appear to be related to river-borne clay. Possibly its origin can be explained by its position which lies just at the center of a vortex of atmospheric circulation, i.e. outside the main dust trajectory, and where the local fine grained material supplied by long distance downslope submarine transport is diluted only by clay and fine-grained silt from marginal — or distal dust falls.

Concentrations of terrigenous silt during the last glacial maximum (18 000 years B.P.)

The abundance and the modal sizes of terrigenous silt have varied considerably during the past 30 000 yrs B.P. To demonstrate this trend, we have selected Valdivia core 13289, 18°W, 18°N (Figure 4), which was retrieved from an isolated, 300-500 m high seamount on the continental rise where lateral sediment supply by bottom transport is rather unlikely. Coarse grain sizes reach a maximum (up to 66 % > 6 μm, modal sizes up to 45 μm) during the glacial stage 2 culminating around 18 000 yrs B.P., and a marked minimum (< 10 % > 6 μm, modal sizes down to < 25 μm) during the climatic optimum of stage 1 around 6 000 yrs B.P. The drastic changes obviously reflect the history of dust influx from the Sahara at a critical position near the source (Figure 5); possibly the changes are also related to marine current activity which temporarily might have somewhat restricted the deposition of fine grains on top of the seamount. Nevertheless at this core position, the pulses of coarse silt should give a good

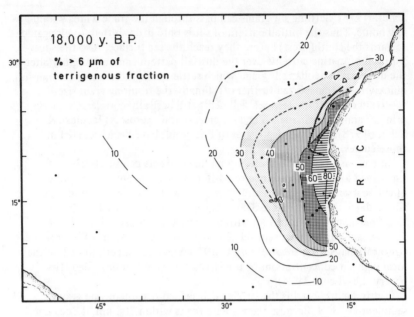

Figure 5. Distribution of percent silt ($> 6\ \mu$m, carbonate- and opal-free) in the sediments of the Last Glacial (18 000 years B.P.) times. Area with supply of unstained (aeolian) sediment (< 15 per cent stained quartz) is surrounded by a dotted line.

idea of the unusually, today unequalled (Bagnold 1954), coarse grain-size distribution (up to 7,4 % $> 62\ \mu$m) of aerosols which were carried to more than 150 km offshore during phases of extreme Glacial wind regimes (also observed by Parkin & Shackleton 1973). These winds coincided with hyper-aridity that is generally accepted for the adjacent part of Mauritania at that time.

During the last Glacial times, the silt proportion ($> 6\ \mu$m) surmounted 40-45 per cent of the terrigenous sediment input as far west as the Cape Verde Isles, from 12°-23°N (Figure 5). This area basically coincides with the distribution of modal grain sizes coarser than 20 μm (Figure 6). As a result, the center of silt deposition and of maximum modal grain sizes remained almost stationary off Africa at 15°-20°N, where extended dune fields then indicated a hyperarid climate (Sarnthein 1978). Since the (lower) Senegal River did not show any activity during glacial times (Strick-Rossignol et al. 1979, Michel 1979), and per analogy to the modern dispersal patterns, we interpret the enrichment of silt as a deposit of (Harmattan) airborne dust. A comparison of the modern (Figures 2, 3) and the Glacial (Figures 5, 6) sediment-distribution patterns clearly reveals that the Saharan dust outbreaks to the Atlantic ocean, i.e. the Harmattan wind circulation, did not undergo noteworthy latitudinal

246

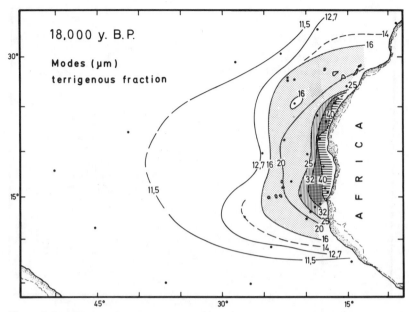

Figure 6. Distribution of modal grain sizes of terrigenous silt (>6 μm, carbonate- and opal-free) in the sediments of the Last Glacial (18 000 years B.P.) times.

shifts in position from Interglacial to Glacial conditions, despite varying climatic zonations on land. This conclusion which is based on a rather dense sample grid, does not corroborate previous deductions concerning a major latitudinal shift of the dust depocenter in the deep sea such as drawn from the concentrations of various clay minerals and stained quartz (Chamley et al. 1977), and from the distribution of quartz percentages in deep-sea sediments (Kolla et al. 1979). However, our available sample grid at 18 000 yrs B.P. is not sufficient to discern details of atmospheric circulation such as the modern vortex pattern.

Three main differences distinguish the Glacial (18 000 yrs B.P.) sediments from the modern ones: (i) Fine grained, i.e. river-borne sediments are completely absent along the whole African continental margin from some 10-12° to over 27°N, i.e. also off the North Sahara, and have only occurred off Morocco at 32°N and off Sierra Leone, 8°N, at the 18 000 yrs B.P. level. (ii) The much expanded coarse-silt area up to 800 km offshore West Africa corresponds to a general doubling of sedimentation rates at 18 000 yrs B.P. (Koopmann 1979). Both features reflect a substantially increased dust output from the South Sahara as compared to today, a dust load which has possibly led to changes of the Earth's albedo. (iii) Almost all surface samples derived from the area influenced by modern dust falls contain abundant stained

247

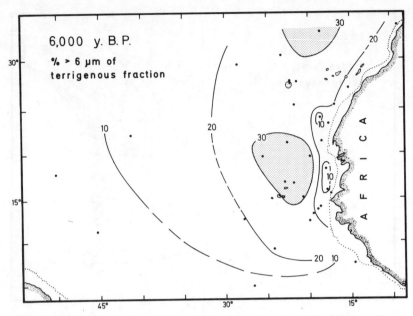

Figure 7. Distribution of percent silt (> 6 μm, carbonate- and opal-free) in the sediments of the Holocene climatic optimum (6 000 years B.P.).

quartz in the fine-sand fraction, except for one small section off-shore 21-23°N, at the southern end of the 'fine-grained eye' region (Figure 2). However, at 18 000 years B.P., this pale-silt region has been greatly elongated parallel to the continental margin up to north of 27°N, as indicated on Figure 5 (and Sarnthein 1979). The stained-quartz dust is obviously related to the dominant south Saharan wind load, whereas the pale-quartz dust can serve as a tracer for a local dust discharge which is supplied by northeasterly (Trade) surface winds from the northwestern Sahara margin such as observed in the modern Baie du Levrier at 21°N (Koopmann et al. 1979). Obviously during the Glacial aerial exposure of the shelf, this local second source as well as the transporting wind regime were much increased, and the distance to the deep sea reduced. Our interpretations are at variance with those of Diester-Haass (1976, 1979), Diester-Haass et al. (1973) and Chamley et al. (1977), who based their conclusions on a rather more limited sampling set.

Concentrations of terrigenous silt during the last climatic optimum (6 000 years B.P.)

Figure 4, 7 and 8 show that the time slice at 6 000 yrs B.P. has differed significantly from both the Glacial and the modern time slices under discussion:

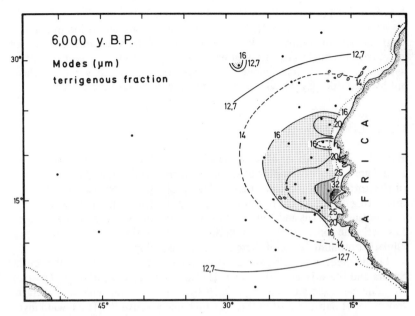

Figure 8. Distribution of modal grain sizes of terrigenous silt (>6 μm, carbonate- and opal-free) in the sediments of the Holocene climatic optimum (6 000 years B.P.).

high clay concentrations (with less than 10-20 per cent silt > 6 μm) occur all along the continental margin from Morocco to Gambia, becoming slightly more prominent to the south. Following the rules of the modern dispersal pattern of grain sizes, and in harmony with a generally humid land record (Sarnthein 1978), we ascribe the continuous high-clay band to dominating river supply close to the continental margin.

Higher silt proportions appear 300-800 km offshore. They would seem to reflect an ongoing large scale dust supply which has remained undiluted by fluvial sediments at this more distal position. The distribution of modal grain sizes can basically confirm this suggestion (Figure 8), but their rather low values (16-25 μm) disclose that the transport capacity of the Harmattan winds then was drastically reduced. However, as with the Glacial period, the center of Saharan dust outbreaks did not shift during the climatic optimum but remained almost stationary at 15-20°N. In addition, the area of dust related coarse modal grain sizes displayed a minor northward expansion up to 23°N (Figure 8), compared with the present. From this we infer either a weakening or possibly a temporary complete breakdown of the (Trade) surface winds from northeast which in the modern situation displace the dust falls from the upper atmospheric levels slightly to southwest. Such a 'quiet' situation at 6 000 yrs B.P. is corroborated by the almost complete lack of locally derived pale-dust deposits off the north Sahara.

At a low seamount (18°N, 18°W), the simultaneous deposition of abundant clay and of comparatively coarse modal grain sizes (23 μm) forms a local contrast. This might provide an interesting insight into the prevailing transport systems, when assuming a model of superposition of directly wind-furnished silt by fluviatile clay which is supplied by long-range suspension transport.

CONCLUSIONS

Both the concentrations of terrigenous sediment fractions > 6 μm and the modal grain sizes of terrigenous silt were mapped in detail in the sediments of the northern subtropical Atlantic. The resulting dispersal patterns reveal interesting details on the modern modes of sediment transport of wind and river borne materials and help to assess the differences of climatology at the East Atlantic continental margin during last Glacial times and the Holocene climatic optimum. Some of these results are as follows:

1) The modern contribution of aeolian dust is concentrated off the southern Sahara and the adjacent Sahel zone. The vortex-shaped maximum of (coarse) silt in deep-sea sediments likely reflects by its position the long-term average trajectory of Saharan dust outbursts which occur above the Trade wind inversion and finally arrive at the Canary Isles. But the extent of North African aridity is not directly indicated by marine dust deposits, a large proportion of which is derived from the semi-arid Sahel zone.

2) In the present center of aridity, 20-23°N, proximal deposits of pale quartz mark a narrow zone of dust supply by (Trade) surface winds from the northern Sahara.

3) High concentrations of clay and fine grained silt (modes < 14 μm, rarely up to 25 μm) immediately adjacent to the continental margin reveal the influence zone of river mouths south and north of the Saharan desert. On the other hand, a narrow continuous belt of very coarse grained silt (modes > 32-40 μm) at the middle and lower continental slope marks the zone of continuous downslope bottom transport.

4) Present marine current activity locally results in a striking superposition of high river-borne clay concentrations over comparatively coarse, wind-borne silt modes such as off north Senegal.

5) The aeolian dust contribution to the east Atlantic was much increased up to 800 km offshore during the last Glacial times, at 18 000 yrs B.P., but its center showed almost no latitudinal shift to the south as compared to today. This agrees with land data from North Africa that suggest no major shift, but an expansion of the synglacial aridity belt both north and south, and enhanced wind speeds (Sarnthein 1978). These are reflected by the massive and unusual transport of coarse grained silt and even of fine sands to the deep sea. The local supply of pale-quartz dust from the North Sahara was also much expanded during the Glacial aerial exposure of the shelf.

250

6) During Glacial times, no fluvial sediment supply is identified from 10° or 12°-28°N, but only offshore Morocco and offshore Sierra Leone.

7) During the climatic optimum at 6 000 yrs B.P., the supply of aeolian dust by the Harmattan winds was strongly reduced, but its center again remained almost stationary at 15-20°N (thus the theories of the latitudinal shift of the upper-wind circulation belts (Rognon 1976) are not corroborated). Abundant river-borne sediments largely diluted the aeolian silt input in a 200 km wide high-clay belt all along the upper rise from Morocco to Gambia and implied continental humidity. No more local dust supply by the Trade winds can be recognized.

Finally, we want to point out that the high concentrations of modern aeolian dust already resemble much more those of a Glacial regime than those of the peak Interglacial which was particularly characterized by ubiquitous river muds at the West African continental margin. Shackleton suggested a similar trend of a massive increase of dust generation prior to the formation of ice caps during a later Interglacial during oxygen isotopic stage 5e (Shackleton 1979).

ACKNOWLEDGEMENTS

We owe sincere thanks to many colleagues of our Kiel university who helped by critical discussion to improve our results, particularly to Drs E.Seibold, D.Fütterer, H.Lange, E.Walger and S.McLean. Stable isotopes of foraminiferal tests were measured by Dr H.Erlenkeuser, Institut für Kernphysik, Kiel University, who kindly made available his unpublished results for this paper. During processing the samples, we received large support from Ms E.Albrecht, Ms Sh.Akester and Ms M.Kiper who also gave valuable help by applying Q-Mode multivariate techniques to further process our grain-size data. Finally, we would like to thank the shipboard parties of R.V.Valdivia and R.V.Meteor, and the core curator of the Lamont-Doherty Geological Observatory, New York, for generously furnishing deep-sea samples. This study is a CLIMAP contribution and part of the thesis of B.Koopmann. It has received extensive support from the Deutsche Forschungsgemeinschaft.

REFERENCES

Abel, N., R.Jaenicke, C.Junge, P.Kauter, R.G.Prieto & W.Seiler 1969. Luftchemische Studien am Observatorium Izaña (Teneriffa). *Met. Rundschau 22:*158-167.

Bagnold, A. 1941/1954. *The physics of blown sand and desert dunes.* Methuen & Co., London.

Bein, A. & D.Fütterer 1977. Texture and composition of continental shelf to rise sediments off the northwestern coast of Africa: An indication for downslope transportation. *'Meteor' Forsch.-Ergebn. C 27:*46-74.

Beltagy, A.I., R.Chester & R.C.Padgham 1972. The particle-size distribution of quartz in some North Atlantic deep-sea sediments. *Marine Geol. 13:* 297-310.

Biscaye, P.E. 1965. Mineralogy and sedimentation of recent deep-sea clay in the Atlantic Ocean and adjacent seas and oceans. *GSA Bull. 76:* 803-832.

Bowles, F.A. 1975. Paleoclimatic significance of quartz/illite variations in cores from the eastern equatorial North Atlantic. *Quater. Res. 5:* 225-235.

Carlson, T.N. & J.M.Prospero 1972. The large-scale movement of Saharan air outbreaks over the Northern Equatorial Atlantic. *J. Appl. Meteor. 11:* 283-297.

Carlson, T. & J.Prospero 1977. Saharan air outbreaks: Meteorology, aerosols, and radiation. *Report US GATE Central Program Workshop* (NCAR), Boulder, Colorado, pp.57-78.

Chamley, H., L.Diester-Haass & H.Lange 1977. Terrigenous material in East Atlantic sediment cores as an indicator of NW African climates. *'Meteor' Forsch.-Ergebn. C 26:* 44-59.

Chester, R., H.Elderfield, J.J.Griffin, L.R.Johnson & R.C.Padgham 1972. Eolian dust along the eastern margins of the Atlantic Ocean. *Marine Geol. 13*(2): 91-106.

Chester, R. & L.R.Johnson 1971. Atmospheric dusts collected off the Atlantic coasts of North Africa and the Iberian Peninsula. *Marine Geol. 11*(4): 251-260.

Cline, R.M. & J.D.Hays (eds.) 1976. Investigation of Late Quaternary paleooceanography and paleoclimatology. *GSA Mem. 145.*

Delany, A.C., A.C.Delany, D.W.Parkin, J.J.Griffin, E.D.Goldberg & B.E.F.Reimann 1967. Airborne dust collected at Barbados. *Geochim. et Cosmochim 31:* 885-909.

Diester-Haass, L. 1975. Late Quaternary sedimentation and climate in the East Atlantic between Senegal and Cape Verde Islands. *'Meteor' Forsch.-Ergebn. C 20:* 1-32.

Diester-Haass, L. 1976. Late Quaternary climatic variations in NW Africa deduced from East Atlantic sediment cores. *Quater. Res. 6:* 299-314.

Diester-Haass, L. & H.-J.Schrader & J.Thiede 1973. Sedimentological and paleoclimatological investigations of two pelagic ooze cores off Cape Barbas, NW Africa. *'Meteor' Forsch.-Ergebn. C 16:* 19-66.

Emiliani, C. 1955. Pleistocene Temperatures. *J. Geol. 63:* 538-578.

Erlenkeuser, H., B.Koopmann & M.Sarnthein 1979. *Oxygen-isotope stratigraphy of 'Meteor' cores at the eastern Atlantic continental margin* (in preparation).

Folger, D.W. 1970. Wind transport of land-derived mineral, biogenic, and industrial matter over the North Atlantic. *Deep-Sea Res. 17:* 337-352.

Jaenicke, R. & L.Schütz 1978. Comprehensive study of physical and chemical properties of the surface aerosols in the Cape Verde Islands region. *J. Geoph. Res. 83*(C7): 3585-3599.

Johnson, L.R. 1979. Mineralogical dispersal patterns of North Atlantic deep-sea sediments with particular reference to eolian dusts. *Marine Geol. 29*(1-4): 335-346.

Kiper, M. 1977. Sedimente und ihre Umwelt im Senegaldelta. Unpubl. Dipl. Arb. Univ. Kiel.

Kolla, V., P.Biscaye & A.Hanley 1979. Distribution of quartz in Late Quaternary Atlantic sediments in relation to climate. *Quater. Res. 11:* 261-277.

Koopmann, B. 1979. Saharastaub in Sedimenten des subtropisch-tropischen Nordatlantik während der letzten 20 000 Jahre. Thesis Univ. Kiel.

Koopmann, B., A.Lees, P.Piessens & M.Sarnthein 1979. Skeletal carbonate sands and wind-derived silty marls off the Saharan Coast: Baie du Levrier, Arguin Platform, Mauritania. *'Meteor' Forsch.-Ergebn. C 30:* 15-57.

Lange, H. 1975. Herkunft und Verteilung von Oberflächensedimenten des westafrikanischen Schelfs und Kontinentalhanges. *'Meteor' Forsch.-Ergebn. C 22:* 61-84.

Lepple, K.F. 1975. Eolian dust over the north Atlantic ocean. Ph.D. Thesis Univ. of Delaware.

Lutze, G.F., M.Sarnthein, B.Koopmann, U.Pflaumann, H.Erlenkeuser & J.Thiede 1979. Meteor cores 12309: Late Pleistocene reference section for interpretation of the Neogene of Site 397. In: U.von Rad & W.B.F.Ryan, et al. *Initial Reports DSDP 47.* US Government Printing Office, Washington D.C. 727-739.

Michel, P. 1979. The southwestern Sahara margin. Sediments and climatic changes during the Recent Quaternary. *Palaeoecology of Africa 12.*

Morales, C. (ed.) 1979. *SCOPE 14. Saharan dust. Mobilization, transport, deposition.* J.Wiley & Sons, Chichester and New York.

Parkin, D.W. 1974. Trade-wind during the glacial cycles. *Proc. R. Soc. London A 337:*73-100.

Parkin, D.W. & N.J.Shackleton 1973. Trade wind and temperature correlations down a deep-sea core off the Saharan Coast. *Nature 245*(5426):455-457.

Parmenter, C. & D.W.Folger 1974. Eolian biogenic detritus in deep sea sediments: A possible index if Equatorial Ice Age aridity. *Science 185*(4152):695-698.

Prospero, J.M., E.Bonatti, F.Schubert & T.N.Carlson 1970. Dust in the Caribbean atmosphere traced to an African dust storm. *Earth Planet. Sci. Lett. 9:*287-293.

Prospero, J.M. & T.N.Carlson 1972. Vertical and aerial distribution of Saharan dust over the Western Equatorial North Atlantic Ocean. *J. Geophys. Res. 77:*5255-5265.

Rognon, P. 1976. Essai d'interprétation des variations climatiques au Sahara depuis 40 000 ans. *Rev. Géogr. phys. Géol. dyn., II, 18*(2-3):251-282.

Rossignol-Strick, M. & D.Duzer 1979. Late Quaternary pollen and dinoflagellate analysis of marine cores off West Africa. *'Meteor' Forsch.-Ergebn. C 30:*1-14.

Sarnthein, M. 1978. Sand deserts during glacial maximum and climatic optimum. *Nature 271*(5648):45-51.

Sarnthein, M. 1979. Indicators of continental climates in marine sediments, a discussion. *'Meteor' Forsch.-Ergebn. C 31:*49-51.

Seibold, E. 1979. Climate indicators in marine sediments off NW Africa: A critical review. *Palaeoecology of Africa 12.*

Shackleton, N.J. 1977. The oxygen isotope stratigraphic record of the Late Pleistocene. *Phil. Trans. R. Soc. Lond. B 280:*169-182.

Shackleton, N.J. 1979. Oxygen isotopes in deep-sea cores: Indicators of continental precipitations in pericontinental areas. Paper presented at the Mainz Symposium on Sahara and surrounding seas – Sediments and climatic changes, 1-4 April 1979.

Tetzlaff, G. & W.Wolter 1979. Meteorological patterns and the transport of mineral dust from the North African continent. *Palaeoecology of Africa 12.*

Windom, H.L. 1975. Eolian contributions to marine sediments. *J. Sed. Petrol. 42*(2):520-529.

REFERENCE SECTIONS ON LAND

THE SAHELIAN ZONE AND THE PROBLEMS OF DESERTIFICATION

Climatic and anthropogenic causes of desert encroachment

HORST G. MENSCHING
Geographisches Institut, Universität Hamburg

1. INTRODUCTION

Since the last great drought disaster (1969-1973) the problems of the degradation and destruction of the Sahelian ecosystem have become of major interest all over the world. In the affected countries in the tropical Sahelian zone, from Senegal to the upper Nile, areas of desert-like conditions expanded changing the natural ecological preconditions of the thorn-scrub savanna and also partly of the northern dry savanna. The different systems of land use dominant in that zone, both animal husbandry and cultivation, have been negatively affected, especially in the drought periods. Consequently large regions failed to subsist their inhabitants, which led to great migrations, especially towards the southern parts of the savanna belt. A part of the population returned north after the drought period of 1973 was over, but they did not find their former basis of living there. The reasons for the advancement of the cultivation boundary northwards are mainly due to the fact that before that major drought, a longer period of wetter years occurred, thus making the land use potential appear more favourable than the long-term climatic ecological conditions of the semi-arid marginal tropical zone allow.

This paper intends to discuss both the climatic and the anthropogenic causes of desertification. The questions of climatic evolution and morphogenesis play an important role in judging the present situation of the Sahelian ecosystem. As a transitional zone between the Saharan and Sudan climatic zones since the young Tertiary, the Sahel belongs to a climatic fluctuation zone in which more arid and more humid phases of the marginal tropical climatic system *more effectively* interchange than in the case of the Sahara itself. This is supported by numerous pedological, morphological and hydrological evidences which are widely spread throughout the Sahelian zone.

2. PRESENT DYNAMICS AND THE GENETIC TRENDS IN THE SAHELIAN CLIMATIC ZONE

The Sahelian climate shows all the characteristics of the transition from the

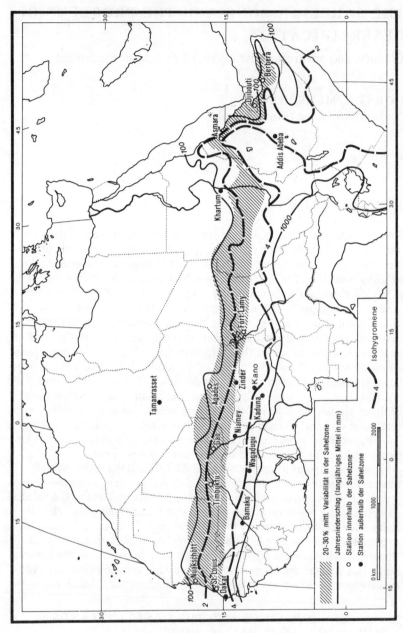

Figure 1. The situation, mean variability of precipitation and the Isohygromenes (2 and 4) in the Sahelian zone.

savanna to the arid zone south of the Sahara. This applies to climate as well
as vegetation, water balance and also to the geomorphological structure with
its current processes. But the genetic aspect of these geofactors must be kept
in mind when considering the regional and local disparities of the vulnerability
of the Sahelian ecosystem to human impact. In the physical geographical
sphere the following analyses are carried out.

2.1 *Climatic and palaeo-climatic analysis*

The climatic boundaries of the Sahelian zone can be drawn according to different criteria. The least convincing of them are the long-term annual means of
precipitation, because the variability is very high. Our most recent studies[1]
have shown that the 1905-70 annual mean of selected stations in West Africa
(dry savanna to thorn-scrub savanna) have an average deviation from the mean
of 20-30 per cent (Figure 1), while the average variability range (positive and
negative) was 30-60 per cent. The analysis of the climatic data shows that
although the positive deviation (precipitation above the mean) is generally
higher, the number of dry years is also higher, while the percentage of mean
negative deviation is lower than the positive deviation.

A typical example of the southern Sahel is the station of Zinder (Republic
of Niger): (56-year mean)
Zinder (453 m above sea level)
490 mm annual precipitation mean
average positive deviation 24,3 per cent in 31 wet years
average negative deviation 19,6 per cent in 40 dry years
average variability range 43,9 per cent

While in Gao/Mali the positive and negative deviation as well as the number
of wet and dry years are well-balanced, an example of the northern Sahel (the
station of Kidal) shows the following:
Kidal (459 m above sea level)
average positive deviation 38,5 per cent in 20 wet years
average negative deviation 25,9 per cent in 30 dry years
average variability range 64,4 per cent

In the central Saharan stations, the mean variability range rises above 100 per
cent (Tamanrasset: 118 per cent and Bilma 134 per cent). Thus the annual
variability is a decisive hygric climatic characteristic, to which one should add
the regional disparities and the high rainfall variability within each summer
rainy season. Thus the distribution of rainfall within the rainy season has
proved to be a standard value for crop abundance, which shows a high
regional fluctuation. Annual means of rainfall can, therefore, only give a
rough judgment of the agroecological situation.

For limiting the Sahelian zone, the number of 'humid' months also gives
a rough idea. The transition from the Sudan zone to the Sahelian zone can be
set (per definitionem) at 4-3 humid months, while the transition to the Sahara

lies at about one humid month. Here, empirical values, especially the distribution and density of the vegetation cover, are also taken into consideration. Closer details on this subject are not intended to be given here.

Attention should be drawn to the fact that a Sahelian zone which is defined according to the isohygromenes is also liable to high variabilities of humidity (whatever criteria are used here, as long as the relationship between precipitation and evaporation is the basis used). In the West African Sahel, for example, the towns Niamey and Zinder, as well as N'Djamena (formerly Fort Lamy) in Chad, lie in the southern margin of the so-defined Sahelian zone, while Kano, for example, clearly belongs to the Sudan climatic zone (Figure 1).

One of the good attempts to display the distribution of aridity and humidity in a manner compatible with the ecosystem is made by D.Henning & H.Flohn (1977)[2]: 'Climate aridity index map, Budyko-Ratio 1:25 Mio'. In that map, the Sahelian zone extends between the aridity indices 2 and 7. The radiation data used are to be preferred to other indices. Owing to the high variability of precipitation, however, the problem of using the annual means remains unsolved. Numerous studies and calculations have proved that the Sahelian zone is climatically nearly identical with the vegetation zone of the thorn-scrub savanna (cf. 1.2.).

Morphogenetically and palaeoclimatically, the following facts should be borne in mind when assessing the ecological resources of land use under desertification hazards:

The present relief provides varying preconditions for the various ecological areas. Apart from highlands and mountain regions (e.g. Aïr and Jebel Marra), one can differentiate the following major relief units which give some proof of the climatic genesis since the Tertiary.

1. The morphogenesis of laterite plateaux covered with ferricrusts indicates also a trend towards climatic fluctuations with periods of increasing aridity already in the Tertiary phases. However, absolute climatic data can hardly be gained from the deeply weathered soils and their sediments. Relatively wetter and drier phases followed in succession, as P.Michel,[3] for instance, has shown by his studies in the further Senegal basin. For the question at hand, this means that the Sahelian zone already possessed in the Tertiary a wide range of climatic variability in the hygric sphere, as can be retraced by the distribution of laterite and ferricrusted rests of the old surfaces. The northern limit of this old relief of various niveaux lies in the southern Saharan zone, indicated by small plateaux rests.

2. In the Quaternary, pediments and glacis have largely replaced the old surfaces forming lower sedimentation plains. During that time, the new Quaternary hydrologic system was gradually formed. These younger surfaces and valley plains are the most important agricultural areas in West Africa today. Climato-genetically, several fluctuations from wetter to drier conditions can also be assumed here, as can be shown in the Sahel since the Tertiary. During the time of enhanced surface erosion of sediments the old Quaternary

and Tertiary horizons of ferrous laterite were often washed off their over-laying sediments. Owing to their high edaphic aridity, they hardened and built a ferricrust.

3. In the young Quaternary (young Pleistocene and Holocene) period, a relatively arid phase from about 21 000 to 15 000 B.P. is assumed on the basis of the different studies made on the dune belt of the Sahel. The Carbon datings vary greatly from region to region. However, a large part of the dune belt between the Senegal and the Nile must have been built in that period (as it is one of the most important periods of aeolian morphodynamics). It must be taken into consideration, however, that these dune complexes, which are 'fixed' with vegetation today, are bound to large sedimentation basins of major fluvial systems whose enormous sand transport is primarily due to the former wetter period.

These old dunes play a decisive role in the spread of cultivation today. Thus they are highly susceptible to anthropogenic desertification, as will be shown below.

Owing to increased humidity fluctuations, the period from 11 000 to 5 000 B.P. is supposed to have been generally wetter than today. This is indi-cated by radio-carbon data of sediments and by the different cultures of the stone age, particularly of the Neolithic Age. How far the stone age cultures already possessed a concentration of population, which might have triggered off the first extensive phenomena of desertification, must yet be studied more closely. A beginning has been made in this field. However, climatic interpretations of prehistoric settlement areas should pay special attention to the particular relief conditions of each of these places (escarpments, wadis, margins of basins and the like) before establishing generalized climate curves.

Data of historical and present climatic development, especially those denoting the fluctuation of precipitation, have been lately collected all over the world. Yet, many questions are still open, especially in the Sahelian zone. A general view over the stand of research is given by F.K.Hare (1977)[4] in his contribution to the Nairobi Conference on Desertification in 'Climate and Desertification'. He shows, for instance, that the question of the periodicity of drought periods is not yet clarified. Though Landsberg (1975) assumes a two to three year and a ten year periodicity from the analysis of the long-term precipitation measurement in Dakar since 1887, other authors studying rain-fall data of the Republic of Niger and northern Nigeria could perceive no periodicity of the dry and wet periods. Many arguments for trends of aridity are often doubtful, especially if such trends are deduced from shorter periods or if stations in rather arid regions (e.g. Agadez/Niger) are chosen. Owing to their different variability rates, stations of 150 mm cannot be compared with those of 500 mm, especially if one takes short climatic periods. The connec-tion between desertification and drought periods will be tackled below.

The present phenomena of desertification are to be seen almost entirely in connection with the climatic development in this century correlated with the anthropogenic impact of overpopulation, overstocking and ignoring the natural limits in the different land use systems thus incurring the degradation and destruction of the ecosystem. Older desertification phases in historical time could have taken place in the Sahel, too. But they must have been different in degree from those of the northern margin of the Sahara, e.g. since the Roman times. The reason for this is the different distribution and density of population in the two marginal zones.

2.2 *Analysis of the ecosystems*

To judge desertification phenomena of different degrees as those occurring at different places in the Sahelian zone, it is indispensable to carry out an analysis of the ecosystems.

The *zonal* ecosystem of the Sahel embraces the thorn-scrub savanna with its various *Acacia* species, each of which acts as a good indicator of a respective soil-water balance and even of groundwater (*A.albida*). This applies also to the different grasses (*Aristida, Cenchrus biflorus* and others) which are very useful resources. The northern transition of the Sahelian zone towards the Saharan semidesert/desert is as indistinct as its southern transition into the dry savanna which lies at about 600 mm annual mean precipitation. The fact that the northern limit of the dry savanna is strongly anthropogenically influenced is shown by the occurrence of Baobab trees (*Adansonia*) far into the southern Sahelian zone. These trees represent a relict within the man-influenced ecosystem of the Sahel.

Surfaces which are free from grasses and scrubs during the dry season (December to May/June) are important for the processes of desertification, for with the beginning of the first summer rainfalls, strong erosion hazards accompanied by an enormous transport of sediments occur. One must, however, consider the different substrates (proportion of clay, silt and sand). It is well-known that a rainfall of 50 mm in the dry climates seeps only 10 cm deep into the clayey substrate, 50 cm into the sandy one and 100 cm into the stoney substrate where it is stored. The measurements I made in Upper Volta (August 1974) showed a seepage of 30 cm (maximum) in clayey soil and 100 cm in sandy sediments which had been washed out at the slope of an escarpment of the laterite plateau. An analysis of the amounts of summerly rainfalls shows that even in wet years after rainfalls of 10-30 mm in August (wettest month) an average of 2.5 rainless days follow. On those days, the soil became so hard (some centimetres deep) that the following rainfall is usually lost to evaporation after being included in surface run-off. Only the sandy soils were an exception, and for this reason, they bore great significance for the spread of millet cultivation, while clayey soils possessed less cultivation value.

A general analysis of the zonal ecosystem of the Sahel shows that the present geographical vegetation boundaries, especially those between the thorn-scrub savanna and the dry savanna have been changed by the impact of man. This happened in the first place through the spread of cultivation and secondly by the increase of the density of the sedentary and semi-sedentary population from the Sudan zone. This was mainly the case only in the last 100 years.

2.3 The ecological differentiation through relief and soil

The following major relief units differentiate the Sahelian ecosystem:

a) The laterite plateaux with a ferricrust. To the north they dissolve into relict surfaces. Here there is a clear difference between the West African Sahel and the Sahel of the Republic of the Sudan in which laterite surfaces are almost completely missing.

b) The younger and lower pediments, glacis and wash-off surfaces with a relatively high sediment transport. How far these morphologically active surfaces have been brought under cultivation depends on the sufficient soil water balance in sufficiently thick sediment layers.

c) The sahelian old dune belt, which extends from the Senegal, over Chad to the Nile. Most of the dune systems are bound to vast depressions with old, partly fossil wadi tributaries which have transported most of these sand masses.

d) The wide, flat-floor valleys (Flachmuldentäler) and wadis which unite to form a vast hydrological system which is ecologically stressed through a higher soil water balance.

e) The highlands and the volcanic mountain massifs (e.g. Aïr and Jebel Marra), which possess hydrological controlling functions for a vast surrounding region. These are the conditions under which the important system of Dallol tributaries of the Niger has been formed in West Africa.

These morphological units differentiate the Sahelian ecosystem through the effect of the differing soil and ground water balance as well as the difference in their susceptibility to desertification.

The *morphodynamics* on the laterite plateaux, as far as these cover a large area, is mainly a surface transport of sediments which occurs mostly in rhythmic waves. This gives the bush-stock, the well-known vegetation form of the 'brousse tigrée' which is widely spread in the West African Sahel. Grass sheets grow only in patches in favourable areas. As the possibility of cultivation is limited, the land is used for grazing.

The younger, lower lying wash-off surfaces are found in the surroundings of highlands and mountain ranges (such as in the further foreland of Jebel Marra) and are formed on pediments and pediplanes. The sediment layer consists of pebbles and debris. The ecosystem is impoverished through the lack of soil water and is formed mainly of tree and scrub vegetation. Grasslands grow only on the sediments of the wash-off surfaces.

A great part of the lower wash-off surfaces which are no longer covered with secondary laterite ferricrust and are partly covered with thick sediment layers possess — under natural conditions — the densest plant cover. Tree and bush stock interchange with dense grass sheets so that an almost closed vegetation carpet is formed regenerating annually as it is adapted to the high variability of climate, especially of precipitation. Therefore, we might consider this ecosystem as 'stable' (adapted). But as already mentioned, particularly these areas have been changed very much through cultivation and overgrazing so that everywhere desertification damages caused by enhanced erosion are to be observed.

Of great significance is the Sahelian old dune belt, whose sand masses mainly originated in the young Pleistocene phase whose climate was drier than the present one. According to pollen analysis of the Meteor bore-samples a specially arid period is postulated between 19 000 and 12 500 B.P. for the West African Sahel, cf. Rossignol-Strick.[5] Under the present climatic conditions of the Sahel, these dunes are covered and 'fixed' by a tree and scrub stock and an annual grass sheet. So they are morphodynamically less active. These dunes appear 'reddish' because the weathering allows iron to be set free and the sand grains of the upper decimetres are thinly covered with clay. The southern limit of these dunes in West Africa corresponds largely to the Sahelian limit ('Extension of the Ergs fixés de l'Orgolien-Kanémien' according to P.Michel, who dates them back to the time between 20 000-13 000 B.P.[6]).

As a result of hoe cultivation and overgrazing, the natural ecosystem of these dunes has been disturbed and partly destroyed. Affected by this destruction especially is the regeneration ability of the vegetation cover which was regularly damaged by misuse. In such areas, the reddish upper layers have been eroded by deflation so that they are now covered with white sand. Ecologically, this means a deterioration of the soil-water balance because the adhesiveness of the soil-water to the less weathered sands is much reduced. This leads to the decrease of crop.

An ecological exception among the Sahelian ecosystems are the major wadis and flat-floor valleys (Flachmuldentäler). Owing to their high water table, they possess very dense stocks of trees and scrubs as well as a concentration of palmtrees (mostly Dum palms, but also Phoenix dact). The Dallol systems in Niger had a special significance. Their formation is due to the new adaptation of the West African river network to the young Tertiary epirogenetic, southerly directed, inclination of the Niger-Aïr-upland, and the filling up of the Continental-Terminal-Basin. These Dallol valleys are incised into the laterite plateaux encroaching upon the old surfaces through their numerous tributaries. Through the interchange of humid and arid phases in Quaternary and Holocene, there were different hydrological phases of periodic and partial drainage into the Niger and other rivers, similar to the present. Through the formation of lateral alluvial fans, partial basins were built in the valleys forming 'mares' after being filled with water in the rainy season.

As a result of the concentration of settlement and land use on the fringes of these Dallol, a drastic deterioration of the original vegetation occurred there. This radical destruction of the natural ecological structure made these areas highly susceptible to erosion and thus increased sediment transport into the valleys. Here we find a concentration of the phenomena of desertification.

Semiarid, marginal tropical ecosystems prevail in the highlands and volcanic mountain ranges (Aïr and Jebel Marra). They can be considered here as independent ecological complexes within the Sahelian zone which will not be closely tackled in this paper. One must observe, however, that the degradation of the natural plant cover varies highly from one place to another. Due to the extreme concentration of population in Jebel Marra ('Fur') this mountain range has been almost completely deforested and so the conditions of erosion have been changed there. We are planning to carry out more detailed research on this problem. The Aïr mountain range, in contrast, shows less population density with a concentration on the major valleys. So the destruction of the mountain ecosystem is less significant than in Jebel Marra.

3. THE EFFECTS OF THE ANTHROPOGENIC IMPACTS ON THE ECOSYSTEM IN THE SAHEL – THE CONSEQUENCES OF DESERTIFICATION

Many reports have been written[7] on the phenomena of desertification as caused by the land use system of the Sahelian population which ultimately lead to the destruction of the ability of the plant cover to regenerate. This implies a degradation of soil productivity, too. Particularly affected by this process is the old dune belt with its original thorn-scrub savanna.

Here, an overview should be given of the effects of the anthropogenic impacts represented by unadapted land use methods practised by the Sahelian population leading to intensified desertification in the Sahelian ecosystem.

From the *climatic* point of view, anthropogenic impacts upon the ecosystem trigger off microclimatic changes added to some indirect changes in the macroclimatic sphere (so far of largely unknown dimensions). The causes are twofold:

1. The complete clearing of the vegetation cover leads to an enormous enlargement of the surfaces which are unprotected. This is especially the case when stalks are uprooted from cultivated land after cropping (as in the case of millet). Through the destruction of the whole tree stock on cultivated land the soil is deprived of the positive effects of shade. The increase of the evaporation ratio makes the soil dry up also involving microclimatic changes. Surface incrustation is clearly enhanced in fine-grained soils. Land surfaces which are cleared of their plant cover effect an increase of wind speed close to the ground and this causes stronger deflation (dust storms).

2. While the regeneration of the natural vegetation of the thorn-scrub savanna takes place even after dry periods and despite the high variability of rainfall (for instance through the adaptability of the seeds to remain conserved in the soil in extreme droughts), drought effects are strengthened and prolonged by the anthropogenic enlargement of the vegetation-free areas. The formerly 'stable' ecosystem becomes unstable. In desertified areas, for instance, the regeneration of the ecosystem does not profit much from the first wet years following a run of dry ones, though the amount of rain may be high. One can conclude, therefore, that drought periods have a lasting impact on the microclimate. Deflation and erosion render these damages permanent.

Pedologically a lasting impoverishment of the soil takes place. This is caused by the deterioration of the soil-water balance and also by intense deflation and fluvial erosion. The result is considerable crop failure. Overgrazing leads to an alteration of the grass and scrub species: palatable grasses and scrubs are replaced by less palatable species. This compels the herdsmen to enlarge their pastures in search of fodder.

Shallow soils lying over rocky pediments or laterite ferricrusts (Tertiary and Quaternary land surfaces) are exposed to erosion thus becoming skeletic. After being laid bare the laterite iron crust can hardly develop a grass cover. Bushes dominate the vegetation scene. In this connection, the hypothesis is yet to be proved, whether the 'brousse tigrée' is an outcome of, or at least influenced by, man's impact through excessive overgrazing. This would, however, mean a process which extended over a few hundred years. Also, one has to find out to what extent the transition from the Neolithic Cattle Period to the present climate of greater aridity should have played a part in that process.

Generally, one can observe that owing to the enhanced wash-off of the higher, old relief (Tertiary and older) a constant transport of young, less weathered, sandy sediments takes place onto the hydrologic-morphologic erosion forms and depressions (partly endoreïc). In the region of the West African Continental Terminal, the proportion of fine sediments from the weathered products of the older Tertiary is quite high. This effects an improvement of the soil in all lower lands in the Sahel. Compared with the higher relief parts, there is a concentration of cultivation in the low lands.

Geomorphologically (morphodynamically) the active processes, both the fluvial and aeolian, have thus been explained. It has become clear that their regional distribution conforms to the Tertiary and Quaternary relief evolution. For surveying and investigating these areas, aerial photographs and satellite imagery were indispensable.[8]

REFERENCES

1. Lück, P. 1978. Die Problematik der Tropengrenze in Westafrika unter besonderer Berücksichtigung der hygrischen Verhältnisse. Staatsexamensarbeit, Hamburg.
2. Henning, D. & H.Flohn 1977. *Climate aridity index map (Budyko-Ratio)*. UNCOD A/Conf. 74/31 (Map 1:25 Mio).
3. Michel, P. 1973. *Les bassins des fleuves Senegal et Gambie*. Mém. Off. Rech. Sci. Techn. Outre-Mer, 63, 3 vols.
4. Hare, F.K. 1977. Climate and desertification. In: *Desertification: Its causes and consequences*. Comp. and Ed. by Secr. UN Conf. on Desertification Nairobi, pp.63-167.
5. Rossignol-Strick, M. & D.Duser 1979. Late Quaternary West African climate inferred from palynology of Atlantic deep-sea cores. *Palaeoecology of Africa 12*.
6. Michel, P. 1977. Recherches sur le Quaternaire en Afrique Occidentale. *Rech. Fr. sur le Quaternaire INQUA 1977. Suppl. au Bull. AFEQ, 1977-1, No. 50*.
7. Mensching, H. & F.Ibrahim 1977. The problem of desertification in and around arid lands. *Appl. Sciences and Development*. Spec. Issue UN Conf. on Desertification Nairobi, pp.8-43.
8. Mensching, H. 1974. Aktuelle Morphodynamik im afrikanischen Sahel. In: *Geomorphologische Prozesse und Prozesskombinationen in der Gegenwart unter verschiedenen Klimabedingungen*. Abh. Akad. Wiss. Göttingen – Rep. of the Comm. on Present-day Geomorph. Proc. Math.-Phys. Kl. 29, pp.22-38.

NORTHWESTERN SAHARA MARGIN: TERRESTRIAL STRATIGRAPHY OF THE UPPER QUATERNARY AND SOME PALEOCLIMATIC IMPLICATIONS

H. ROHDENBURG / U. SABELBERG

Aiming at geomorphological problems, our research concentrated on terrestrial sediments in the western Mediterranean area, that is, on fluvial and eolian sediments (loess, wind-blown sand), which have been deposited on slopes, on pediments, and on alluvial fans. As in Central Europe, fossil soils turned out to be especially helpful for a stratigraphic classification of these sediments. Consequently, our research included the analysis of all pedogenetic alterations.

Our research concentrated on coastal areas of Catalonia, the Balearic Isles, and on Southwest Morocco, as in these areas a relatively reliable chronology is provided by the interfingering of terrestrial and marine sediments. Especially useful were the marine sediments of the last widespread transgression before the Late-Pleistocene regression (i.e. sediments of Tyrrhenian II corresponding to the Moroccan Ouljian age), which served as a most important key horizon. Primarily, terrestrial sediments on top of these marine deposits were analyzed, because their stratigraphic position in the Last Glacial and Holocene can be taken as certain. For this period, highly differentiated soil-sediment-series can be presented.

However, the analysed coastal profiles do not include as a rule the last Interglacial, and presumably also not parts of the Early Glacial (Early Würm). Nevertheless, we know, from earlier analyses of Middle-and Early-Pleistocene sediment-soil-series, about the general structure of the Interglacial-Early-Glacial sequence (Rohdenburg & Sabelberg 1973): it corresponds with the 'Untere Serie' in the sense of Kukla (1963), and includes relatively thin, and in general fine-clastic sediments, which have been strongly altered in most cases by pedogenetic processes, and can thus be subdivided using several well-developed soil horizons.

Research in the area of Cala Conta, Northwest of Ibiza, confirmed these results for the Eemian-Early-Würm period. In contrast to all other Eemian-Early-Würm-profiles analysed, in this area not only beach-profiles, but also profiles of the transitional zone to inland conditions are exposed (Figure 5). Here, rather close to the beach, three well-developed lime crusts are separated by reddish soil-sediments, which are intermixed with marine material indicating sea-level oscillations during that time. More inland, due to the decreasing thickness of the intercalated sediments, the three lime crusts converge, and finally form a single, very thick lime crust, which thus represents several individual crusts from the Eemian-Early-Würm-period.* The area around Cape Rhir (Southwest Morocco) provides the most detailed stratigraphy for the period of the last-Glacial marine regression (Figure 1, cf. Sabelberg 1978). There, over long distances, the recent cliff-line cuts Late Quaternary sediment-series, which are built up either by alluvial fan aggradations, or by sand accumulations. All these series were deposited over marine sediments of the last transgression, which preceded the Late-Pleistocene marine regression (Ouljian). Based on $^{230}Th/^{234}U$-datings (Stearns & Thurber 1967, 95 000-75 000 years), the Ouljian has long been regarded as an equivalent of the Eemian. The assumption that the Ouljian not only corresponds to the Eemian, but also includes, in part, the Early Würm (cf. Sabelberg 1978), seems to be confirmed by more recent datings of samples from Cape Rhir (Hoang et al. 1978).

Remnants of very well-developed, formerly humic soils, which cover, or have developed in Ouljian sediments, are frequently found, even in locations of high fluvial potentials (e.g. alluvial fans). These soils substantiate for the early phase of the Glacial (Early Soltanian) very stable morphodynamic conditions of high pedoclimatic humidity, and dense vegetational cover. The same is indicated by the presence of pedogenetic lime crusts, which have developed contemporaneously with the soils. In an earlier publication (Rohdenburg & Sabelberg 1973), we tentatively correlated this sequence with the Early Würm.

The soils of this Early Glacial section were frequently reworked in the time following their formation. Then, reddish alluvial sediments (soil sediments) occur instead, which, as a characteristic, are almost free of any coarse fractions. At the same time, or a short time later, thick sand accumulations were deposited locally, particularly in the proximity of alluvial fans. These accumulations include in the beginning, numerous pedogenetic intercalations, later on increasing evidence of fluvial redeposition, and in the end they appear sometimes as pure dune-accumulations. Due to their geomorphological situation, these sands must originate from the shelf. Consequently, they are an indication of relatively close proximity to the beach. However, the marine transgression must have already set in. In addition, the formation of pure eolian landforms (dunes) seems to indicate a first, short arid phase (cf. Figure 1, Cycle S1).

After the accumulation of these series, humic soils developed repeatedly

* This is another example supporting Sabelberg's (1978) theory that thick lime crusts have probably developed additively, that is, in several phases which were favourable to such soil formation. The fact that Early-Pleistocene (Villafranchian) lime crusts are always much thicker than more recent ones (among others: Ruellan 1967) has not yet been explained satisfactorily. But now the theory of additive development of complex lime crusts by a sequence of many periods of soil formation offers a convenient explanation.

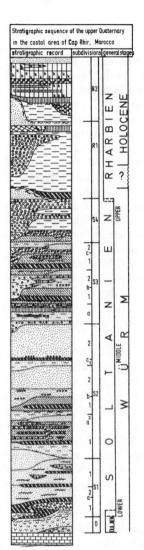

Fig.1

in the entire coastal area, but distinctly weaker than the previous ones. Especially striking is the decrease in the intensity of pedogenetic lime crusts. Only remnants of these soils can be found, as they were in general reworked by later fluvial processes. Characteristic for this profile section are the increasingly stronger intercalation and the intermingling of soil-sediment with coarse materials. This indicates a noticeable growth of morphodynamic activity,

269

even if the alluvial fans do not yet show full fluvial activity. The accumulation of reddish fine sediments still remains dominant (Figure 1, S2a).

After an obviously short period of transition, this phase is followed again by the accumulation of eolian sands of marine origin. In their basal parts, they show signs of strong fluvial transport. Outside the larger alluvial fans, the formation of massive and polycyclic dunes set in. In almost all Late Quaternary profiles of the coastal area (with the general exception of typical alluvial fan sites), this dune complex represents the largest sand accumulation of the entire Soltanian. It is present along the entire Moroccan Atlantic coast and is known as 'Oulja Wall'. It can be interpreted as the result of an efficient combination of marine sand supply and eolian sand transport, which took place during an already highly active arid phase (Figure 1, S2b-c).

Later on, the profiles frequently show reddish sediments, often separated by several weakly developed soil horizons. A very short time later, especially on the alluvial fans, the transport of coarse materials increases considerably. For a short period, even lateral planation occurs on the alluvial fans. Consequently, the rather weakly developed fossil soils of this phase are found only outside the alluvial fans, especially in areas of prevailing dune formation. Datings of shells from fossil Kjökkenmöddingers from this section of the profile (cf. Sabelberg 1978) provided Middle Würm ages.

For this period again, repeated indications of strong arid phases can be found. But these are, in general, only well-recognisable distant from the alluvial fans, as in the meantime, a considerable lowering of the sea level obviously had taken place (thus reducing the supply of sand) (Figure 1, S3).

After the cycle S3, indications of very weak soil formations are found only in a few places. Also the reddish sediments disappear almost completely. In contrast, removal and wide-spread accumulation of coarse debris on the alluvial fans demonstrate a considerable increase in morphodynamics, which must have been active in wide areas of the hinterland. Possibly this phase did not last for very long. The alluvial fan-activity soon waned, because of growing aridity. However, dune-formation is lacking almost completely in the profiles; obviously, the sand supply from the then remote coast was too low (maximum of the regression; Figure 1, S4).

On top of these Late Pleistocene accumulation series, a very intensive red soil with an underlying lime crust occurs, which is partly decalcified even in the presently semi-arid area of Cape Rhir. This soil demonstrates a basic change of ecological conditions: strong pedoclimatic humidity and a complete stabilisation of the relief by a protecting vegetational cover.

Contrary to former practice, we now regard this soil as the reasonable paleo-ecological limit between Soltanian and Rharbian (Sabelberg 1978).

Also the Rharbian ($\sim$ Holocene) represents a highly differentiated period with regard to its morphodynamic development. The well-developed soil of the Earliest Rharbian was eroded to a great extent soon after its formation, so that in most cases only the lime crust remained. On top of this, thick fine-sediment series accumulated (including the arid areas), which can be

270

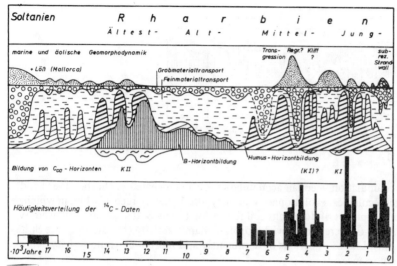

Fig. 2

stratified by sinter layers and weak humic horizons. In general, these series
still characterise a very stable morphodynamic period. Only after the Middle
Holocene marine transgression (Mellahian), a short phase of increased morphodynamic activity is noticeable. A repeated change between development
of humic soils and shorter periods of increased transport of coarse materials
is apparent on alluvial fans as in run-off channels. Some parts of smaller ter-
raced valleys between Agadir and Tamri particularly suggest finer subdivisions
of the Holocene (cf. Rohdenburg 1977).

In a first step of evaluation, we attempted to characterize the geoecologi-
cal conditions of the entire catchment area by considering individual exposures
in relation to the entire catchment area. For that purpose, we estimated in two
chronological/spatial diagrams (Figures 2, 3) the proportions of areas with
soil formation (and dense vegetation), of areas with stronger transport of fine
material (and open vegetation), and of areas with transport of coarse material
(and sparse vegetation). This procedure seemed to us admissible, due to the
great number of exposures in the research area. In Figure 3, drawn by U.Sabel-
berg, the estimated areas of dominant eolian activity are also included. This
has not been possible in H.Rohdenburg's Figure 2, as he considered more
terrestrial areas with generally steep slopes, which did not allow any state-
ments on eolian activities, as larger areas were devoid of eolian sediments.
Consequently, the relative importance of the eolian activity has been represented only in the form of an additional column.

From Figure 3 can be seen that the geo-ecological development of the
research area in the Late Pleistocene possesses a distinct periodicity. In the
course of the development, the sequence: soil formation — fluvial erosion and

271

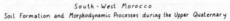

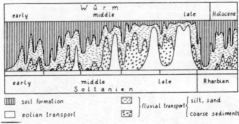

South - West Morocco

Soil Formation and Morphodynamic Processes during the Upper Quaternary

Fig.3

accumulation — eolian accumulation occurs repeatedly. On the other hand, a geo-ecological evolutionary tendency also is clearly recognisable, in which initial phases of dominant soil formation (Early Soltanian) are followed by phases of increasing fluvial morphodynamic activity (Middle and Late Soltanian). Likewise, periods of dominantly arid morphodynamic activity gain importance.

Both the highly differentiated geo-ecological subdivision of the last Glacial, as deduced from field studies, and its close similarities with the geo-ecological development of Central Europe in the Würm Glacial are remarkable. This is true for the periodic decrease of the soil formation potential, as well as for the periodic increase of slope formation and arid morphodynamic activity, even though the processes may differ individually. A direct chronological correlation of individual phases seems not yet possible, because available datings are still too scarce.

This is not the case for the Holocene, as a greater number of datings has been published recently (Figure 2, cf. Rohdenburg 1977, Sabelberg 1978). Figure 2 shows that the Holocene of the research area is characterised by a geo-ecological development, which resembles, in general, the development of an Early Soltanian cycle. However, the much more detailed chronology of Figure 2, as compared with Figure 3, has to be taken into account. This chronology resulted from an abundance of datings. As already demonstrated for the Eemian-Early Würm period (v.s.), the Holocene of the research area also possesses characteristics of an 'Untere Serie' in the sense of Kukla (1969), or at least of a part of the series. Contrary to e.g. the mid-latitude Holocene, the generally high soil formation potential was repeatedly weakened by a revival of geomorphodynamic activity. The actual state of the landscape with partly open vegetation and a tendency towards increased erosion of fine sediments is, however, strongly influenced by human interference. Thus, the natural soil formation potential appears too small.

Frequently, characteristics of the sediment have led to immediate conclusions concerning the paleoclimatic conditions, especially the annual amounts of precipitation. Our advancing geo-ecological studies, in which we attempted to include at least the most important factors of the ecosystem, make a

272

simple relation between sediment characteristics and paleoclimatic conditions seem rather doubtful. In the following, our theory will be explained in detail by means of Figure 4, then further conclusions will be drawn.

The precipitation hitting the surface of the soil — interception will be neglected in a first approximation — can be divided into surface run-off (Figure 4: y-axis) and infiltration (Figure 4: x-axis). The total precipitation results from the addition of respective axial distances in the first quadrant of the coordinate system. Precipitation increases in direction of the bisector. Perpendicular to it run the lines of equal precipitation (isohyetes).

At a given temperature and an annual distribution of precipitation, the amount of infiltrating water determines the vegetation potential, and thus the soil formation potential.

Figure 4 shows further that the density of vegetation grows with increasing infiltration, but decreases with simultaneously growing surface run-off, as the surface run-off influences directly and indirectly the potential of the vegetation (mechanical damages — intensified by transported materials, soil erosion, loss of nutrient, etc.). The density of the vegetation influences, via friction, to a great extent, the speed, and thus the competence of the surface run-off. Consequently, coarse gravels will only be transported in the case of missing, or very incomplete, vegetation. On the other hand, a dense vegetation, which completely protects the soil, will only exist in the case of a distinct dominance of infiltration. Between both extremes, there is an abundance of intermediate stages of discontinuous (with patchy relics of dense vegetation) or gradual thinning of vegetation — depending on plant associations.

The diagram demonstrates that the total amount of precipitation is of only minor importance (with the exception of extremely arid areas) in determining the qualities of a site, with regard to surface protection by vegetation, soil erosion, or transportable material. A given amount of precipitation will, e.g. in the case of missing vegetation transport coarse material by surface run-off, or, in the case of dense vegetation, be completely unable to erode the soil. Which factors distinguish between these different ecological stages? First of all, of course, the primary site properties, such as substratum and slope angle, but, on the other hand, also consequential effects on the ecosystem itself, as for example the amount of continuous coarse pores in the soil, which exert a great influence on infiltration. The most important climatic factors influencing the relation between surface run-off and infiltration, however, are the intensity and the distribution of precipitation occurrences. These can have both direct and indirect effects (density of the vegetation, soil structure, etc.). The indirect effects will result from a positive feedback effect of the ecosystem itself, so that even little variations of the rainfall distribution may cause distinct shifts of the surface run-off — infiltration ratio, and of the erosion- and transport-qualities of a certain site. Thus, a lower precipitation intensity with a generally more balanced distribution of the precipitation will lead to a higher vegetation density, the infiltration capacity will improve, and the surface run-off will be further reduced. On the other hand, in the case of rainfall

273

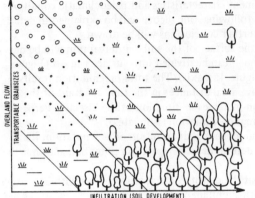

Fig. 4

distribution, characterised by short, high-intensity events, the increased surface run-off will damage the vegetation, e.g. by baring of the roots, eluviation of the nutrients, etc.). As a consequence, for example, the soil structure will become less favourable. This will lead to a reduced infiltration capacity, the surface run-off will continue to increase, and so will its competence, due to reduced friction (thinning of the vegetation).

Because primary site-factors, such as slope angle and substratum, have not changed to any great extent during the Late Quaternary, the demonstrated changes of the geomorphodynamics must result, more or less, from variations of the rainfall distribution. Excepting the periods of soil formation, the Late Pleistocene rainfall distribution was much more intermittent than today. Probably also the total amount of the precipitation has changed, which can, however, only be deduced by means of additional findings. Also the Glacial lowering of the temperatures, which led to a reduction of the total evaporation, has to be considered. Decalcified soils in certain stratigraphic positions (Eemian, Earliest Rharbian) are indications of high pedoclimatic humidity. A strong eolian geomorphodynamic is certainly an indication of a low total precipitation. Periods, in which much coarse material was transported, were certainly not too extremely dry, but probably characterised by a lower total precipitation, as compared with today. Fine sediments with intercalated sinter layers, humic soils, and a rich malacologic fauna are an indication of a rather high precipitation (e.g. Early Rharbian), other fine sediments might also originate from very arid periods. In the case of fine sediments, the paleoclimatic classification is most difficult, unless there are additional criteria.

The Figure 4 demonstrates quite clearly that the frequently used term 'Pluvial' is rather unsuitable for a characterisation of observed paleoclimatic changes. Many observations have been summarised by this term. However, some of these observations indicate a high pedoclimatic humidity (dense vegetation, soil formation), whereas others indicate strong fluvial geomorpho-

274

dynamics. Thus opposite ecological stages are both considered as pluvial. We now know the reasons for this discrepancy: it is the too simple correlation between sediment and soil properties with a single paleoclimatic quality, i.e. the amount of the precipitation. For all further paleoclimatic discussions — and this we want to stress — it is absolutely necessary to differentiate between the antagonistic elements of the water balance, namely surface run-off and infiltration.

REFERENCES

Hoang, C.T., L.Ortlieb & A.Weisrock 1978. Nouvelles datationes ^{230}Th/^{234}U de terrasses marines 'ouljiennes' du sud-ouest du Maroc et leurs significations stratigraphique et tectonique. *Cr. Acad. Sci. Paris, 286:*1759-1762.

Kukla, J. 1969. Die zyklische Entwicklung und die absolute Datierung der Lössserien. *Periglazialzone, Löss und Paldolithikum der Tschedioslowakei* (J. Demek & J. Kukla, eds.), 75-95, Bino.

Rohdenburg, H. 1977. Neue ^{14}C-Daten aus Marokko und Spanien und ihre Aussagen für die Relief- und Bodenentwicklung im Holozän und Jungpleistozän. *Catena 4:*215-228.

Sabelberg, U. 1977. The stratigraphic record of late Quaternary accumulation series in South West Morocco and its consequences concerning the Pluvial hypothesis. *Catena 4:*209-214.

Sabelberg, U. 1978. *Jungquartäre Relief- und Bodenentwicklung im Küstenbereich Südwestmarokkos.* Landschaftsgenese und Landschaftsökologie 1. Catena-Verlag, Cremlingen-Destedt.

Stearns, C.E. & D.L.Thurber 1967. Th^{230}/U^{234} dates of late Pleistocene marine fossils the Mediterranean and Moroccan littorals. *Prog. in Oceanogr. 4:*293-305.

THE LITTORAL DEPOSITS OF THE SAHARIAN ATLANTIC COAST SINCE 150 000 YEARS B.P.

A. WEISROCK

Geographical Institute, University of Nancy II

The study of the littoral deposits of the Saharian Atlantic coast since 150 000 y. B.P. offers an extremely interesting but complex field. In this text, they are compared to those of northwest (Moroccan) and southwest (Mauritanian) borders of the Atlantic Sahara.*

1. THE MARINE DEPOSITS AND THE PROBLEM OF THE SEA-LEVEL SINCE 150 000 Y. B.P.

The last two known world-wide transgressions (Riss-Würm and Holocene) are called in Morocco Ouljian (Gigout 1949) and Mellahian (Biberson 1958). In Mauritania, if the Holocene transgression, called Nouakchottian (Elouard 1966) is well-known, the local Inchirian (Faure & Elouard 1967) corresponds to an intermediate stage of the Würm. The observations made in the western Sahara (Ortlieb 1975) do not yet allow indisputable correlations.

1.1 *The marine deposits are always found at a low altitude, but their extents are quite variable.*(Figure 1)

1.1.1 *In the Atlantic Atlas and the Anti-Atlas.* The *Ouljian* deposits remain on a platform of abrasion which cannot be found above 4 or 5 m high. They are made of shelly sands or shelly limestones which are more or less coarse and generally very hardened and crusted on their surface (laminated crust). Their thickness is less than 1 m, and they are partly dismantled, but still very continuous, except in the Essaouira Basin and in the Sous, where they pass under the present level. So, the Ouljian shore constitutes only a very narrow band along the coast.

The *Mellahian* deposits are nearly non-existent, even if traces of a late

* Acknowledgement: I am indebted to L.Ortlieb and N.Petit-Maire for much of the early work on which this paper is based.

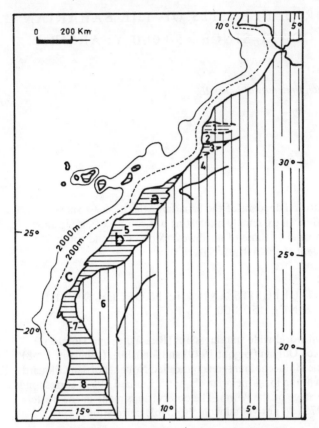

Figure 1. The marine deposits of the Saharian Atlantic coast.
1. Essaouira Basin; 2. Atlantic Atlas; 3. Sous; 4. Anti-Atlas; 5. Western Sahara Basin: (a) Tarfaya Province; (b) Seguia el Hamra; (c) Rio de Oro; 6. Reguibat Arch; 7. Mauritanian Basin; 8. Senegal Basin.

marine standing a little bit higher than the Present can be recognized. For example, an abandoned shore bar at 12 km north of Essaouira shows pebbles which had been shaped by the sea and are mixed with various shells (*Patella* sp., *Thaïs, Cymatium* . . .). One of them has given an age of 2 770 y. B.P.

1.1.2 *In the West-Saharian Basin.* The *Ouljian* deposits are very comparable with those of the Atlas. They are always situated at below the 5 m level, but they are less continuous. For example, the El Khebita shore (Tarfaya Province) shows ouljian deposits with a sub-horizontal stratification. They are grey-natural beach-rocks, whose beds are made of remains of shells and of whole shells. Towards the top, these sandstones progressively become dune

278

Table 1. Datings now available

Ages in thousand years B.P.	Number of datings Th/U	^{14}C	Places	Authors	Names
148-120	3		Tarfaya, S. el Hamra	Hoang et al. 78	Ouljian 1
97-60	6		id., + Atl. Atlas	id.	Ouljian 2
	2		Agadir	Stearns 65	
46-43	3		Atl. Atlas, Tarfaya	Hoang et al. 78	Ouljian 3
39-28,7		9	Mauritania	Hébrard 78	Inchirian
32-30		2	Tarfaya, Atlas	Rohdenburg 77	
6,3-4,2		11	Mauritania	Hébrard 78	Nouakchott
6,1-3		13	West-Sahara	Delibrias et al. 75	id.
7,3-2,7		5*	Atlantic, Atlas	Rohdenburg 77 Weisrock, unpublished	Mellahian

* Indirect evidences; a single dating of shell *in situ* at 2 770 B.P.

sandstones, azoïc, with an oblique stratification. L.Ortlieb (1975) does not find any trace of these deposits beyond the 26th parallel, that is to say, South of Bojador Cape.

The *Nouakchottian* deposits: along Seguia el Hamra and Rio de Oro, evidence of the last transgression is exclusively located in littoral sebkhas, which were once gulfs. Thus, the fossiliferous beds of the Nouakchottian sea may be found mainly through excavating some of these coastal plains. Littoral sebkhas, covering a number of square miles, are cut off from the present sea by sand bars. The sebkha level is even, and varies from −1 to +2 m relative to present mean sea level. Superficial sand-deposits of the sebkhas, as those of the Sebkha Mahariat (Barbas Cape, Rio de Oro), are protected from deflation by an accumulation of various shells, which have been given an age near 5 000 y. B.P. Contrary to Ouljian deposits, the Nouakchottian deposits have not been found in the north of Bojador Cape. However, there do exist here, too, some sebkhas, but they are either still connected with the Ocean or they have supplied no fossiliferous level.

1.1.3 *Comparison with the Northern Mauritania.* The late fluctuations of the shorelines have affected very large stretches, since the corresponding marine deposits (Inchirian and Nouakchottian) can be respectively found up to 150 and 90 km from the East of the present shoreline, but always at a low level (+4 m). The marine invasion of the sea is deep inland, but the water-level remains low. So, they appear to be essentially made of beach-sands and accumulation of shells, as L.Hebrard (1978) described them in the Gulf of Tafoli, north of Nouakchott.

279

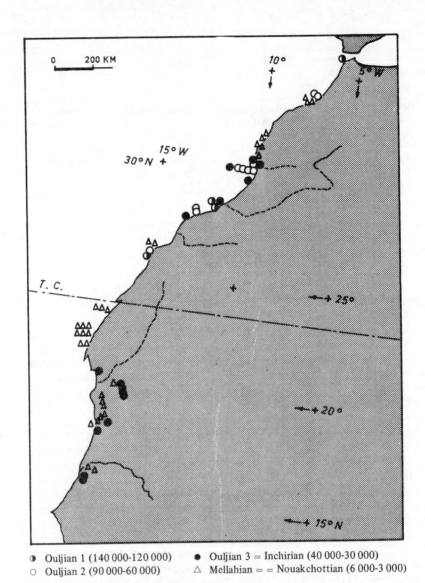

◐ Ouljian 1 (140 000-120 000)   ● Ouljian 3 = Inchirian (40 000-30 000)
○ Ouljian 2 (90 000-60 000)   △ Mellahian = = Nouakchottian (6 000-3 000)

Fig. 2. Distribution of dates (*in situ* shells).

1.2 *Isotopic datings and the problem of correlations*

1.2.1 *Table of the datings now available.* This table was established with the study of sea-shells between Safi and Nouakchott (Table 1 and Figure 2).

1.2.2 *Points to be noted.*
— The Ouljians 1 and 2 of the Atlantic Atlas and of the Northern West-Saharian Basin have been found, until now, neither in the Rio de Oro nor in Mauritania, for lack of datings ^{230}Th/^{234}U in these regions. Apparently, some authors want to link up the Ouljian (1 and 2) with the Aïoujian of Mauritania, but this is just an assumption. Aïoujian seems to be older (Ortlieb 1975).
— There should be a rather good equivalence between the Ouljian 3 and the Inchirian. However, the number of datings Th/U of the Ouljian 3 is still low and needs to be corroborated. Furthermore, datings between 40 000 and 30 000 y. B.P. are situated at the limit of possibilities of the ^{14}C method, and so are not admitted by everyone unreservedly. Finally, deposits of this period have not been found again between Tarfaya and Mauritania.
— The less arguable correlation appears between the Mellahian and the Nouakchottian. However, the Mellahian is very badly represented throughout the North of the studied coastline, while the Nouakchottian from the south (Seguia el Hamra, Rio de Oro, Mauritania) is well-developed.

1.2.3 *Conclusions.*
1. The distribution of marine deposits in the Western Sahara (area of the Ouljian in the north, area of the Nouakchottian in the south) seems to correspond grossly to the late morpho-tectonic evolution of this region:
— Subsidence of Mauritania and, in a more located way, of certain areas of the Rio de Oro, Sous and Essaouira Basin.
— Bulging of the Regueibat Arch and the Anti-Atlas which are two elements of the Saharian Plate.
— Strong upheaval of the Atlas Chain.
2. Along the coast, some general remarks can be verified about the oceanic levels:
— Levels of different ages are found at equal height or at neighbouring height. The Ouljian platform which is so continuous around +4 m in the Atlantic Atlas has been dated between 97 000 and 43 000 y. B.P. (Hoang et al. 1978). In the cutting of Hasseï Gaboun, L.Hebrard (1978) has shown that the Nouakchottian covers the Inchirian and can be found only at around 2 m under the present zero.
— During the Ouljian time, the level of the sea has witnessed many oscillations: three in the Atlantic Morocco, around 120 000, 90 000 and 70 000 y. B.P.
— A level close to that at present is revealed about 40 000-30 000 y. B.P.
— There is a certain shifting of the transgressive maximum, according to the regions: the Ouljian three seems to be a little earlier than the Inchirian,

281

Table 2. Faunas of the Ouljian *sensu lato*. (For Mauritania, Aïoujian, then Inchirian)

Regions		Whole nb. of species	Tropical species Nb	%	Temperate species Nb	%	Cold species Nb	Nb. of sites
N								
Atlantic Atlas	G	102	26	25,5	23	22,5	1	23
	B	33	5	15	2	6	0	19
Anti-Atlas	G	31	9	29	5	16	0	7
Tarfaya Province	G	34	14	41	9	26	0	13
	B	7	2	28,5	0		0	5
Seguia el Hamra	G	32	8	25	3	9	0	5
	B	15	1	6,7	0		0	5
Mauritania, Aïoujian								
in Lecointre	G	14	5	35,7	1	7	0	3
in Hébrard	G	24	6	25	0		0	?
Mauritania, Inchirian	G	16	6	37,5	0		0	?
	B	27	4	14,8	0		0	?
S								

Table 3. Faunas of the Nouakchottian (For the Atlantic Atlas, Present)

Regions		Whole nb. of species	Tropical species Nb	%	Temperate species Nb	%	Cold species Nb	Nb. of sites
S								
Mauritania	G	28	7	25	0		0	?
	B	31	3	9,6	0		0	?
Rio de Oro	G	27	5	22	3	11	0	6
	B	18	3	16,6	0		0	6
S. el Hamra	G	26	7	27	1	3,8	0	3
	B	12	3	25	0		0	3
Atlantic Atlas, Present	G	39	8	20,5	6	15,4	1 (2,5)	8
N								

G = Gastropods.   B = Bivalves

while the situation is reversed in the case of the Mellahian and Nouakchottian but no explanation can be given to this phenomenon.

2. MARINE DEPOSITS AND CONTINENTAL DEPOSITS: CLIMATIC PROBLEMS

It is impossible to reconstruct the climatic evolution of the west Saharian coast since 150 000 y. B.P. because our observations are still much too sparse. We present here just an evaluation of our knowledge about the west Saharian coast paleoclimates according to two themes which have been recently studied: the paleoclimatic conditions concerning the littoral waters using the fossil faunas (Ouljian, Inchirian and Nouakchottian), and the paleoclimatic interpretations of the sedimentology of littoral deposits according to the different authors. This second part can only concern the last 40 000 years which is the reach of the ^{14}C dating method.

2.1 The climatic indications of the marine faunas

2.1.1 *The working method.* The fossils have been found in the marine shores of the Atlantic Atlas by A.Weisrock, of the Anti-Atlas by P.Oliva, and of the West Saharian Basin by G.Lecointre, then L.Ortlieb. Determinations have been made by P.Brébion, J.Meco and A.Rage-Lauriat. For comparisons with Mauritania, we use the works of L.Hebrard (1978) and P.Elouard (1976).

Some well-known species have a significant distribution according to the climatic point of view. From their present area, we can assign to them a tropical character (shells from the Guinean and Senegalese Provinces), a temperate character (Atlanto-mediterranean shells), or a cold character (shells from the North Atlantic). The lists of these species have been established by P.Brébion for the Gastropods and by A.Rage-Lauriat for the Bivalves. We will not forget the incompleteness of this method, the results being done on the basis of very unequal gathering. Furthermore, they take just a few species into account (about half have no precise climatic meaning), neglect the local conditions (bathymetry and nature of the bottom), and are based upon a current classification. So the results are indicative and must be cautiously considered.

2.1.2 *The results.* For the results, see Tables 2 and 3.

2.1.3 *Conclusions* (Figure 3).
1. We have only a general view for the Ouljian stage (Inchirian for Mauritania), with a hiatus at the level of Rio de Oro.
2. There are more species of Gasteropods than species of Bivalves, except for the Mauritanian-Inchirian. It must be perhaps attributed to the conditions of the deposit in itself, with sandy beaches which are larger in Mauritania than in other regions. At the Ouljian stage, the graphs which have been made for the Gasteropods on the one hand and for the Bivalves on the other, hare parallel. So Gasteropods and Bivalves present the same local variations.
3. At the Ouljian stage, we can notice that:
— the percentage of tropical species increases from Rhir Cape to Juby Cape, constantly lesses towards Seguia el Hamra and increases again towards Mauritania:
— the percentage of temperate species remains rather steady from Rhir Cape to Juby Cape (slight increase), then constantly diminishes, without completely disappearing, towards Mauritania;
— the cold species only exist in the Atlantic Atlas.

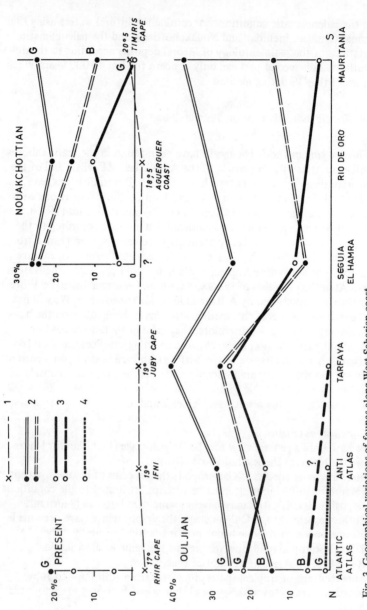

Fig. 3. Geographical variations of faunas along West-Saharian coast.
1. Annual temperatures of ocean; 2. Tropical faunas; 3. 'Atlanto-mediterranean' faunas; 4. Cold faunas.

So, in the aggregate, we can notice, as we can nowadays, a warming from the north towards the south: Rhir Cape has a cool aspect and Mauritania a warm aspect. But the area between Juby Cape and Bojador Cape appears also cool, with a maximum mixing of species. This character seems to be a good indicator of upwelling: capes are regarded to be more sensitive to upwelling than bays. Is that also to say that upwelling was extended as far north as 26°N?

4. In the Nouakchottian, and for the southern part of the coast only, we have the corroboration of the cool aspect (important mixing and temperate species always present). The same remark can be made about the present fauna of Atlantic Atlas. This shows the cooling and the global impoverishment of the fauna between the Ouljian and the Northern Present, between the Inchirian and the Nouakchottian in Mauritania.

2.2 The climatic evolution since 40 000 years B.P. according to the littoral deposits

2.2.1 The observations of the different authors. These observations are listed in Table 4.

2.2.2 Conclusions. The climatic evolution is grossly parallel in the different studied areas:

1. From 40 000 to 28 000 years B.P., there is a spread of a very thick littoral dune which marks the proximity of the ocean level, and does not mean a general dryness; in the Atlas, a good number of kjökkenmödding shows that this dune was frequented by man.

2. A damping is indicated in the north from 28 000 to 26 000 years B.P., with pink alluvial deposits and Helix. Around 24 000-23 000 years B.P., there was in Mauritania an old lacustrian period (P.C.Chamard 1973).

3. The period from 20 000-15 000 years B.P. is a very dry period in the South and in the West Sahara, with the expansion of the great Ogolian ergs. In the North, only thin, silty, pink deposits of that period can be found. Is this an indication of a region becoming a desert or being still damp? We can then observe a big new dune in the North; but this dune, which is highly developed in the Essaouira Basin, has not yet been dated precisely: before 5 700 years B.P. at the Ksob Wadi. Perhaps it simply may be linked with the onset of the Mellahian, and devoid of climatic meaning. Furthermore, we have no precise dating of the coarse base of one alluvial terrace which could be situated in that time too. We do not know if we must consider a period of strong erosion — and scarceness of deposits — as an indication of aridity.

4. From 11 000 years B.P. in the South, and 7 300 years B.P. in the North, the increase in moisture is general. There are the well-known episodes of the Tchadian of Mauritania, and old Rharbian of Morocco. However, we notice that rainfall seems to occur earlier and be strongly increased in the South. In

Table 4. Chronology of the west-Saharian littoral deposits. (-) = number of datings.

Datings B.P.	Atlantic Atlas A.W. & P.R. 77	Atlantic Atlas H.R. 77 - U.S. 78	Anti-Atlas P.O. 76 - W.A. 77	West-Saharian basin L.O. 75	Mauritania L.H. 78
880-Present	Low historic terrace. Historic dune Kjökken (5)	= jüngste terrasse = subrezent. Strandwall = Kjökken, Humusboden (5)	= sub-present and present terrace of Draa wadi	Dunes	Dunes (15)
2 000-800	1 850: running wild of Mogador Island, after roman occupation. Hiatus		Pedogenesis	aridification	return of the desert — 2 000 (52)
2 200-2 000	Dune on the Mogador Island	Kjökken, Humusboden thin grey-brown 'Grünschwarzserie', summit (3)	thin grey-brown alluvial fragments 2 500		
2 770	Shore-bar (1)				
3 700-3 400	Alluvial terrace (1) Clayed soil of Phoenician occupation in the Mogador Island	= Terrasse and 'Grünschwarzserie'(2)	'Saharian Neolithic'	lagoon deposits kjökken reddish silts (2) kjökken — 3 000	'debadeb' = montmorillonitic gypsum clay kjökken — 3 000
4 300		Terrasse + Humusboden (2)	grey and grown soils Vegetation 5 500	lagoon/marine deposits nouakchottian deposits, diatomaceous lakes deposits (12) — 4 000	nouakchottian deposits, sahelo-soudan. flora, kjökken (36) — 4 000
5 700-4 800	Indications of the oceanic climb: Alluvial terrace on rehandled dunes, kjökken (2)	= Kjökken in Strandsand, Holzk. unter Küstendüne (2)	7 000 silty terrace		arid episode? — 7 000
7 300-6 300	'Feinschuttakkumulat' soils silty terrace	'Feinschuttakkumulat' Humusboden (4)		reddish silts (2) — 8 000 / 11 000	tropical ferruginous soils, diatomaceous lake deposits, spreading of gravels (10)
13 000-7 000	Coastal dune crust	Kalklamelle, 'graue Schotter'?	11 400 calcareous tufa ? alluvial cones		
14 000-13 000	Thin slope deposits (2)	Hiatus		Hiatus 15 000-11 000 great funded dune	
26 000-14 000	Pink silts (2)	'obere rote Serie'	brown reddish soils dune		big 'ogolian' erg — 15 000 / 20 000
28 000-26 000	Big coastal dune, kjökken (1)	'Haupt-Düne' (2)			S.Chemchane lacustrine deposits (3) — 21 000 / 23 000
34 000-28 000	Ouljian 3 (3)	high sea level (2)			
about 40 000				Ouljian 3 (1)	Inchirian (13)

286

the same way, this damp period ends somewhat earlier in the South (around 3 000) than in the North (about 2 000 years B.P.).

5. The process of becoming arid occurs earlier in the South, harder and continuous up till today, while the North still has various oscillations, with man certainly accelerating the process of aridity.

REFERENCES

Andres, W. 1977. Studien zur jungquartären Reliefentwicklung des südwestlichen Anti-Atlas und seines saharischen Vorlandes (Marokko). *Mainzer Geogr. Stud. 9.*

Biberson, P. 1958. Essai de classification du Quaternaire marin du Maroc. *C.r. somm. Soc. Géol. France, 4:*67-69.

Brebion, P. & L.Ortlieb 1976. Nouvelles recherches géologiques et malacologiques sur le Quaternaire de la Province de Tarfaya. *Géobios 9:*529-550.

Brebion, P. & A.Weisrock 1976. Faunes de Gastéropodes et morphologie des plages étagées plio-quaternaires de l'Atlas atlantique marocain. *C.r. Acad. Sci. Paris, 283:* 1145-1148.

Chamard, P.C. 1973. Monographie d'une sebkra continentale du SW saharien: la sebkra de Chemchane. *Bull. Inst. Fond. Afr. Noire, 35:*207-243.

Delibrias, G., L.Ortlieb & N.Petit-Maire 1976. New ^{14}C Data for the Atlantic Sahara (Holocene): tentative interpretations. *J. Hum. Evol. 5:*535-546.

Elouard, P. 1966. Définition du Nouakchottien. *Bull. Liais. Ass. Sénégal Et. Quatern., Dakar 10:*9.

Elouard, P. 1976. Application de la paléoécologie des Mollusques à un problème de stratigraphie: différenciation de deux étages du Quaternaire marin de Mauritanie. *Notes afric. 151;*65-73.

Faure, H. & P.Elouard 1967. Schéma des variations du niveau de l'Océan Atlantique sur la côte de l'Ouest de l'Afrique depuis 40 000 ans. *C.r. Acad. Sci. Paris 265:*784-787.

Gigout, M. 1949. Définition d'un étage ouljien. *C.r. Acad. Sci. Paris 229:*551-552.

Hebrard, L. 1978. Contribution à l'étude géologique du Quaternaire du littoral mauritanien entre Nouakchott et Nouadhibou. *Doc. Lab. Géol. Fac. Sci. Lyon 71.*

Hoang, C.T., L.Ortlieb & A.Weisrock 1978. Nouvelles datations ^{230}Th/^{234}U de terrasses marines 'ouljiennes' du sud-ouest du Maroc et leurs significations stratigraphique et tectonique. *C.r. Acad. Sci. Paris 286:*1759-1762.

Lecointre, G. 1965. Le Quaternaire marin de l'Afrique du Nord-Ouest. *Quaternaria, 7:* 9-28.

Oliva, P. 1976. La plate-forme moghrébienne: néotectonique et eustatisme sur le littoral de l'Anti-Atlas. *Méditerranée 2:*73-91.

Ortlieb, L. 1975. *Recherches sur les formations plio-quaternaires du littoral ouest-saharien.* Trav. et Doc. 48. ORSTOM, Bondy, France.

Rohdenburg, H. 1977. Neue ^{14}C Daten aus Marokko und Spanien und ihre Aussagen für die Relief- und Bodenentwicklung im Holozän und Jungpleistozän. *Catena 4:*215-228.

Sabelberg, U. 1978. *Junquartäre Relief- und Bodenentwicklung im Küstenbereich Südwestmarokkos.* Landschaftgenese und Landschaftsökologie. Catena-Verlag, Cremlingen-Destedt.

Stearns, C.C. & D.L.Thurber 1965. ^{230}Th/^{234}U Dates of late Pleistocene marine fossils from the mediterranean and moroccan littorals. *Quaternaria 7:*29-42.

Weisrock, A. & P.Rognon 1977. Evolution morphologique des basses vallées de l'Atlas atlantique marocain. *Géol. médit. 4:*313-334.

PLEISTOCENE LAKES IN THE SHATI AREA, FEZZAN (27°30'N)

N. PETIT-MAIRE / G. DELIBRIAS / C. GAVEN
Laboratoire de Géologie du Quaternaire, C.N.R.S., Marseille-Luminy.
Centre des Faibles Radioactivités, C.N.R.S., Gif.

On the northern border of Murzuk basin in central Libya, the Shati valley lies east to west, from 13°20' to 14°40'E, along the southern slope of the anti-clinal Gargaf structure. The Gargaf arch limits it to the north, the Ubari sand sea to the south. The Cambrian, Ordovician, Devonian and Carboniferous beds successively outcrop from north to south and plunge into Murzuk basin; at places to the north, discordant tertiary continental limestones cover the palaeozoic layers (G.Goudarzi 1968). Quaternary alluvial deposits fill the depression; three to five kilometers wide sebkhas, separated by Carboniferous rocky outcrops, cover its bottom.

About half-slope up between these sebkhas and the tertiary limestones, a 125 km long line of oases marks the position, between Ashkeda and Adri, of about 800 wells and springs from the 'Wadi' Shati artesian or rising waters (Muller-Feuga 1954).

The difference in level between the oasis and the bottom of present sebkhas, which is covered with a saline crust of a 'ploughed soil' type, is to 40 m.

Presently, mean annual rainfall is less than 10 mm in the whole area (8 mm at Ubari, 7.3 mm at Sebha, Dubief 1963). However, evidence for existence, during the Quaternary, of wide free water expanses is found in this hyper-arid region. All along the Shati valley, extensive lacustrine shell deposits mark the Quaternary sediments. The results of the preliminary field trip (Petit-Maire 1977) and first laboratory research (Petit-Maire, Casta & Delibrias, in press, Gaillard & Testud, in press) will be summarized in this paper and completed with some isotopic dates shedding new light upon Sahara palaeoclimatology in a still poorly known span of time in the central Sahara lowlands.

1. *At about the altitude of the oases* and on both sides of present sebkhas, 0.3 to 10 m thick deposits outcrop at many localities. Their facies vary considerably, sometimes within only a few metres distance. At times, one finds hard lumachelles more or less rich in sand, in which shells may be fragmented to very different degrees (entire valves to 2-10 mm fragments). Elsewhere, well-preserved shells (sometimes still bivalved) are found in sandy layers without any cohesion (Plate 1).

289

Plate 1. Lacustrine shells deposits highest above selekha level.

Some of these 'Cardium' localities were mentioned by Lelubre (1946), Muller-Feuga (1954) and Desio (1968), and attributed to an undetermined 'Quaternary' age. In fact, *Cerastoderma glaucum,* the Mediterranean cardium (determination Gaillard) is dominant at all localities, but always together with *Melania tuberculata;* at times *Hydrobia* sp., *Bullinus truncatus,* Ostracoda *(Cyprideis* sp.) and Foraminifera (*Ammonia beccarii* (Linné), var. *tepida* and *Protelphidium paralium* (Tintant), det. Blanc Vernet and Carbonel) are also associated. Some of these associations are probably due to reworking.

These beds, lying unconformably upon the fossiliferous palaeozoic layers, are often eroded and gullied, resulting in frequent destruction of the shell layers; this (together with the facies variations) probably indicates important local differences in life and/or sedimentation conditions.

The lumachelle exposures have clear shore-line traits: sand grain morphology shows heavy water movement, ebb and flow effects are indicated by typical beach-type breaking up of shells and by frequent sorting of cardium valves nesting one into the other, according to size. In areas of Gargaf wadis deltas, gravels, either mixed or interbedded with fossiliferous layers, indicate runoff and erosion of Mollusc littoral colonies. The continuity of that type of exposures indicates a past lake shoreline about 100 km between the villages of Ashkida and Wenzerik. The large extent of the lake is confirmed by magnitude of water movement, not possible in lakes of small extent.

Gypsum micro-crystals are always present and they sometimes encrust the shells and indicate evaporation processes related perhaps to short-term lake oscillations, to capillary rise of brackish water during lower level lake phases, after deposition or even to recent phenomena.

If association of *M.tuberculata* with *C.glaucum* is not due to *post-mortem* transport, one can estimate the lake(s) salinity to have been at least three per thousand (lower Cardium tolerance) but not much more, Melania being a typical fresh water Mollusc. However, according to Rosso & Monteillet 1977, Melania occasionally has been found alive in 29 per thousand salinity Senegal delta water.

Cardium shells from eight localities were dated through Goldberg & Koide (1962) method, based upon ionium disequilibrium (valid up to 200 000 B.P. with a 5 per cent error). Little or no thorium contamination was indicated by low ^{232}Th/^{230}Th ratios. The ^{234}U/^{238}U ratios (mean: 1.36) are characteristic of continental waters (Cherdyntsev 1962). The amount of uranium is low (a few tenths $\mu g/g$) and about equal in all samples.

Recrystallisation is unimportant in two samples (VII[6] and X, less than 3 per cent calcite) but high in the others, therefore suggesting discretion in the interpretation of the results listed below (sample numbers referring to localities between Ashkida and Quttah, cf. Petit-Maire, Casta & Delibrias (in press).

In the agricultural areas recently set up, through aquifer pumping, in the Quaternary alluvium south and east of Brak, and approximately at the same altitude, other deposits were observed. They are frequently eroded by past runoff or present irrigation processes. The shell beds are generally 0,10 to

291

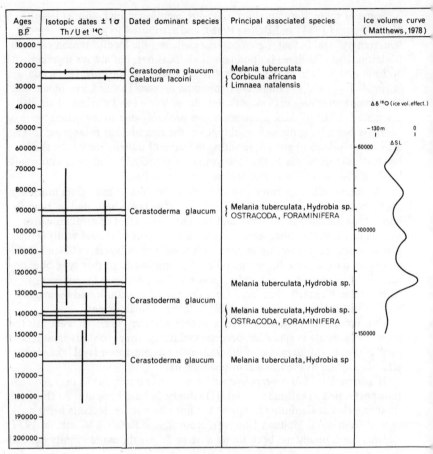

Figure 1. Lacustrine shell deposits in the Shati area (Fezzan, Libya).

0,50 m thick, but one exceptional *C.glaucum* 4 m mound was also found at about the same level.

The Invertebrate fauna is either characterized by the same associations as in the oases area, i.e. *C.glaucum* and *M.tuberculata* (localities V, VI, west of Brak; IV, south of Brak) or by a combination of fresh water species, *Caelatura lacoini, Corbicula africana, Limnaea natalensis, Bullinus truncatus,* with very few *M.tuberculata* and *C.glaucum* fragments (determinations Rosso & Testud, localities III, southeast of Brak). In the present state of research, one cannot tell whether such a difference corresponds to lenticular formations (local colonies around springs?) or to significant levels in lake evolution.

The same Ostracoda and Foraminifera are also found in all these sites.

292

Locality	Isotopic age	
XIV	90 000	± 20 000
VII d	92 500	± 7 000
I	125 000	± 11 000
VII b	127 000	$\left\{ \begin{array}{l} + 13\,000 \\ - 12\,000 \end{array} \right.$
VII e	139 000	$\left\{ \begin{array}{l} + 14\,000 \\ - 13\,000 \end{array} \right.$
XIII	141 000	$\left\{ \begin{array}{l} + 12\,000 \\ - 11\,000 \end{array} \right.$
X	143 000	$\left\{ \begin{array}{l} + 12\,000 \\ - 11\,000 \end{array} \right.$
VII c	163 000	$\left\{ \begin{array}{l} + 20\,000 \\ - 15\,000 \end{array} \right.$

Cyprideis smooth shells indicate low salinity. The layers with *C. lacoini,* etc. also testify to a salinity less than or equal to 2 per cent.

The association of fresh and brackish water species is probably due to post-mortem displacement of isolated shells (this could account for the scarcity of the rarer species) or to local, maybe seasonal, variations in water salinity.

C. lacoini at explosure III has been radiocarbon dated 26 100 ± 600 B.P. (Gif 4603).

3. *The bottom of the sebkhas* in covered, under the 'ploughed soil' crust, with a compact 0,15-0,20 m thick gypsum layer. This indicates evaporation of a wide expanse of water. No core could yet be made in sebkhas.

4. *Slightly above the present oasis* level and quite independently from the Pleistocene lake deposits, marshy soils mixed with calcified *Phragmites* nodes are found northeast of Brak in association with Neolithic sites. The organic matter was dated 4 590 ± 50 B.P. (Gif 4574) and a fire-place with Neolithic artefacts, in the same area was dated 3 650 ± 70 B.P. (Gif 4575). It brings evidence for possibility of human life at that age.

CONCLUSIONS

The shell deposits in the Shati area, Libya, testify to Pleistocene lacustrine periods in central Sahara lowlands.

Shell-carbonates isotopic ages, when considered in the frame of their error range, indicate a mid-Pleistocene lake episode around 140 000- 90 000 B.P.

Radiocarbon age at 26 000 correlates with known high levels in Tchad (Servant 1973). Between 4 500 and 5 500 B.P., groundwater rose a few metres higher than the present level.

Archaeological data seem to corroborate these ages: late Acheulean, Mousterian-Levalloisian and Neolithic artefacts have been observed in surface positions. They may be seen *in situ* in the Wadi Hmuda terraces, but could not yet be collected.*

However, one must not forget the high calcite ratio in most dated shells. This calcite cannot be original: recent *C.glaucum* shells from a closed Camargue lagoon and Gabes gulf were tested and resulted in pure aragonite (Lafont). Thus, the isotopic ages here recorded must be considered with caution. However, absence of (continental) [232]Th and lack of variation in uranium content indicate an early opening of the system in an aqueous environment, without any pollution. Moreover, three ages obtained from shells with low calcite ratios are concordant with ages from three other high ratio samples. Anyhow, the liability of all these isotopic ages is being tested.

The Shati lake(s) may have originated from groundwater rise and/or from runoff contributions from the northern djebels. Both are probably involved.

Runoff is proven by presence of fluvial deltas, by terraces in dry wadis coming down from Gargaf, by gravels interbedded or transported into shell layers, and by presence of volcanic material from northern djebels (Casta).

The Shati lakes fossiliferous beds have great informative value to Sahara palaeoclimatic history.

Further field and laboratory work, including stratigraphy, topography and precise altimetry, coring, Ostracoda-Foraminifera ecology, further radiochronology and stable isotopes studies, should now complete these preliminary observations.

REFERENCES

Chavaillon, J. 1964. *Les formations quaternaires du Sahara nord occidental*. CRZA, Paris, Série géologie 5.

Cherdyntsev, V.V. 1969. Uranium 234 atomizdat, *IPST* Cat. 2313, Moscou.

Conrad, G. 1968. L'évolution continentale post-hercynienne du Sahara algérien. Thesis, Univ. Paris.

Coque, R. 1962. *La Tunisie présaharienne,* Armand Colin, Paris.

Desio, A. 1971. Outlines and problems in the geomorphological evolution of Libya, from the Tertiary to the present day. *Symp. Geol. Libya 1968,* Tripoli, pp.11-35.

Dubief, J. 1963. *Le climat du Sahara,* Vol. 2. Inst. Rech. Sahariennes, Alger.

Gaillard, J. (in press). Comparative study of Quaternary and present lagoons Populations of *Cardium glaucum* Bruguière (Mollusca, Bivalvia). *2nd Symp. Geol. Libya.* Tripoli 1978.

Goldberg, E.D. & M.Koide 1962. Geochronological studies of deep-sea sediments by the Io/Th method. *Geochim. Cosmoch. Acta 26:*417-449.

Goudarzi, G. 1971. Geology of the Shati valley Area Iron Deposit, Fezzan. *Symp. Geol. Libya, 1968,* Tripoli, pp.489-500.

* Authorization for archaeological sampling has not yet been given by Libyan authorities.

Lelubre, M. 1946. A propos des calcaires de Murzuk. *C.r. Acad. Sci. Paris, 222:*359-361.

Matthews, R.K. 1978. Barbados high stands of sea-level and the deep sea [18]O ice volume curve. *Litoralia Newsletter,* 11.5.1978.

Monod, Th. 1931. L'Adrar Ahnet. *Rev. Géogr. phys. Géol. dyn. 4:*222-261.

Monteillet, J. & J.C.Rosso 1977. Répartition de la faune testacée actuelle (Mollusques et Crustacés Cirripèdes) dans la basse vallée et le delta du Sénégal. *Bull. IFAN 39*(4): 788-820.

Muller-Feuga, R. 1954. Contribution à l'étude de la géologie,de la pétrographie et des ressources hydrauliques et minérales du Fezzan. *Ann. Mines et Géol., Tunis 12.*

Petit-Maire, N. 1977. Rapport sur la mission de prospection effectuée en Libye occidentale du 10.12.76 au 10.2.77. *Doc. Lab. Géol. du Quat. CNRS.* Paris.

Petit-Maire, N., L.Casta & G.Delibrias (in press). Quaternary palaeolacustrine deposits in the Wadi Shati area, Libya. Preliminary data. *2nd Symp. Geol., Libya,* Tripoli, 1978.

Servant, M. 1973. *Séquences continentales et variations climatiques: Evolution du bassin du Tchad au Cénozoïque supérieur.* ORSTOM, Paris.

THE SOUTHWESTERN SAHARA MARGIN: SEDIMENTS AND CLIMATIC CHANGES DURING THE RECENT QUATERNARY

P. MICHEL
Institute of Geography, University of Strasbourg

My research work has been based mainly on the Senegal valley and its neighbouring region, an area that has not lost its interest for me since my first visit there in 1954. Thorough research work on eustatic variations and marine deposits has been carried out by P.Elouard and L.Hébrard, especially in the vicinity of Nouakchott. The observations and conclusions presented in this paper are based mainly on data obtained from 12 borings carried out at Bogué in the Senegal valley (Figure 1), and from which I have established a general profile (Figure 6).

1. CLIMATIC VARIATIONS AND SEA-LEVEL FLUCTUATIONS

A convenient starting point, as far as the Late Quaternary evolution of the area is concerned, would be the end of a long humid period during which the Senegal probably flowed northwest and fed vast lakes in the Ferlo and Trarza areas. Lake deposits of limestone, for instance, can be observed in the Aftout ech Chergui Depression, situated northeast of Lake Rkiz. These deposits lie topographically below partly cemented ancient Quaternary quartz gravels and duricrust debris which are underlain by sandstones of the 'Continental terminal' (Figure 2). Borings in different sites reveal lake deposits of limestone overlain by dunes. Some lakes could also have resulted from an upward migration of the watertable. The end of this humid period occurred about 100 000 years ago and was probably contemporaneous with the Aïoujian transgression which corresponds with the Riss-Würm interglacial. Fossiliferous deposits, dating from this period, occur in Mauritania, especially in the region of Nouadhibou and northeast of Nouakchott.

During the next regression, a series of tectonic movements led to a westward diversion of the lower course of the Senegal which then became entrenched in its present channel downstream from Bogué (Figure 1). The delta region is still subject to subsidence. In the ensuing arid climate, diffuse surface run-off led to the formation of the low glacis on the edge of the valley and an erg evolved in the northern Ferlo south of the lower valley.

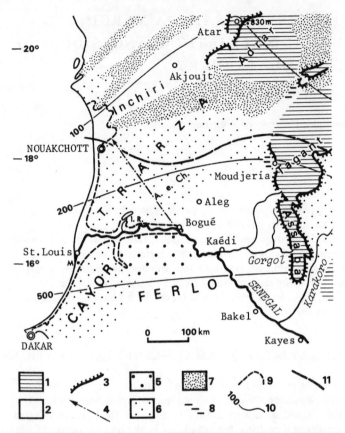

Figure 1. Southern Mauritania and Northern Senegal. The southwestern part of the Sahara and the Sahel.
1. Great sandstone plateaus; 2. Plains and lower plateaus; 3. Main cuestas; 4. Probable ancient course of the Senegal River; 5. Subdued 'ancient erg' dunes; 6. Fixed Ogolian dunes; 7. Mobile or revivified dunes; 8. Holocene lake deposits; 9. Limit of the Nouakchottian transgression; 10. Annual isohyet (in mm); 11. Limit of the Sahel and the Sahara.
A.e.Ch. = Aftout ech Chergui; L.R. = Lake Rkiz; M = Mouit.

A succeeding humid period occurred between 40 000 and 30 000 B.P. Dunes constituting the ancient Ferlo erg became subdued in form and underwent soil formation. They are presently characterized by well-developed ferruginous tropical soils. This humid period is synchronous with the Inchirian transgression during which the sea formed a gulf in the Inchiri area, northeast of Nouakchott (Figure 1). Shells obtained from this area have enabled a partial ^{14}C dating of this transgression at between 35 000 to 31 000 B.P. Borings at Mouit, 18 km south of St Louis, near the present mouth of the Senegal,

298

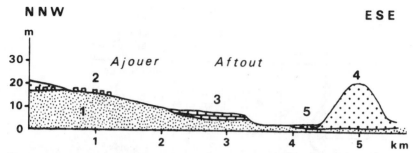

Figure 2. Cross-section of the Aftout ech Chergui Depression (Southern Mauritania).
1. Continental terminal sandstones; 2. Duricrust debris and quartz gravels partially cemented; 3. Hardened lake limestone *(calcaire du Rkiz)*; 4. Ogolian dune; 5. Friable clayed limestone *(calcaire des gouds)*.

reveal shell horizons at depths of about −20 m and −13 m. They have been dated by ^{14}C at 34 300 and 31 300 B.P. respectively (Michel 1973). These deposits have been lowered by subsequent subsidence.

2. THE LAST MAJOR ARID PERIOD

With the post-Inchirian regression, the climate became dry once more (Figure 3). It was probably characterized by major contrasts, with long periods of drought punctuated by heavy showers. The Senegal incised deeply into its bed. Borings reveal an incision that extended right down to the Eocene basement. A depth of −20 m is recorded at Bogué and −27 m at Rosso. The borings at Bogué were carried out on the banks and the bed of the river which is marked by a big meander at this point (Figure 5). The Eocene sediments are overlain by well-sorted medium to coarse sands as well as by gravel and shingle horizons of quartz and jasper. These ancient overlying horizons, taken together, vary in thickness from 10 to 15 m while the diameter of the gravels is in the vicinity of 35 mm (Figure 6). This major regression was probably characterized by slight sea-level oscillations and the ancient (post-Inchirian) sediments could have been deposited during a slight rise in sea level.

The climate became arid. The massive Ogolian dunes which presently occupy southern Mauritania and western Senegal in the region of Cayor (Figure 1) were formed at the height of this arid period. The great northeast-southwest trending dunefields gradually blocked the course of the river up to the vicinity of Kaédi (Figure 4). These dunes consist of fine to medium sands. Grain shape analyses indicate that these sands have been transported over only short distances by the wind. The dunes presently occur as relicts in the lower Senegal valley and in the Delta area where they are surrounded by recent alluvium.

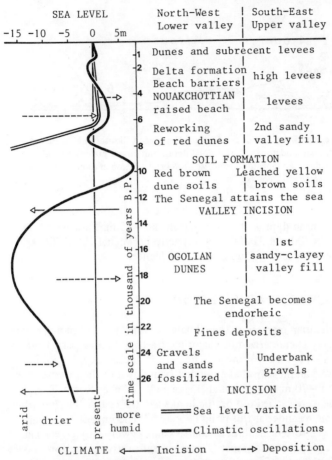

MORPHOGENESIS

	North-West Lower valley	South-East Upper valley
-1	Dunes and subrecent levees	
-2	Delta formation Beach barriers	high levees
-4	NOUAKCHOTTIAN raised beach	levees
-6		
-8	Reworking of red dunes	2nd sandy valley fill
	SOIL FORMATION	
-10	Red brown dune soils	Leached yellow brown soils
-12	The Senegal attains the sea	
	VALLEY INCISION	
-14		
-16	OGOLIAN	1st sandy-clayey
-18	DUNES	valley fill
-20	The Senegal becomes endorheic	
-22	Fines deposits	
-24	Gravels and sands	Underbank gravels
-26	fossilized	
	INCISION	

SEA LEVEL
-15 -10 -5 0 5m

Time scale in thousand of years B.P.

═══ Sea level variations
▬▬▬ Climatic oscillations

arid drier present more humid

CLIMATE ◄——— Incision ----► Deposition

Figure 3. The geomorphological evolution of the Senegal valley and its margins since 26 000 years B.P. (before present).

Upstream from the dune barrier, the Senegal accumulated sandy-clayey deposits that thus built up the 'first valley fill'. These are badly sorted sediments deposited by a river rendered endorheic by the dunefields lying across its channel. These sandy-clayey deposits overlie the ancient sediments and, in turn, are overlain by well-sorted aeolian sands (Figure 6). The borings were, in fact, carried out close to the extreme limit of a series of dunes of Ogolian age (Figure 5). This major arid period probably occurred between 20 000 and 14 000 B.P., and was accompanied by a southward shift of desert conditions over a distance of about 4 to 5° latitude.

300

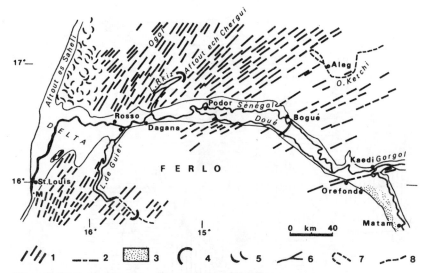

Figure 4. Ogolian ergs on either side of the Senegal valley.
1. Longitudinal red dune barriers; 2. Main dune barring the valley; 3. First valley fill
terrace; 4. Extreme limits of the Nouakchottian marine gulf; 5. Reworked red dunes;
6. Lake fed by the Senegal River; 7. Temporary Lake; 8. Wadi or dry valley. M = Mouit.

3. THE HUMID PHASE AND THE HOLOCENE
   TRANSGRESSION

When the climate became less dry, the Senegal incised into the 'first valley
fill' deposits, forming a terrace in the process. It also cut through the dune
barrier in order to regain access to the sea. Sea level was then still low. Downstream from Dagana, borings indicate an incision depth of −28 m. The river
gradually regained its former valley. The now-dry stream network in the Ferlo
Plateau was active during this period and carved out the depression now
occupied by Lake Guier, before finally joining the Senegal (Figure 4). Several
small coastal streams have incised into and cut through the Ogolian dunes
between St Louis and Dakar (Figure 1).

The last transgression was attended by a humid phase that lasted from
11 000 to 8 000 B.P. During this phase, traces of which can be observed
throughout the southern Shara and which is known as the Tchadian, the
Senegal deposited clays which thus overlie the aeolian sands (Figure 6). They
contain pollen belonging to a flora, notably ferns, that could only have
thrived in a humid environment (Michel & Assemien 1969). The Ogolian
dunes have been stabilized by savanna vegetation and are characterized by
reddish sands: hence the name red dunes. On them have developed red-brown
soils. The interdune depressions are often occupied by temporary pools of

301

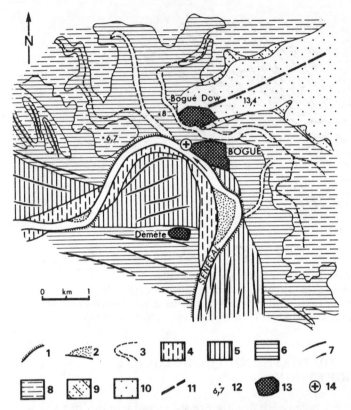

Figure 5. Geomorphological map of Bogué and its surrounding region.
1. Eroded bank; 2. Sand bar; 3. Defluent; 4. Present levees; 5. Sub-recent levees; 6. Fluvio-deltaic deposits; 7. Main levees; 8. Clay basins; 9. Sandy terraces; 10. Ogolian dunes; 11. Dune alignment; 12. Trig. Point; 13. Settlement; 14. Bore hole sites.

stagnant water and contain deposits of friable clayey limestones known as the 'calcaire des gouds'. Such deposits occur for instance in the Aftout ech Chergui Depression between an Ogolian dune and the terrace formed from the 'calcaire du Rkiz' (Figure 2). Lake deposits can also be observed at the foot of the sandstone plateaux, as in the neighbourhood of Moudjéria, west of the Tagant (Figure 1).

During a minor arid phase (Figure 3), the Ogolian dunes became partially reworked, especially near the coastline. This resulted in the formation of small north-south oriented dunes. The Senegal deposited sands upstream, thus building up the 'second sandy valley fill' terrace. In the borings at Bogué, the clays give way to sandy deposits towards the top (Figure 6).

The rising sea invaded the Delta region and the lower Senegal valley, thus

302

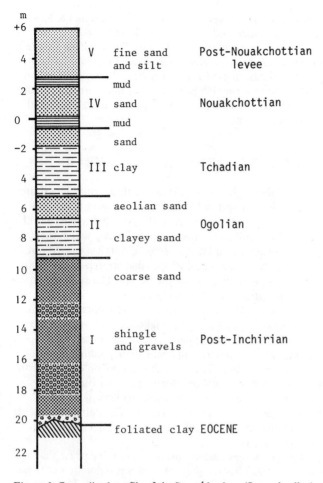

Figure 6. Generalized profile of the Bogué borings (Senegal valley).

occupying the Lake Rkiz and Lake Guier depressions on either side of the valley (Figure 4). At the height of this transgression, locally known as the Nouakchottian, the sea attained Bogué, 230 km from the present coastline, thus transforming the lower valley into a narrow elongated gulf. This transgressive phase also resulted in the formation of a wide gulf northeast of Nouakchott (Figure 1). The raised beach resulting from this transgression contains shell horizons principally characterized by *Arca Senilis.*. The first detailed observations on this raised beach were carried out in the region of Nouakchott, from where samples dated by ^{14}C gave ages ranging from 6 000 to 5 000 B.P. In the Senegal Delta, two samples gave a ^{14}C age of 5 470 B.P. The summit of this raised beach is situated at between +1.5 and 2 m. Monteillet & Rosso are presently carrying out detailed studies on its fauna. At the edge of the lower valley can be observed a narrow stretch of sandy terrace often situated at the foot of the Ogolian dunes. It is underlain in places by a mud horizon. The lower mud horizon in the borings at Bogué contains *Rhizophora* pollen, and is overlain by white sands over which lies the upper mud horizon (Figure 6). During the Nouakchottian, the climate became quite humid, and faunal studies indicate that the sea was calm.

4. THE EVOLUTION DURING THE LAST 5 000 YEARS

The sea subsequently became rough. An important north-to-south longshore current built up a series of barrier spits which gradually closed off the gulf northeast of Nouakchott forming, in the process, the great Ndramcha Sebkha. The spits also dammed off the gulf in the lower Senegal valley which then became transformed into an elongated lagoon. Ten ^{14}C datings, carried out on the barrier spits in the vicinity of St Louis give ages that stretch from 4 000 to 1 900 B.P., ages decreasing from inland toward the sea. The spit sands have been reworked by winds to form small yellow dunes.

The Senegal brought down a significant amount of sediments during this period and it built up a network of high post-Nouakchottian levees of fine sand and silt. These deposits overlie the upper mud horizon observed at Bogué, and their base contains *Elaeis* pollen which indicate a fairly humid environment while the upper section contains Mimosa and grass pollen characteristic of the present vegetation. Downstream from Bogué, the river built up an elongated delta that gradually filled up the then existing lagoon. During the process, the river became split up into several channels. The deltaic sediments that engendered this splitting up contain salt deposits as well, especially in the area west of Rosso (Figure 4).

The formation of the high levees ceased during a minor dry phase that probably occurred around 2 000 B.P. (Figure 3). The Senegal undercut these high levees and traced out a number of meanders in the process. The river however constructed a series of smaller 'sub-recent' levees within the major meander bends (Figure 5). The ancient distributary mouths of the delta became dammed off by littoral dunes which were formed as a result of this dry phase. The river thus abandoned its delta, shifted south-southwest and joined the sea through an unstable estuary south of St Louis. North of the delta region, the Ogolian dunes were reworked by northwest maritime winds (Figure 4).

The climate was apparently still relatively humid around 2 900 B.P. in southern Mauritania. This is indicated by the presence of sediments deposited in a former lake that then extended along the foot of the Tichit cuesta.

Generally, however, the available evidence indicates dry climatic conditions since 2 500 B.P., but this period has been punctuated by minor variations. The Middle Age was characterized, for instance, by relatively humid conditions. The post-Nouakchottian littoral dunes that were formed during the minor dry phase around 2 000 B.P. have been partially fixed by vegetation. Since the Nouakchottian, sea level has similarly undergone slight oscillations before assuming its present stand.

5. CONCLUSION

The sedimentary record and geomorphic evolution of the Senegal valley are now fairly well-known. They are both closely linked to climatic variations and sea level fluctuations. However, more work needs to be done on the stratigraphy of neighbouring regions especially southern Mauritania. Lappartient is presently carrying out new research work on the Aftout ech Chergui Depression while Barbey is engaged in a study of the ergs in western Mauritania, where several generations of dunes probably exist as in northern Senegal.

New ^{14}C dates are absolutely essential if a more detailed chronology is to be established, especially of those periods predating the Nouakchottian. Unfortunately, present research work indicates an abysmal dearth of datable material. Meanwhile, it is at least necessary to redate sediments deposited during the Inchirian transgression since the ages ranging from 35 000 to 31 000 B.P. are close to the absolute limit of ^{14}C dating and can thus be doubted. The necessity to use another method which can give fairly accurate ages beyond the limit of ^{14}C dating is thus evident.

REFERENCES

Barbey, C., J.P.Carbonnel, S.Duplaix, L.le Ribault & J.Tourenq 1975. Etude sédimentologique de formations dunaires en Mauritanie occidentale. *Bull. Inst. fond. Afr. noire, A, Sénégal, 37:*255-281.
Beaudet, G., P.Michel, D.Nahon, P.Oliva, J.Riser & A.Ruellan 1976. Formes, formations superficielles et variations climatiques récentes du Sahara occidental. *Rev. Géogr. phys. Géol. dyn., Fr. 18:*157-174.
Daveau, S. 1965. Dunes ravinées et dépôts du Quaternaire récent dans le Sahel mauritanien. *Rev. Géogr. Afr. occ., Dakar 1(2):*7-47.
Daveau, S. 1970. L'évolution géomorphologique quaternaire au sud-ouest du Sahara (Mauritanie). *Ann. Géogr., Fr. 431:*20-38.
Einsele, G., D.Herm & U.Schwarz 1977. Variations du niveau de la mer sur la plateforme continentale et la côte mauritanienne vers la fin de la glaciation du Würm et à l'Holocène. *Bull. Ass. Sénégal. Et. Quat. afr. Dakar 51:*35-48.
Elouard, P. 1968. Le Nouakchottien, étage du Quaternaire de Mauritanie. *Ann. Fac. Sci., Dakar 22:*121-137.
Elouard, P. 1975. Formations sédimentaires de Mauritanie atlantique. In: *Notice Carte Géol. Mauritanie au 1/1 000 000.* Bur. Rech. géol. min., Orléans, pp.171-233.
Elouard, P. & P.Michel 1958. Le Quaternaire du lac Rkiz et de l'Aftout de Boutilimit, *C.r. Soc. Géol., France 12:*245-247.

Faure, H. & P.Elouard 1967. Schéma des variations du niveau de l'Océan Atlantique sur la côte de l'Ouest de l'Afrique depuis 40 000 ans. *C.r. Acad. Sci., D, Fr.265:*784-787.

Grove, A.T. & A.Warren 1968. Quaternary landforms and climate on the south side of the Sahara. *Geogr. Journal, London 134:*194-208.

Hébrard, L. 1968. Contribution à l'étude géologique des formations quaternaires la bordure de la sebkha Ndramcha près de Nouakchott (Mauritanie). *Rapp. Lab. Géol. Fac. Sci., Dakar 27.* 28p.

Hébrard, L. 1978. Contribution à l'étude géologique du Quaternaire du littoral mauritanien entre Nouakchott et Nouadhibou, 18-21° latitude Nord. Thèse Lyon 1973. *Docum. Lab. Géol. Fac. Sci., Lyon 71.* 210p.

Lappartient, J.R. 1976. Le Quaternaire continental de Mauritanie: bordure nord de l'Aftout ech Chergui. In: *Palaeoecology of Africa 9.* p. 7.

Michel, P. 1957. Rapport préliminaire sur la géomorphologie de la vallée alluviale du Sénégal et de sa bordure. *Bull. Miss. Aménag. Sénégal, Saint-Louis 111.* 85p.

Michel, P. 1968. Genèse et évolution de la vallée du Sénégal, de Bakel à l'embouchure (Afrique occidentale). *Z. Geomorphol., Stuttgart 12:*318-349.

Michel, P. 1973. Les bassins des fleuves Sénégal et Gambie. Etude géomorphologique. Thèse Strasbourg 1970. *Mém. Off. Rech. Sci. Tech. Outre-Mer., 63,* 3 vols. 752p.

Michel, P. 1977a. Geomorphologische Forschungen in Süd- und Zentral Mauritanien: Probleme der Lateritmäntel, Eisenpanzer, Kalkkrusten; der Einfluss von Klimaschwankungen. *Mitt. Basler Afrika Bibliogr. 19:*81-108.

Michel, P. 1977b. Recherches sur le Quaternaire en Afrique occidentale. In: Recherches françaises sur le Quaternaire. *Supplément au Bull. AFEQ 50:*143-153.

Michel, P. 1979. Les modelés et dépôts du Sahara méridional et Sahel et du Sud-Ouest Africain. Essai de comparaison. *Rech. Géogr., Strasbourg 5:*5-39.

Michel, P. & P.Assémien 1969. Etude sédimentologique et palynologique des sondages de Bogué (basse vallée du Sénégal) et leur interprétation morphoclimatique. *Rev. Géorm. dyn., Fr. 19:*97-113.

Michel, P., P.Elouard & H.Faure 1968. Nouvelles recherches sur le Quaternaire récent de la région de Saint-Louis (Sénégal). *Bull. Inst. fond. Afr. Noire, A. Sénégal 30:*1-38.

Monteillet, J. 1977. Tourbes de l'Holocène inférieur (Tchadien) dans le Nord du Delta du Sénégal. *Bull. Ass. Sénégal. Et. Quat. afr., Dakar 50:*23-28.

N., 1978. Fichier des âges radiométriques du Quaternaire au nord de l'équateur (10e série). *Bull. Ass. Sénégal. Et. Quat., afr., Dakar 52-53:*11-177.

Rognon, P. 1976. Essai d'interprétation des variations climatiques au Sahara depuis 40 000 ans. *Rev. Géogr. phys. Géol. dyn. 18:*251-282.

Rosso, J.C., P.Elouard & J.Monteillet 1977. Mollusques du Nouakchottien (Mauritanie et Sénégal septentrional): Inventaire systématique et esquisse paléoécologique. *Bull. Inst. fond. Afr. noire, A., Sénégal 39:*465-486.

Toupet, Ch. 1976. L'évolution du climat de la Mauritanie du Moyen-Age jusqu'à nos jours. In: *Colloque Nouakchott sur la désertification du Sud du Sahara.* Nouv. Edit. Afric., Dakar, pp.56-63.

Tricart, J. & M.Brochu 1955. Le grand erg ancien du Trarza et du Cayor (Sud-Ouest de la Mauritanie et Nord du Sénégal). *Rev. Géom. dyn., Fr. 6:*145-176.

MODELLING OF CLIMATE AND PLANT COVER IN THE SAHARA FOR 5 500 B.P. AND 18 000 B.P.

W. LAUER / P. FRANKENBERG
Geographisches Institut, Universität Bonn

In the last decades, questions arose concerning past climates as to a general demand for modelling future climate changes. The Sahel drought disaster led to investigating the change of climatological and ecological conditions in the North African dry belt.

According to our quantitative studies about the relationships between plant cover and climate in the Sahara and its adjacent areas (Lauer & Frankenberg 1977, Frankenberg 1978), mainly based on floristic lists, we present a model of past plant cover by reconstructing past climatic conditions with the help of literature (cf. Lauer & Frankenberg 1979).

The first point to stress is the actual correlation between plant cover and climate. The extension of the Sahara desert and its forming part of the holarctic or palaeotropic seem to be of major scientific importance (Figure 1). As to the absolute or relative domination of floristic elements which can be correlated with isotherms or isohyets by means of Chi^2-Tests, one may assume that the northern and southern floristic boundaries of the Sahara correlate well with the 100 mm isohyets and the floristic boundary between palaeotropic and holarctic with the mean annual temperature of 24,5°C.

Secondly, the absolute number of plant species (Figure 2) was analysed in a square grid of 80 km. The number of plant species reflects the net productivity of plant cover (1 species = 1 gr net production per square metre per year) and the percentage of soil covered by plants (log y = 1,08 log x − 0,134 1; correlation coefficient = 0,7 log = log 10; y = percentage of soil covered by plants, x = absolute number of plant species).

The absolute number of plant species shows quite a strong correlation with the annual mean of the water balance (ETP − P: y = 1,72x + 7,38, all values = log 10; correlation coefficient = 0,7; number of climatic stations: 48; y = absolute number of plant species; x = ETP − P). Using these correlations between the actual plant cover and climate, it is possible to calculate past plant cover by modelling past climate.

We have chosen two marked climatic dates for giving a scenario of their plant covers as compared with the actual ecological situation in the Sahara desert, i.e. 18 000 and 5 500 B.P. Around 18 000 B.P. climate was generally

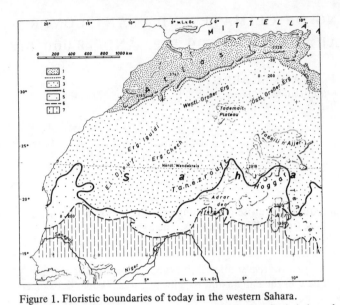

Figure 1. Floristic boundaries of today in the western Sahara.
1. Mediterranean holarctic floristic region; 2. Northern floristic boundary of the Sahara;
3. Saharo-arabic holarctic floristic region; 4. Floristic boundary between the holarctic and palaeotropic; 5. Saharo-arabic palaeotropic florstic region; 6. Southern floristic boundary of the Sahara; 7. Sahelian palaeotropic floristic region.

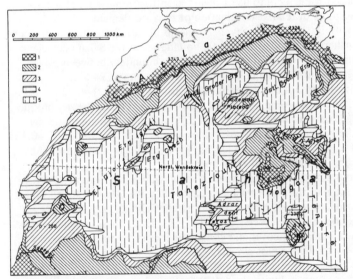

Figure 2. Actual absolute number of plant species in the western Sahara (1: above 161;
2: 81 to 160; 3: 41 to 80; 4: 21 to 40; 5: below 21).

308

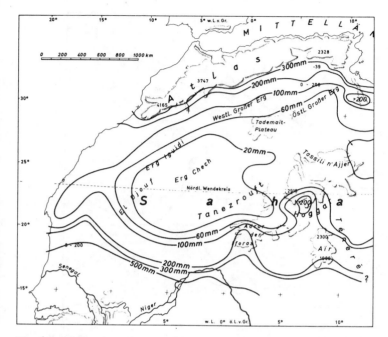

Figure 3. Mean annual precipitation in the western Sahara, modelled for 5 500 B.P.

very cold and dry, showing a maximum extension of the Würm or Wisconsin glaciation with a general southward shift of climatic zones. Compared with the present, the mean annual temperatures were about 4-7° lower in the area under investigation.

In the Mediterranean region south of the Atlas-mountains, precipitation seems to have been gradually higher at 18 000 B.P. than today (Alayne Street-Grove 1976, Conrad 1969, Fontes-Perthuisot 1973, Nicholson 1976, Diester-Haass 1976). Sarnthein (1978) indicates valid proves of more arid environmental conditions in the Mediterranean regions concerned. He states a general onset of wetness after 17 700 B.P. in the Algerian North Sahara, so that our estimation of more humid conditions in the North Sahara may be valid mainly for the period after 17 700 B.P.

At about 5 500 B.P., warm and relatively wet conditions prevailed on the globe (altithermal, climatic optimum). Temperatures were about 1 to 2°C higher than today. At that time, the Sahara seemed to have been wetter both in the south and in the north, which is difficult to explain by a simple shift of climatic zones.

Sarnthein (1978) dates the time-span of that climatic optimum from 6 000 B.P. up to 5 500 or 5 000 B.P. The overall wetter climatic conditions during the climatic optimum are indicated by most of the scientists having practised

309

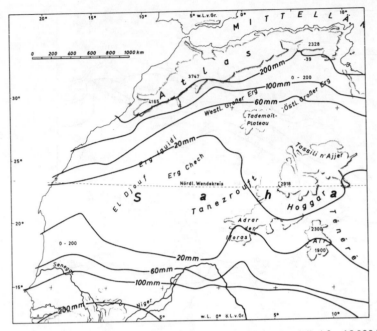

Figure 4. Mean annual precipitation in the western Sahara, modelled for 18 000 B.P.

studies in the area under investigation (Sarnthein 1978, Nicholson 1976, Suzuki 1976, Alayne Street-Grove 1976, Rognon 1976, Mauny 1956, World Climate 1966).

We tried to reconstruct climatic conditions for both climatic epochs mentioned above according to the literature, fully cited in Lauer & Frankenberg 1979. Figures 3 and 4 show the results of modelling isohyets for 5 500 B.P. and 18 000 B.P. Potential evapotranspiration (ETP) for 5 500 B.P. was calculated to be nearly the same as today. The ETP for the cold and dry climatic conditions at about 18 000 B.P. was 60 per cent of its present values in the north and 90 per cent in the south in the area under investigation.

According to the isohyets and isotherms assumed for 5 500 B.P. and 18 000 B.P., one may give a scenario of the floristic boundaries of the Sahara and the position of the boundary dividing holarctic and palaeotropic at that time (Figures 5 and 6). Figure 5 presents the shrinking desert during the climatic optimum and the far northward domination of tropical plant species in comparison with the actual patterns of the present. Figure 6 gives quite an opposite picture. In the north, compared with the present, we can assume a certain southward shift of the steppe vegetation not only due to higher precipitation, which may be doubtful but perhaps as a result of reduced potential evapotranspiration and cooler temperatures. A pronounced southward shift

310

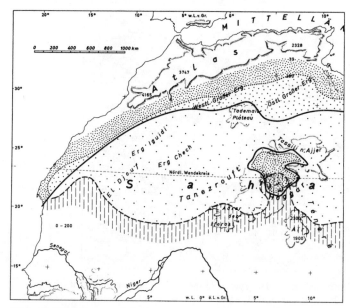

Figure 5. Floristic boundaries in the western Sahara, modelled for 5 500 B.P. (for legend, see Figure 1).

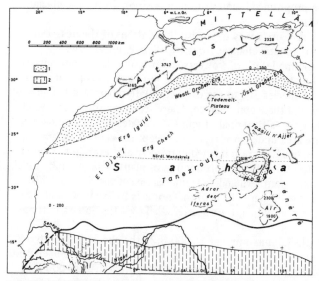

Figure 6. Floristic boundaries in the western Sahara, modelled for 18 000 B.P.
1. Northern floristic margin of the Sahara; 2. Southern floristic margin of the Sahara;
3. Floristic boundary between the holarctic and palaeotropic.

311

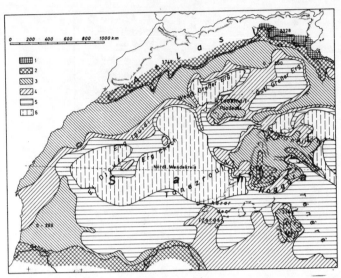

Figure 7. Absolute number of plant species in the western Sahara, modelled for 5 500 B.P. (1: above 300; 2: 161 to 300; 3: 81 to 160; 4: 41 to 80; 5: 21 to 40; 6: below 21).

of the Sahara is less questionable. The boundary between holarctic and palaeotropic also shows a pronounced more southern position during the glacial maximum than today.

In summary, we recognize very extensive fluctuations of the desert in North Africa due to climatic changes from 18 000 B.P. up to now.

With regard to the modelling of isohyets and of potential evapotranspiration for the two climatic epochs concerned, we were able to calculate past absolute numbers of species due to the close correlation between the absolute numbers of species and the water balance. These scenarios are presented in Figures 7 and 8. For the 'altithermal', Figure 7 indicates higher numbers of species in the entire area under investigation, except the central Sahara. There the climate seems to have remained extremely arid even in the 'climatic optimum'. At that time, the net production of plants was generally much higher at the northern, southern and western margins of the Sahara. This seems to have been a good basis for neolithic nomads to live there and even for many tropical animals which were scattered all over the Sahara in those times.

During the last glacial maximum at 18 000 B.P., the chrological patterns of absolute numbers of species in the western Sahara seem to have been quite different from those of 5 500 B.P. and of today (Figure 8). At 18 000 B.P. the region with very low values was situated far to the south, whereas some parts of the actual central Sahara showed a higher number of species, mainly because of the lower potential evapotranspiration.

312

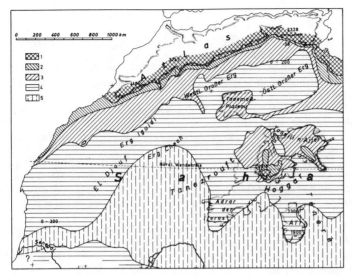

Figure 8. Absolute number of plant species in the western Sahara, modelled for 18 000 B.P. (1: above 300; 2: 161 to 300; 3: 81 to 160; 4: 41 to 80; 5: 21 to 40).

The different maps of the number of species (Figures 7 and 8) indicate plant cover density and net production for marked climatic epochs and, thus, quantitatively express the changes of ecological conditions within the North African dry belt.

Changing climate does not only mean changing vegetation cover but also changing soil formation and fluctuating coast-lines. The drop of the sea level at about 18 000 B.P. means for example a broad part of the actual shelf is above sea level (coastal desert?), from where sand could have been blown into the interior to form dunes even under relatively wet conditions. Changing soil formation would alter the infiltration capacity and could also lead to more humid conditions for plant cover.

Some general trends of climatic change seemed to have prevailed in the area under investigation. The northern parts showed more stable climatic conditions than the southern ones. The Atlas mountains seem to have orographically fixed the climatic zones, whereas no such barriers exist in the south.

Climate — over long periods — shows relatively stable conditions and extreme fluctuations within short periods. It seems that climate is subject to increasing entropy and then returns to more stability at a different temperature level. Cyclic changes of entropy may explain climatic change.

313

REFERENCES

Alayne Street, F. & A.T.Grove 1976. Environmental and climatic implications of late quaternary lake-level fluctuations in Africa. *Nature 261:*385-390.

Beaudet, G., P.Michel, D.Nahon, P.Oliva, J.Riser & A.Ruellan 1976. Formes, formations supeficielles et variations climatiques récentes du Sahara occidental. *Rev. Géogr. phys. Géol. dyn. 18:*157-174.

Beucher, F. 1971. Etude palynologique de formations néogènes et quaternaires au Sahara Nord-occidental. Thèse, Univ. Paris.

Capderou, M. & A.Verdet 1977. *Gravures rupestres de l'Atlas Saharien de 6 000 a.v. à 100 a.p. J.C.* Toulouse.

CLIMAP Project Memebers 1976. The surface of the Ice-Age earth. *Science 191:*1131-1137.

Conrad, G. 1969. Contribution à l'histoire continentale posthercynienne du Sahara algérien. *Palaeoecology of Africa 4:*37-47.

Cour, P. & D.Duzer 1976. Persistence d'un climat hyperaride au Sahara central et méridional au cours de l'holocène. *Rev. Géogr. phys. Géol. dyn. 18:*175-198.

Diester-Haass, L. 1976. Late quaternary climatic variations in Northwest Africa deduced from East Atlantic sediment cores. *Quatern. Res. 6:*299-314.

Fontes, J.Ch. & J.P.Perthuisot 1973. Climatic recorder for the past 50 000 years in Tunisia. *Nature Phys. Sci. 244:*74-75.

Frankenberg, P. 1978. Florengeograpische Untersuchungen im Raume der Sahara . . ., *Bonner Geogr. Abh. 58.*

Gabriel, B. 1977. Zum ökologischen Wandel im Neolithikum der östlichen Zentralsahara. *Berliner Geogr. Abh. 27.*

Geyh, M.A. & D.Jäkel 1974. Spätpleistozäne und holozäne Klimageschichte der Sahara aufgrund zugänglicher 14C Daten. *Z. Geomorphol. Stuttgart. N.F. 18:*82-88.

Jäkel, D. 1977. Abfluss und fluviatile Formungsvorgänge im Tibesti-Gebirge als Indikatoren zur Rekonstruktion einer Klimageschichte der Zentralsahara im Spätpleistozän und Holozän, Vortrag 10. INQUA Congress, Birmingham (20.8.77).

Lauer, W. & P.Frankenberg 1977. Zum Problem der Tropengrenze in der Sahara. *Erdkunde 31:*1-15.

Lauer, W. & P.Frankenberg 1979. Zur Klima- und Vegetationsgeschichte der westlichen Sahara. *Akad. Wiss. Lit. Mainz., Math. Naturw. K. 1.*

Mauny, R. 1956. Préhistoire et zoologie: la grande 'faune éthiopienne' du Nord-Ouest Africain du paléolithique à nos jours. *Bull. IFAN, A 1:*246-279.

Nicholson, S.E. 1976. A climatic chronology of Africa: Synthesis of geological, historical and meteorological information and data. Diss. Univ. of Wisconsin.

Rognon, P. 1976. Essai d'interprétation des variations climatiques au Sahara depuis 40 000 ans. *Rev. Géogr. phys. Géol. dyn. 18:*251-282.

Saltzmann, B. 1975. A solution for the northern hemisphere climatic zonation during a glacial maximum. *Quartern. Res. 5:*307-320.

Sarnthein, M. 1978. Sand desert during glacial maximum and climatic optimum. *Nature 272*(5648):43-46.

Suzuki, H. 1976. World precipitation, present and hypsithermal. *Japanese Progress in Climatology, Nov:*63-79.

World Climate 1966. World climate from 8 000 to 0 B.C. *R. Meteorol. Soc. Proc. Int. Symp. London,* 19.4.1966.

CLIMATOLOGICAL ASPECTS OF THE SPATIAL AND TEMPORAL VARIATIONS OF THE SOUTHERN SAHARA MARGIN

DIETER KLAUS
Geographisches Institut, Universität Bonn

SPATIAL-TEMPORAL PRECIPITATION FLUCTUATIONS AT THE SOUTHERN SAHARA MARGIN

All precipitation data available for the Sahel area as well as the long time series of other West African countries were comprised of area mean values for the period 1934-1973 (3 x 3 degrees of latitude). The time series of the annual precipitation anomalies of the area mean values were subject to an eigenvector analysis. Eigenvector analysis, also referred to as principal components or empirical orthogonal function analysis enables fields of highly correlated data to be represented adequately by a small number of orthogonal eigenvectors and corresponding orthogonal time coefficients. These coefficients when combined with the associated eigenvectors, would provide for an exact reproduction of the original data field. Reference is made to Kutzbach (1970) for discussion of eigenvector analysis techniques, including the application to climatological problems.

The eigenvectors only depend upon the interrelationship within the data analyzed. Simply stated, the first eigenvector represents that linear combination of the original variables which, when used as a linear predictor of the original variables, explains the greatest fraction of the total variance. The subsequent eigenvectors are required to account for the largest parts of the remaining variances.

The spatial pattern of the eigenvector components shows maximum values of variance explained by the first eigenvector at 13-19°N. The time series of the eigenvector coefficients is marked by high positive coefficients in the years 1935, 1936 as well as during 1950-1958. Around the year 1940 and particularly after 1967, negative coefficients were predominant. In the case of positive eigenvector coefficients and positive eigenvector components high positive precipitation anomalies correspond with the eigenvector components and in the case of negative coefficients the anomalies are reversed.

Thirty-one per cent of the variance of all time series considered are explained by the first eigenvector. The spatial pattern of the eigenvector components shows that the precipitation fluctuations of West Africa are marked by strong

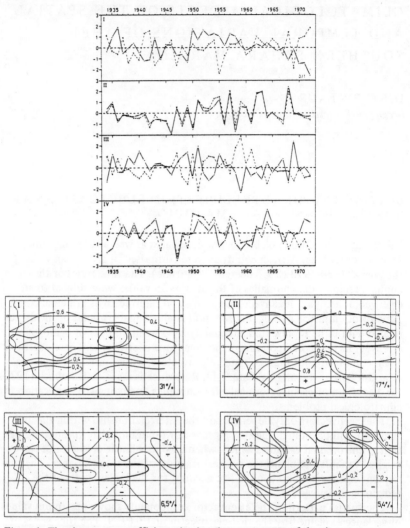

Figure 1: The eigenvector coefficients (top) and space patterns of the eigenvector components determined on the basis of the time series of the anomalies of the area mean values of the annual precipitation for 1934-1973.

in-phase changes in the entire West African region. A general decline of the annual precipitation totals can be noticed since 1955.

The second eigenvector still explains 17 per cent of total variance and is marked by positive components at the southern margin of the Sahara and by negative components between 13-16°N while a positive sign of the compo-

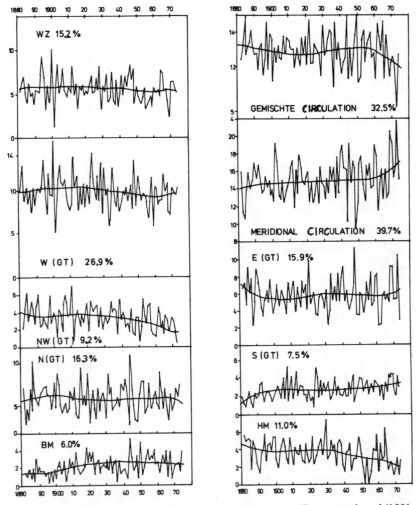

Figure 2. Temporal frequency development of some European 'Grosswetterlagen' (1881-1973) and their approximation by 4th degree polynomes.

pents south of 13°N is predominant. The time series of the coefficients shows that there is no dominant trend. The forties and the years after 1965 – except for 1969 – are marked by negative anomalies at the southern Sahara margin as well as north of the coast of Guinea due to negative coefficients. A zonally oriented reversed behavior becomes evident.

The third and fourth eigenvector together explain another 12 per cent of total variance. Strong meridionally oriented anomaly patterns are predomi-

317

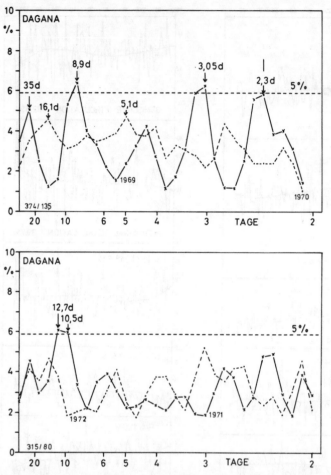

Figure 3. Powerspectra of the daily precipitation totals for the rainy season of 1969, 1970 (top) and 1971, 1972 (bottom) calculated for Dagana, Senegal.

nant. In this connection, the west coast shows an inverse relationship vis-à-vis the eastern Sahel area (third eigenvector) on the one hand, and inverse vis-à-vis the area between 6°W to 3°E on the other hand (fourth eigenvector).

Almost 50 per cent of total variance of the West African precipitation time series are explained by the first two spatial eigenvector patterns: 1) Spatial coherence of the precipitation decreases and/or precipitation increases; 2) Precipitation decreases in the northern Sahel area are linked with precipitation increases in the Sahel-Sudan zone and with precipitation decreases in the Guinea coast area

318

CHANGES IN THE CIRCULATION STRUCTURE

The spatial circulation structure of the European Atlantic sector can be described by the Grosswetterlagen frequency (Hess, Brezowski 1969). Figure 2 contains the temporal frequency developments from 1881-1973 for some 'Grosswetterlagen'. The trend was determined by fourth degree polynomials and show that from 1954 onward the frequency of the half meridional 'Grosswetterlagen' decreases whereas the frequency of the meridional 'Grosswetterlagen' increases. The latter is conditioned particularly by the increased occurrence of northern and eastern 'Grosswetterlagen' during the summer which leads to an equatorward extension and westward shift of the trough at the 200 mb level existing at approximately $25°E$ in the long-term mean (Thompson 1965). The Tropical Easterly Jet (Flohn 1971) is thereby weakened and shifted to the south. At the same time, the southern Algerian anticyclone at the 200 mb level becomes isolated and intensified. Large areas of the Sahel thus come under the influence of a strong North West wind at the 200 mb level which, according to Osman & Hastenrath (1969), leads to precipitation deficits in the Sahel. Similar conditions can be noticed over Mexico during the summer months when the 200 mb trough at $60°W$ is extended meridionally.

The quasi stationary trough existing to an increasing extent over northern Africa since 1955 reduces the frequency of the 'lignes de grain' as was shown by Hammer (1976) for the years 1968-1973 for the eastern part of the Sahel and as is presented in Figure 3 for a typical western Sahelian station. The power spectrum of the daily precipitation totals of Dagana ($16°31'N$, $15°30'W$) under mean conditions shows an approximately 3-5 day period exceeding the level of significance as can be observed in 1969. In the exceptionally dry years, the spectrum of the daily precipitation totals is similar to the continuum of the white noise. Even period lengths exceeding 35 days which reflect the ITC migration cannot be observed in these dry years.

SEASONAL CHANGE OF THE DISTRIBUTION OF THE PRECIPITATION IN THE SAHEL

A change of the seasonal distribution of precipitation (Figure 4) is connected with the precipitation deficits in the years of drought in 1968, 1972 and 1973. In the latitudinal sector $16-19°N$, the shares of September precipitation in the annual total increase particularly at the west coast. In the sector $13-16°N$, the September precipitation increases only near the west coast in the remaining areas only the July and June precipitation increase. This trend is still recognizable in the latitudinal sector $10-13°N$.

The reason for this feature must be seen in the increased occurrence of polar troughs at high altitudes which may lead to Sudano-Saharan depressions (Morell 1973, Maley 1977) as well as to rare but very intensive 'ligne de grain'

319

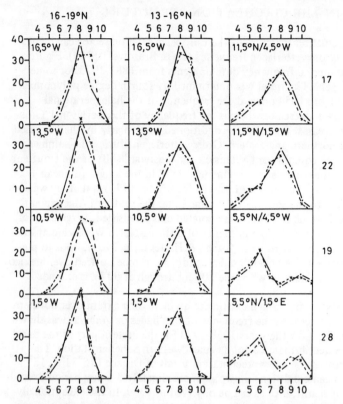

Figure 4. Share in percent of the monthly precipitation totals in the annual total for the long-term mean conditions (solid line) and for the years of drought 1968, 1972 and 1973 (dotted line).

(Riehl 1954, Dhonneur 1974). Particularly in the west coast area, these phenomena occur in a very intensive manner.

CHANGES OF PRECIPITATION IN THE SAHEL AS A FUNCTION OF THE FREQUENCY OF MERIDIONAL CIRCULATION STRUCTURES

Figure 5 shows the precipitation totals in percent of the mean value resulting from the formation of the mean value (1921-1973) for all those years in which, during the summer months, the frequency of meridional 'Grosswetterlagen' exceeds the long-term mean. A T-test shows that in the Sahel area precipitation deviations by more than five to seven per cent from the mean value can be regarded as significant. In the entire northern region of Western Africa,

320

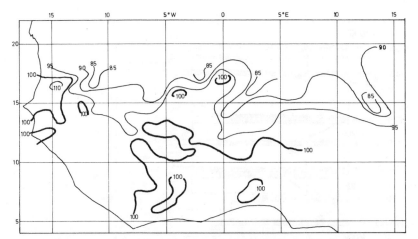

Figure 5. Mean precipitation totals for the months of May-October in percent of the mean (1921-1973) occurring during these months in West Africa connected with the prevailance of meridional large scale weather patterns ('Grosswetterlagen') over Europe.

considerable negative precipitation anomalies occur (Figure 5) when, during the summer months, meridional circulation structures within the Atlantic-European sector prevail.

The historically proven droughts in the Sudan-Sahel area (Plote 1974) are shown in Figure 6 in relation to the temperature changes in Northwest Greenland (Dansgaard et al. 1969). In this connection, a close relationship between the cooling-off phases and the droughts becomes clearly visible.

Cooling-off phases in Greenland are conditioned by an increased frequency of meridional circulation structures which in turn imply precipitation deficits in the Sahel. This relationship apparently is valid beyond the short period of observation analyzed here.

These results correspond very well with the conclusions of Meehl (1978) concerning worldwide circulation patterns linked with temperatures below normal over Greenland: 1) Stronger trade winds in the Atlantic. 2) Highly significant correlation of northeast and southeast trade wind strengths. 3) ITC in Africa, defined by the belt of heaviest precipitation, shifts south. 4) Position of the high altitude extratropical windbelt (300 mb) shifts south over Africa. Most of these results are even true in the two seasons prior to below normal temperatures over Greenland connected with the so-called 'Greenland Seesaw'.

THE RELATIONSHIP TO THE CIRCULATION PATTERN OF THE SOUTHERN HEMISPHERE

Dorize (1974), Leroux (1972) and Vittory (1973) have indicated an influence

321

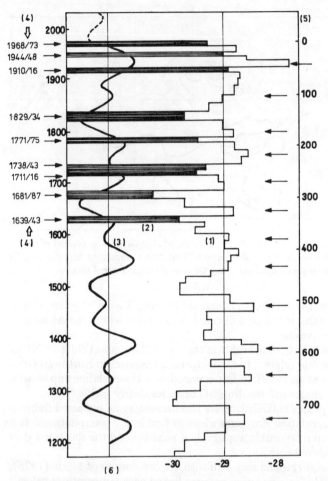

Figure 6. Relationship between cooling phases in Greenland and the occurrence of droughts in the Sahel.
1. $^{18}O/^{16}O$-share of the Camp Century ice core; 2. Cooling phases in Greenland (Johnson et al. 1970); 3. 180-year cycle with 80-year harmonic; 4. Years of occurrence of droughts in the Sahel-Sudan area; 5. Years B.P.; 6. ^{18}O-share in parts per mille.

of the Sahelian precipitation fluctuations by the southern hemisphere circulation. It was Fletcher (1970) who first compiled indicators which show that under long-term conditions a strong southern hemisphere circulation leads to a strong, zonally oriented circulation on the northern hemisphere, whereas a weakened southern hemisphere circulation conditions a meridional circulation on the northern hemisphere. Lamb (1977) and Krishnamurti (et al. 1975)

322

as well, furnish strong proof that the intensity of circulation of both hemispheres undergo almost simultaneous changes. In this connection, a warming-up of the southern hemisphere seems to be linked with a reduction of the pack ice belt around Antarctica and a weakened southern hemisphere circulation. This would result in an increased meridional circulation pattern on the northern hemisphere as can be increasingly noticed since 1955. A global warming due to the CO_2-effect will thus trigger a southern hemisphere temperature increase and a weakening of the southern hemisphere circulation. This will result in an increase of the northern hemisphere meridional circulation pattern which in turn will condition a cooling of the Arctic as is recognizable since 1945. Precipitation deficits in the Sahel due to increased northern hemisphere meridional circulation patterns have already been described.

Simultaneous changes in northern and southern hemisphere circulation patterns could explain, why 31 per cent of the West African precipitation variances are marked by uniform precipitation changes all over the West African continent. In spite of a strong link between the two hemispheres, the variations of trough positions and centers of action on the northern and southern hemisphere as well as the sea surface temperatures in the Gulf of Guinea and at the African west coast (Dorot 1973) imply many modifications of this coherence as can be seen in the eigenvector patterns II-IV.

INDICATIONS FOR THE PAST 500 YEARS

In Figure 7, the sea-level variations of Lake Chad (Teuchebeuf de Lussigny 1969, Maley 1973) are correlated with the precipitation totals in the Santiago District (Taulis 1935, Lamb 1972) and with the ice cover at the coast of Iceland (Koch 1945, Lamb 1972). Dryness in Santiago is linked with a southward shift of the extratropical cyclon tracks, and the enlarged icecover near Iceland with a reduced northern hemisphere circulation. Figure 7 shows that the reduced circulation of both hemispheres coincides with a drastic reduction of the Chad lake-level. The lake-level fluctuation around 1630 which is not very well proved also coincides with a reduced circulation and/or a southward shift of the climatic zones on both hemispheres.

The number of years of drought and famine (Nicholson 1976, Lamb 1977) in Algeria is also quite high at that time which, however, may be conditioned by hunger rather than drought, for bad harvests, due to low temperatures and excessive precipitation, are imaginable. The temporary parallelity between the number of years of famine and drought in Algeria and Senegambia apparently only applies to the west coast region, as the comparison to the Chad lake-level fluctuations shows. For years of predominant meridional circulation patterns positive precipitation anomalies in Senegambia are shown in Figure 5 whereas the entire Sahel to the east is marked by negative precipitation anomalies.

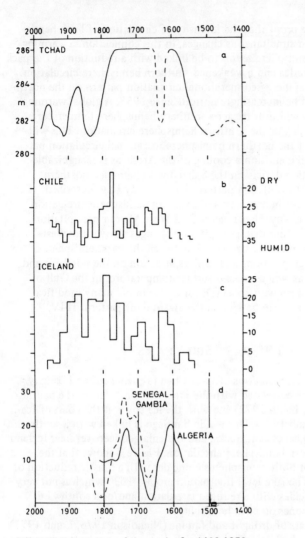

Figure 7. Comparison of time series for 1400-1970:
(a) Lake Chad lake-level variations in 5-year means after Touchebeuf de Lussigny (1969),
   reproduced from Maley (1973);
(b) Precipitation in the Santiago district as compared with today. Each year was weighted:
   (1) very dry; (2) dry; (3) normal; (4) humid; and (5) very humid. The ten year totals
   of these weights are shown from Taulis (1974) and Lamb (1972);
- (c) Polar ice at the coast of Iceland expressed in weeks per year as a 20-year mean after
   Koch (1945) from Lamb (1972);
(d) Frequency of droughts and years of famine in Algeria and in Senegambia as 50-year
   current means shown at ten year intervals from Nicholson (1976) and Lamb (1977).

324

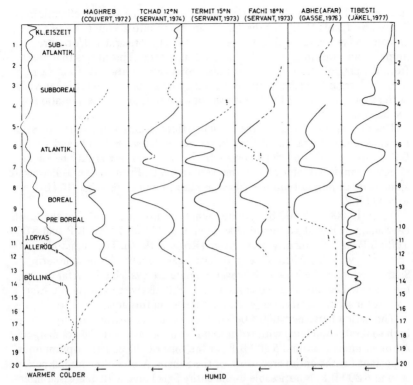

MAGHREB   TCHAD 12°N   TERMIT 15°N   FACHI 18°N   ABHE(AFAR)   TIBESTI
(COUVERT,1972)   (SERVANT,1974) (SERVANT,1973) (SERVANT,1973) (GASSE,1976)   (JÄKEL,1977)

WARMER COLDER   ←   ←   ←   ←   ←
                                              HUMID

Figure 8. Comparison of time series from 0-20 000 B.P. from left to right:
1. Temperature changes of the northern hemisphere (Daansgard et al. 1969, Lamb 1977, Schönwiese 1978);
2. Precipitation totals north of the Sahara according to plant cover in the Maghreb area (Servant 1972, Rognon 1976);
3. Lake Chad lake-level fluctuations (Servant 1972; Rognon 1976);
4. Lake Termi lake-level fluctuations (15°N, 12°W; Servant 1973);
5. Lake Fachi lake-level fluctuations (18°N, 12°W; Servant 1973);
6. Lake Abhé lake-level fluctuations in Afar (Gasse 1976);
7. Humid and arid climatic phases in the Tibesti and the Central Sahara (Jäkel 1977).

THE SOUTHERN SAHARA MARGIN IN THE COURSE OF THE PAST 20 000 YEARS

Figure 8 shows the northern hemisphere temperature fluctuations (Schön-wiese 1978, Johnson et al. 1972, Dansgaard et al. 1969) in comparison to the precipitation fluctuations in the Maghreb area (Couvert 1972, Rognon 1976), the Chad lake-level fluctuations (Servant 1974) as well as the two lakes Termi and Fachi located further north (Servant 1973). Furthermore, lake-level fluc-

325

tuations of Lake Abhé in the Afar region (Gasse 1976) and the climatic curve derived by Jäkel (1978) for the Tibesti are taken into account. All of the climatic curves, compared to the present, reflect more arid and/or more humid conditions in a little substantiated time scale. This means that apart from the precipitation, the temperature trough evaporation is effective as well. With regard to lake-level fluctuations, topographic pecularities as well as the source regions of the tributaries may complicate an interpretation of the curves.

Despite these restrictions, the humidity fluctuations at the southern Sahara margin show a general parallelism. In the period between 20 000 and 12 000 B.P., extreme arid conditions prevail. In the Afar area, the aridity becomes effective somewhat later than in the western Sahel. From 16 000 B.P. onward an increase of humidity in the Tibesti is marked, which remains effective up to 8 000 B.P.

From 12 000 to 8 000 B.P., more humid conditions prevailed at the southern Sahara margin. Two humid phases between 11 000-12 000 B.P. and 8 000-9 000 B.P. have frequently been shown (Rognon 1976). The humid phase coinciding with the Alleröd seems to have been effective in the southern Sahara including the Tibesti; however, no data are available for Afar during this period. The humid phase around 8 500 B.P. likewise exists in the Afar, however it seems to start earlier and last longer in this area.

The period between 8 000-4 000 B.P. is marked by a strong aridity increase in the Tibesti coinciding with reduced humidity at southern Sahara margin. A humid interval around 5 500 B.P., at the time of the climatic optimum, is important for the entire area south of the Sahara. A preceding humid period around 6 800 B.P. is noticeable in the entire Sahel area with decreased intensity towards the North, however, it coincides with great aridity in the Tibesti (Figure 8). The Chad lake-levels around 5 500 B.P. are very high and are followed by an arid period after 4 500 B.P. which is strongly marked particularly in the Tibesti.

During the Period between 4 000-1 500 B.P., a general aridity increase in the Sahel and Tibesti is modulated by minor fluctuations. A drastic aridity increase in the Tibesti introduces the conditions prevailing today.

AN ATTEMPT TO EXPLAIN THE CLIMATIC FLUCTUATIONS AT THE SOUTHERN SAHARA MARGIN

Rognon (1976), Maley (1973) and Leroux (1975) have deduced mechanisms explaining the spatial and temporal changes of the southern Sahara margin. As Maley (1973) was able to show northern hemisphere temperature decreases in the Holocene are linked with increased aridity in the Sahel. This relationship was also confirmed in this study for the most recent droughts.

Figure 9 shows the results of the icecore analyses in a simplified manner for northwest Greenland and the Byrd station/Antarctica as presented by

326

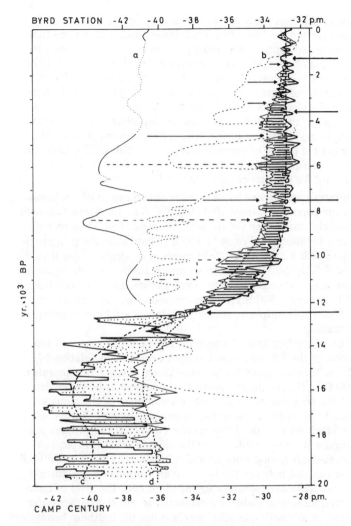

Figure 9. Comparison of time series for 0-20 000 B.P.:
(a) Lake Chad lake-level fluctuations (Servant 1974);
(b) Climatic phases in the Tibesti and the Central Sahara based on morphologic findings
    (Jäkel 1974);
(c) $^{18}O/^{16}O$-share of the Camp Century, Greenland ice core (Johnson et al.1972);
(d) $^{18}O/^{16}O$-share of the Byrd Station, Antarctica ice core (Johnson et al. 1972).

327

Johnson (et al. 1972). This figure shows the envelope of the positive and negative deviations from the long term trend. While smaller deviations from the general trend are still speculative, more pronounced features as well as the general trend may be regarded as relatively certain (Johnson et al. 1972). The two curves have been correlated in Figure 9 in such a manner that the two approximated trends intersect in the present. Taking the isotopic composition of the polar glacier ice ($^{18}O/^{16}O$ ratio) as an indicator for the temperature of formation of the snow at the time of deposition on the surface of the ice sheet, increasing values of this ratio reflect a warming trend of the climate during the period of deposition (Johnson et al. 1972). Thus, the present temperature relationship between the Arctic and the Antarctic becomes a standard for the past epochs presented.

Five time intervals are marked by changed northern and southern hemisphere temperature conditions: from 20 000-12 500 B.P., the Arctic (curve c) is almost exclusively cooler than the Antarctic; compared with the present, the reverse is true for the period from 12 500 B.P. until now. From approximately 12 500-7 500 B.P., the Antarctic becomes increasingly colder than the Arctic and from 7 500-3 500 B.P., the temperature difference — disregarding smaller fluctuations — remains almost the same and steadily decreases from 3 500-1 500 B.P. Around 1 500 B.P., the temperature contrast between the northern and southern hemisphere, still existing today, established, disregarding minor fluctuations.

A comparison of the lake-level fluctuations of Lake Chad (curve a) and the climatic curve of the Tibesti (curve b) shows a remarkable relationship which is similar to the one shown for the most recent climatic development.

Until 12 500 B.P., the Arctic, compared to the present, is by far colder than the Antarctic, resulting in a year-round displacement of the climatic zones to the south; in other words, Saharan conditions prevail throughout the year in the Sahel. As a result of this southward shift of the climatic zones, the Tibesti temporarily comes under the influence of the extratropical circulation and is relatively humid since 16 000 B.P. onwards. From 12 500 B.P. the temperature contrast between the northern and southern hemisphere is reversed due to northern hemisphere warming (Figure 9). The level of Lake Chad reaches maximum values when the temperature contrast between the Arctic and Antarctic intensifies, in other words, when the southern hemisphere climatic zones are displaced equatorward. The seasonal changes of position of the southern hemisphere centres of action are very likely to have been smaller at that time than today. As a result, the southwest monsoon, even during the winter months, penetrated deeper into the African continent than today. According to McIntyre (1974), the July temperatures around 17 000 B.P. in the Gulf of Guinea were about the same as the present, only in the winter months the temperatures were by 4°C lower than in the present. During the summer months of this period, heavy monsoon rainfall can be assumed, whereas in the transitional and winter months, rainfall generated by Sudano-Saharan depressions, as a result of the southwest-monsoon's deep penetration into the North African continent, can be expected.

328

Around 7 500 B.P., the temperature contrast between the northern and southern hemisphere is reduced. More arid conditions occur in the Sahel and in the Tibesti and continue during the following periods. The southern hemisphere circulation is displaced poleward, in other words, the monsoon circulation is less effective. The resulting lack of tropical rainfall cannot be compensated in the Tibesti by extratropical rainfall. As a result of the continuous reduction of the temperature contrast between the northern and southern hemisphere one can observe extremely arid conditions in the Tibesti. Although a short-term southern hemisphere cooling around 7 000 B.P. is still effective in the Chad, which can be seen by increased humidity, it does not affect the more northern regions, as can be seen in the curve of Tibesti. Only around 6 000 B.P., a drastic cooling of the southern hemisphere and a simultaneous warming of the northern hemisphere brings about an intensification of the temperature contrast between the northern and southern hemisphere. The result is an exceptionally high humidity in both the Tibesti and the Chad. The general temperature increase is very likely to have kept the lake-level lower than in the period before 7 000 B.P. which is attributable to increased evaporation.

Around 4 500 B.P., the Antarctic warming results in extreme aridity in the Chad and the Tibesti, and an intensification of the temperature contrast around 4 000 B.P. between the northern and southern hemisphere again leads to more humid conditions in the Tibesti. The Chad-curve is poorly dated in this period. The humidity maximum around 3 000 B.P. in this curve may possibly be dated around 3 500 B.P., as it applies to the Tibesti. All following humidity declines in the Tibesti curve show a temporal coincidence with an Antarctic warming and an Arctic cooling to within the accuracy of the dating.

The explanation attempt presented for discussion here provides indications that not only the short term but also the long-term humidity fluctuations at the southern Sahara margin are conditioned by the changes of the temperature contrast between the Arctic and the Antarctic. By reversing this contrast related to present conditions, as for example prior to 12 500 B.P., extremely arid conditions at the southern Sahara margin are producted. Under present conditions, this area is a region of great climatic instability.

CONCLUSION FOR THE PRESENT

If the relationships presented here are correct, it is quite possible that the CO_2-effect becomes, in the future, responsible for an increased occurrence of droughts in the Sahel. With regard to the future Arctic temperature development, the steady decrease of the temperature contrast between the northern and southern hemisphere, which can be observed since 12 500 B.P., deserves greater attention than until now.

REFERENCES

Couvert, M. 1972. Variations paléoclimatiques en Algérie. *Libyca 20:*45-48.

Dansgaard, W., S.J.Johnson, J.Moller & C.C.Langway 1969. One thousand centuries of climatic record from Camp Century on the Greenland ice sheet. *Science 166:*377-381.

Dhonneur, G. 1974. Essai de synthese sur les theories des lignes de grain en afrique occidentale et centrale. *Publ. Dir. Exploit. Météorol. Dakar, 38.*

Dorize, L. 1974. Oscillation Pluviométrique récente sur le Bassin du Lac Tchad et al circulation atmosphérique générale. *Rev. Géogr. phys. Géolog. 16:*393-420.

Dorot, M. 1973. Contribution à l'étude des interactions océan-atmosphère sur les cotes de l'ouest Africain. *Publ. Dir. Exploit. Météorol. Dakar, 24.*

Flohn, H. 1971. Tropical circulation pattern. *Bonner Meteorol. Abh. 15.*

Gasse, F. 1976. Utilisation des milieux lacustres en milieu desertique pour la reconstitution des oscillations climatiques. *Holocenes. Bull. Ass. Géogr. France 433:*69-75.

Hammer, R.M. 1976. Rainfall characteristics in eastern Sahel. *Nature 263:*48-49.

Hastenrath, S.L. 1967. Rainfall distribution and regime in Central America. *Arch. Meteor. Geophys. Biokl., B 15:*201-241.

Hess, P. & H.Brezowski 1969. Katalog der Groswetterlagen Europas. *Ber. Dt. Wetterdienstes, Offenbach 113*(5).

Jäkel, D. 1977. Eine Klimakurve für die Zentralsahara. In: Museum der Stadt Köln, *Sahara, 10 000 Jahre zwischen Weide and Wüste.* Köln, pp.382-396. (cf. also *Palaeoecology of Africa 11,* 1979).

Johnson, S.J., W.Dansgaard & H.B.Clausen 1972. Oxygen isotope profiles through the Antarctic and Greenland ice sheets. *Nature 235:*429-434.

Klaus, D. 1975. Niederschlagsgenese und Niederschlagsverteilung im Hochbecken von Puebla-Tlaxcala. *Bonner Geogr. Abh. 53.*

Koch, L. 1945. The east Greenland ice. *Medd. om Grönland, 130*(3).

Krishnamurti, T.N., C.E.Levy & H.L.Pau 1975. On simultaneous surges in the Trades. *J. Atm. Sci. 32:*2367-770.

Kutzbach, J.E. 1970. Large-scale features of monthly mean Northern Hemisphere Anomaly Maps of sea level pressure. *Month. Weath. Rev. 98*(9):708-716.

Lamb, H. 1972. *Climate: Present, past and future,* Vol I. London.

Lamp, H. 1977. *Climate: Present, past and future,* Vol. II. London.

Leroux, M. 1972. La dynamique des precipitation en Afrique occidentale. *Publ. Dir. Exploit. Météorol., Dakar 23.*

Leroux, M. 1975. La circulation générale de l'atmosphére et les oscillations tropicales. *Climatologie tropical, Dyon R.C.P.269.*

Maley, J. 1973. Un nouveau mécanisme des changements climatiques aux basses latitudes. *Bull. Liais. Ass. Sénégal. Et. Quatern., Dakar 37-38:*31-40.

Maley, J. 1977. Palaeoclimates of Central Sahara during the early Holocene. *Nature 269:* 573-577.

McIntyre, A. 1974. The CLIMAP 17 000 years B.P. North Atlantic Map, Research Publication, CRU RP2, Norwich, pp.41-47.

Meehl, G.A. 1978. Tropical teleconnections to the seesaw in winter temperatures between Greenland and Northern Europe. *Inst. Arctic and Alpine Research, Univ. Colorado, Occ. Paper 28.*

Morell, M. 1973. Notes sur deux situations Météorologiques remarquables observées en Tchad. *Publ. Dir. Exploit. Météorol., Dakar.*

Nicholson, S.E. 1976. A climate chronology for Africa: Synthesis of geological, historical, and meteorological information and data. Thesis, Univ. Wisconsin.

Plote, H. 1974. L'Afrique sahélienne se desseche-t-elle? Bur. Rech. Géol. et minières, Serv. géolog. national., Orleans, France.

Osman, O.E. & S.Hastenrath 1969. On the synoptic climatology of summer rainfall over central Sudan. *Arch. Meteor. Geophys. Biokl. B 17:*297-324.

Riehl, H. 1954. *Tropical meteorology.* New York, Toronto, London.

Rognon, P. 1976. Essai d'interprétation des variationes climatiques au Sahara depuis 40 000 ans. *Rev. Géogr. phys. Géol. dyn. 18:*251-282.

Schönwiese, C.D. 1978. Zum aktuellen Stand rezenter Klimaschwankungen. *Meteor. Rundschau 31:*73-84.

Servant, M. & S.Servant 1973. Le Plio-quaternaire du bassin Tchad. In: *Le Quaternaire, 9th Congr. Int. Union Quat. Res. Comité* national Français de l'INQUA, 169-175.

Servant, M. 1974. Les variations climatiques des regions intertropicales du continent africain depuis la fin du Pleistocène. *J. hydraul. Paris, Rapport 8.*

Taulis, E. 1974. *De la distribution des pluies au Chile.* Materiaux pour l'étude des calamités, 33. Soc. Géogr., Geneva.

Thompson, B.W. 1965. *The climate of Africa.* Nairobi, London, New York.

Touchebeuf de Lussigny, P. 1969. *Monographie hydrologique du Lac Tchad.* Serv. Hydr. ORSTOM, Paris.

Vittori, A. 1973. Notes sur la secheresse au Sénégal en été 1972. Publ. Dir. Exploit. Météorol., Dakar 28.

331

LATE QUATERNARY CHANGES IN LAKE-LEVELS AND DIATOM ASSEMBLAGES ON THE SOUTH-EASTERN MARGIN OF THE SAHARA

F. GASSE

Ecole Normale Supérieure, Fontenay-aux-Roses

1. INTRODUCTION

This paper reviews recent research into the late Quaternary fluctuations in lake environments along the southeastern margin of the Sahara. The interpretation is based upon several continuous and well-dated biostratigraphic sequences. We have attempted to compare the evolution of the Ethiopian lakes (7-14°N, 38-43°N) with that of some equatorial lakes in East Africa which lie between 2°N-4°S and 29-36°E. Special emphasis is given to changes in diatom assemblages which provide evidence of major hydrologic and climatic fluctuations.

2. THE DIVERSITY OF THE LAKES UNDER DISCUSSION

The lakes selected here are located on Figure 1. Some present-day characteristics of major importance for the diatom flora are summarized in Table 1.

2.1 *Lakes recharged from the Ethiopian uplands*

2.1.1 *Lake Abhé, in a desert basin of low elevation.* Lake Abhé occupies a closed tectonic basin in the Central Afar. It is situated at the downstream end of the permanent Awash River which flows from the Ethiopian uplands. The present lake is hyperalkaline and currently shrinking. The results derive from the analysis of a 50 m core taken from the present shore, supplemented by the study of numerous strandlines and stratigraphic sequences on the margin of the basin (Gasse 1975, 1977a, Gasse & Delibrias 1976). Based upon 80 ^{14}C dates, the complex history of Lake Abhé is one of the most complete for East Africa.

A major problem in interpreting the environmental changes is the origin of the water. How far are the fluctuations of Lake Abhé a result of climatic

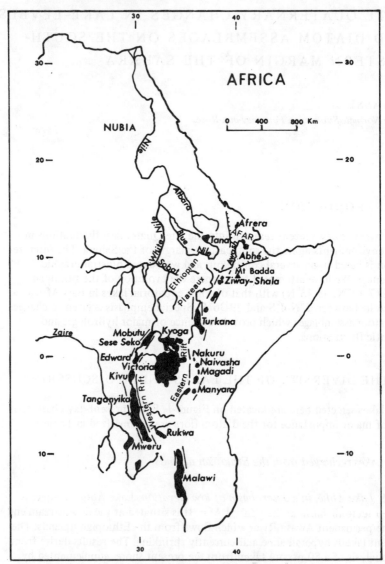

Figure 1. Location of the East African lakes under discussion here.

changes in the Ethiopian plateaux? The water budget of Lake Abhé depends on many factors. The inputs are the Awash inflow, the local runoff, the precipitation (P) on the lake surface and the groundwater influx. The outputs are the evaporation (E) from the lake surface and seepage through the fractured lake substratum. Despite the complexity of these hydrological factors,

334

there is evidence to indicate that Lake Abhé recorded the climatic fluctuations within the Afar itself as well as in the Ethiopian uplands. The synchroneity in the major lake-level oscillations between Lake Abhé and Lake Ziway-Shala in the Ethiopian Rift (Figure 2) marks the immediate effects of such factors as precipitation and runoff rather than the more gradual effect of underground circulation from the Ethiopian highlands. But the Awash inflow into Lake Abhé depends mainly on its nearest tributaries flowing from the Ethiopian escarpment and these present conditions can be extrapolated to the late Quaternary (Gasse et al. 1974, 1979). An increase in the Awash inflow therefore reflects a greater P/E ratio in low-lying areas. A similar inference may be drawn from the Holocene history of Lake Afrera (Gasse 1975), situated in Northern Afar and fed almost entirely by groundwater infiltrated along the foot of the Ethiopian escarpment. Numerous fossil channels of Early Holocene age indicate considerable fluvial activity in the Afar at this time.

2.1.2 *The Ziway-Shala system, in a semi-arid basin of medium elevation.*
The Ziway-Shala basin lies in The Main Ethiopian Rift. It is mainly fed by rivers which flow from the Arussi Mountains and the Ethiopian plateaux. This now closed basin is occupied by a chain of four lakes (Ziway, Langano, Abiyata, Shala) of increasing salinity and alkalinity (Table 1). Its limnological evolution has been studied in detail by Sreet (1977, 1978). The four lakes were united during the wet episodes to form a single water body, Lake Ziway-Sahal, controlled by overflow into the Awash. Diatom analysis from a core

Table 1. Some present-day characteristics of some East African lakes, for which fossil diatom assemblages have been studied.

Lake	Surface altitude (m.a.s.l.)	Total solids (mg/l)	Alkalinity (meq/l)	Silica (mg/l)	Drainage
Ziway	1 636	354	3.9	47	open
Langano	1 582	1 644	15	54	open
Abiyata	1 578	8 358	3 240	55.6	closed
Shala	1 558	16 770	3 300	130	closed
Abhé	240	160 000	1 059	416	closed
Afrera	−80	159 000			closed
Victoria	1 134	118	1.4	9	open
Mobutu	619	580	9.1	3.4	open
George	916	264	2.2	32	open
Kivu	1 460	1 020	18.1	71	open
Naivasha	1 890	152	4.4	63	closed
Manyara	960	34 600	806	1.2	closed

Sources: concerning the Ethiopian lakes, see Gasse & Street 1978, other lakes: Talling & Talling 1965.

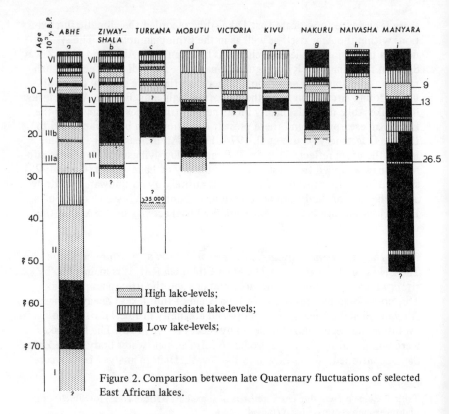

Figure 2. Comparison between late Quaternary fluctuations of selected
East African lakes.

taken at the edge of the present Lake Abiyata adds to our knowledge of the
ecological evolution from about 26 500 to 6 000 B.P. (Gasse & Descourtieux
1979).

2.1.3 *An Afro-alpine peat-bog on Mt Badda*. The Holocene environments of
the high Ethiopian mountains are deduced from the diatoms of a peat-core
(Gasse 1978) collected by A.Street at 4 000 m from a small glacial basin on
Mt Badda. The 3 m long peat-core is dated from 11 500 B.P. to present. The
diatom flora registers a cyclic alternation of truly aquatic, alkaline conditions,
and terrestrial, acidic environments. In comparison to the European peat-bogs,
the acidic stages could record periods of stable or dry climate, and the alkaline peaks may reflect episodes of climatic change or wetter conditions.

2.1.4 *Southern Lake Turkana*. Further south, runoff from the Ethiopian plateaux drains towards the Turkana Basin which is fed by the Omo River. Good
evidence for a complex series of lake-level changes has been demonstrated.
The Early Holocene, and probably the Middle Holocene highstands induced

336

overflow into the White Nile (Bützer 1971, 1976, Bützer et al 1972). At these times, Lake Turkana was linked to Lake Chew Bahir (Grove et al. 1975) and Lake Suguta (Truckle 1976).

2.2 The equatorial lakes

2.2.1 *The lakes of the White Nile system.* Lake Victoria is fed by rivers draining the well-watered Mt Elgon, the Rwanda-Burundi highlands and the Mau escarpment. Water-level and environmental fluctuations are deduced from the sediments and pollen analyses of a core collected in Pilhington Bay (Kendall 1969), supplemented by later investigations by Livingstone (1976). The Victoria Nile flows from Lake Victoria to Lakes Kioga and Mobutu.

Major tributaries of Lake Mobutu are the Victoria Nile, and the Semliki River which originates from Lake Edward. Lake Mobutu overflows into the White Nile. Diatom analysis of a 10.6 m core spanning the last 28 000 years provides a record of the ecological evolution of the lake (Harvey 1976).

Lake Edward and Lake George are little-known, although diatom evidence indicates Late Holocene environmental changes in the later (Haworth 1977).

2.2.2 *Lake Kivu.* Lake Kivu is an open lake belonging to the Tanganyika system. A nearly continuous 14 000 year record is discussed by Degens & Hecky (1973) and Hecky (1978).

2.2.3 *Lakes situated in the Eastern Rift Valley.* The saline-alkaline Lake Manyara occupies a closed basin. Holdship (1976) studied a 56 m core and established a continuous 55 000 year climatic sequence. Interpretation is based upon diatom flora and mineralogic analysis.

Holocene history of Lake Naivasha is deduced from diatom and sedimentologic study of a 28 m core, obtained near the centre of the small crater lake connected to the main lake (Richardson & Richardson 1972).

Dating of the shorelines in the Nakuru basin shows the Holocene lake-level oscillations (Butzer et al. 1972).

Comparison between the evolution of these different lakes is difficult. Their water-level fluctuations are summarized in Figure 2. Interpretation must take into account their altitudinal, topographical and geological setting. Some lake basins lack surface outlets but could overflow during the highstands. Other lakes overflow, but fall in the P/E ratio could bring lake level below the outlet. Water chemistry largely depends on the drainage.

The ecological data presented here are mainly based on the diatom assemblages in the sediments. Interpretation is founded upon comparisons with the diatom communities of existing African lakes, summarized in Richardson (1968, 1969), Hecky & Kilham (1973), Gasse (1975), Harvey (1976), Holdship (1976) and Richardson et al. (1978). Diatom assemblages are sensitive indicators of local conditions and an overview of the fossil flora is therefore

337

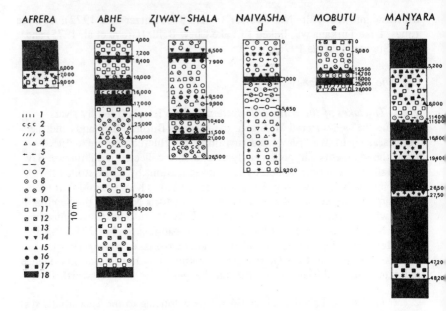

Sources: a, b: Gasse (1975); c: Gasse & Descourtieux (1979); d: Richardson & Richardson (1972); e: Harvey (1976); f: Holdship (1976).

Legend: 1-17: Characteristic diatom species.

1-4: Group I, indicators of cool, or oligotrophic, or acidic conditions (1. *Melosira distans;* 2. *Cocconeis thumensis, C.diminuta, Achnanthes clevei, Pinnularia* spp., etc.; 3. *Stephanodiscus carconensis;* 4. *Cyclotella ocellata*);

5-9: Group II, indicators of dilute water, low alkalinity and/or high silica content (5. *Synedra acus;* 6. *S.berolinensis;* 7. *Melosira ambigua;* 8. *M.granulata;* 9. *M.agassizii, M.granulata* v. *valida, M.nyassensis*);

10-13: Group III, indicators of moderately dilute water, medium alkalinity (at least for 11: low silica content) (10. *Nitzschia fonticola;* 11. *Stephanodiscus astraea* and varieties; 12. *S. hantzschii;* 13. *Nitzschia bacata, N.aequalis, N.stricta*);

14-17: Group IV, indicators of water with medium to high alkalinity and salinity (14. *Cyclotella meneghiniana, C.striata;* 15. *Thalassiosira rudolfi, T.faurii;* 16. *Rhopalodia gibberula, Anomoeoneis sphaerophora, Campylodiscus clypeus,* etc.; 17. *Navicula elkab, Nitschia* spp.);

18: No diatoms.

For each lake, the dates of the major changes are mentioned on the right. Most of these dates are based on interpolation of mean sedimentation rates measured in the dated parts of the sections.

Figure 3. Gross diatom stratigraphy of some East African lakes.

difficult. We have, however, attempted to draw some conclusions about the regional paleoenvironments and climatology. The discussion is based on Figure 3, which only takes into account some points of major interest.

3. MAIN STAGES OF THE LATE PLEISTOCENE LIMNOLOGICAL EVOLUTION

3.1 *The period prior to 30 000 B.P.*

Owing to the diminished amounts of ^{14}C, the problems of establishing a time-scale are great during this period and correlations between the different basins remain hypothetical.

In Lake Abhé, two lacustrine phases, Abhé I and Abhé II, are evident. During Abhé I (70 000-60 000 B.P.?), the diatom flora (*Stephanodiscus astraea, Nitzschia* spp.) indicates a warm and moderately alkaline environment. Abhé II (55 000?-39 000, 30 000 B.P.) was an episode of biochemical sedimentation. The lake (with tropical *Nitzschia* spp., *Stephanodiscus hantzschii* v. *pusilla, Synedra rumpens* v. *neogena*) was warm, with high alkalinity and productivity. Synchronous with a minor regression towards 39 000-30 000 B.P., a change to clastic sedimentation is accompanied by a decrease in productivity which, during a regressive stage, must be attributed to a fall in water temperature. It is registered by the spread of temperate-oligotrophic species (such as *Cyclotella ocellata*). This cooling has been recently confirmed by pollen evidence from the Abhé core (Bonnefille, personal communication 1979).

Two Late Pleistocene lacustrine episodes (Ziway-Shala I, Ziway-Shala II) (Street, personal communication 1979) also occurred before 27 000 B.P. in the Ethiopian Rift. Ziway-Shala II ended around 27 000 B.P. It is characterised, in the Abiyata core, by the abundance of *Stephanodiscus dubius* (Descourtieux personal communication 1979), which suggests a relatively high salinity during the highstand.

Between 48 200 and 47 200 B.P., Lake Manyara was less saline than it is now. The development of a diatom flora (with *Nitzschia fonticola* and *Cyclotella meneghiniana* dominant) reflects a moderate alkalinity (10 to 100 meq/l) (Holdship 1976). This water dilution indicates a humid phase which could be partly equivalent to Abhé II.

3.2 *30 000 - 21 000 B.P.*

3.2.1 *Lake-level oscillations.* In Lake Abhé, the third lacustrine phase (Abhé III) extends from 30 000-29 000 B.P. to 17 000 B.P., but a minor regression towards 20 800 B.P. separates Abhé IIIa and Abhé IIIb. The Abhé IIIa lake attained the highest level reached during the Late Quaternary around 24 000-23 000 B.P. (Gasse 1977a). Equivalence between Abhé IIIa and Ziway-Shala III

(Figure 2) is firmly established, although Ziway-Shala III seems to have been shorter (26 500 to 21 000 B.P.) (Street 1978). In the Turkana basin, no lacustrine highstands are known between > 35 000 B.P. and 9 500 B.P., but non-lacustrine deposits from this period have so far been dated (Butzer 1972).

Lake Mobutu was an open and dilute lake between about 28 180 and 25 000 B.P., except during a brief interval towards 26 500 B.P., such as in Lake Ziway-Shala. The following episode of closure extends from 25 000 to 17 800 B.P. (Harvey 1976). Lake Manyara supported moderately saline conditions and was close to or above its modern level between 29 000 and 24 500 B.P., with a least saline phase dated from 27 000 to 26 500 B.P. (Holdship 1976). The Manyara chronology is consistent with limnological evidence from the crater Lake Ngorongoro, which was at least 2,5 m above its present level between 27 900 and 24 400 B.P. (Hay 1976).

Pronounced parallelism is apparent between the waterlevel fluctuations of the lakes in question. At least between 26 500 and 25 000 B.P., all these lakes were high or even very high. Some differences in the ^{14}C ages of the transgressive and regressive stages could be due to the uncertainties of ^{14}C dating. But they could also betray the influence of the latitude, the basin topography and different inflows. For instance, in Lake Abhé, the lowstand from 20 800 B.P. did not result in complete desiccation, although a marked ecological change reflecting regional environmental fluctuations is recorded at this time. This may be the result of groundwater inflow into this very low basin.

The generalized highstand around 26 500-25 000 B.P. suggests an increase in the P/E ration in Ethiopia as well as in equatorial East Africa. The remarkable amplitude of the water-level rise undoubtedly implies higher precipitation. Hypothetical temperature-evaporation changes may be deduced from diatom studies.

3.2.2 *Diatom record*. In Lake Abhé, the temperate conditions which took place around 39 000-30 000 B.P. persisted until 25 000 B.P. But tropical diatoms (*Melosira granulata* v. *valida*, *M.nyassensis*, *M.agassizii...*), which could support cooler conditions than those of today, multiplied between 25 000 and 21 000 B.P. Ecological conditions deteriorated rapidly around 21 000 B.P., and a diatom community with higher salinity requirements (*Cyclotella meneghiniana, Rhopalodia gibberula, Nitzschia frustulum*) colonized the lake during a short-lived interval only registered in a few centimeters of the Abhé core (Gasse 1975).

The extensive lakes Abhé IIIa and Ziway-Shala III had closely comparable water chemistry recorded by the same phytoplanktonic type, in spite of the relative abundance of the temperate-oligotrophic elements which partly reflect the altitudinal gradient in the highest basin (Gasse & Descourtieux 1979). The abundance of *Melosira* (99 per cent in Lake Abhé) would indicate a low alkalinity (< 10 meq/l) and a high silica content (> 7 mg/l) (Richardson et al. 1978). Ziway-Shala III and Abhé IIIa correspond to the most dilute stage recorded for the whole Late Quaternary in these Ethiopian basins. They also

340

contained the highest frequencies of cool-water species. This lacustrine stage would reflect lower temperatures involving lower evaporation. During the Ziway-Shala III regression, increasingly cold conditions are inferred from the appearance of three palearctic mollusca (Gasse & Street 1978) and a peak (6 per cent) of cold-water diatoms (*Cocconeis diminuta, C. thumensis, Achnanthes clevei*...) in the uppermost Ziway-Shala beds. This regression may result from a parallel decrease in precipitation and in evaporation associated with increasingly cool temperatures.

In Lake Mobutu, the highstands dated from 28 180 to 26 500 B.P. and 26 500 to 25 000 B.P. are characterized by rich diatom assemblages (Harvey 1976). *Stephanodiscus astraea* v. *minutula* is the most numerous species, becoming secondary to *S. carconensis* between 26 500 and 25 000 B.P. *Melosira agassizii,* replaced by *M. granulata* v. *angustissima* after 26 500 B.P., is also of major importance. The surface waters had low to moderate salinity and alkalinity. Harvey (1976) concludes from its diatom study that 'it is doubtful that low temperatures prevailed'. However, the striking spread (up to 49 per cent) of *S. carconensis* suggests to me cooler conditions than those of today if we take into account its modern distribution (Gasse 1977b). In Africa, *Melosira distans* is sometimes abundant in montane lakes but is scarce elsewhere. It is an oligotrophic, oligohalobe to halophobous species (Hustedt 1957, Cholnoky 1968, Abbott & van Landingham 1973) and is considered as a stenothermic cold-water species by Foged (1964). Its frequencies observed here (5,4 to 9,7 per cent) seem to me high enough to be significant. I, therefore, attribute the Mobutu assemblages, between 28 000-25 000 B.P., to a climatic tendency towards temperate conditions.

In Lake Manyara, the water-level rise which began around 29 000 B.P. resulted in renewed development of diatoms between 27 000 to 26 500 B.P. But the alkalinity did not fall below 10-100 meq/l (Holdship 1976), and salinity remains the dominant factor.

When the salinity did not mask the effects of temperature on the diatom flora, the 26 500-25 000 B.P. highstand seems to have been associated both to higher precipitation and lower temperatures. Such conditions could be favourable to the development of glaciers on the surrounding highlands. Small Late Pleistocene glaciers capped the highest Ethiopian summits (Potter 1976, Hastenrath 1977, Messerli et al. 1977), and their advance could correspond to the Abhé III episode, especially to the 30 000-25 000 B.P. interval (Gasse et al. 1979). Such a hypothesis can also be suggested for the last African equatorial glaciation, even if the closing phase was rather dry (Livingstone 1962, 1979). In this case, the development of the equatorial glaciers would involve a temperature lowering of 4-6°C (Osmaston 1965). But it would require a decrease in temperature of 7-10°C if it took place during the arid phase dated from 25 000 to 18 000 B.P. in Lake Mobutu (Livingstone 1979), for which a 29 per cent rainfall reduction has been estimated (Harvey 1976). Further investigation in mountainous environments would be necessary to provide reliable information on the age of the glacial advance.

341

3.3 21 000 - 12 500 B.P.

3.3.1 A marked arid phase in Ethiopia. In Lake Abhé, the water-level rose again after 20 800 B.P. (Abhé IIIb) but did not attain the high level of Abhé IIIa. Detrital components played an increasingly important role in the sedimentation and the diatom content remained very low. The diatom assemblage was dominated by *Stephanodiscus astraea* which presumably required higher alkalinity than the preceding *Melosira* flora, and littoral species were diversified. The lacustrine environment rapidly deteriorated and Lake Abhé was dry, or nearly so, from 17 000 to 10 000 B.P. In the Abhé core, this arid episode is recorded by a deep soil dated from 16 400-16 100 B.P. Nevertheless, this soil contains some diatomaceous levels. The littoral and saline-water assemblages are characterized by the development of forms which are only living, at the present-day in Ethiopia, in mountainous biotopes. They could also indicate a parallel decrease in precipitation and in temperature (Gasse & Delibrias 1976), as mentioned above for the Abhé II and Ziway-Shala III regressions.

This terminal Pleistocene arid phase also occurred in the Main Ethiopian Rift, where the lakes of the Ziway-Shala basin have shrunk to about their present level between 21 000 and 12 000 B.P. (Street 1978).

3.3.2 Climatic fluctuations at lower latitudes. Lake Mobutu was open between 17 800 and 14 000 B.P. The diatom assemblage (Harvey 1976) was dominated by *Melosira*, with *M.agassizii* (47 to 71 per cent), *M.granulata* and its variety *angustissima*. This flora resembled that of Ziway-Shala III and Abhé III. Alkalinity was sufficiently low to cause *Melosira* to become dominant. The succeeding portion of the core contains few or no diatoms. It represents a period of closure which extends from about 14 000 to 12 000 B.P., except during a short-lived interval dated from 12 900-12 800 B.P.

Lake Victoria was also a basin of internal drainage from at least 14 700 until 12 500 B.P., and this implies a ⩾ 10 per cent reduction in the P/E ratio in the area (Kendall 1969). *Stephanodiscus astraea* was more important in the phytoplankton community than it is in the modern one and would indicate higher alkalinity. Livingstone (1976) has demonstrated a lake-level fall of at least 75 m below its modern level. A synchronous low-stand also occurred in Lake Kivu (Degens & Hecky 1973).

Lake Manyara experienced higher dilution than today between 22 000 and 16 000 B.P., with a most dilute phase from 19 400 to 16 600 B.P. (Holdship 1976). The sudden appearance of *Cyclotella meneghiniana* and its associated low alkalinity, without any intervening stages of dilution, suggests a remarkably rapid rise in lake-level around 19 400 B.P. But the biotope rapidly deteriorated. The following diatom community (with *Nitzschia fonticola* and *N.* spp. dominant) reflects salinity similar to that of today, and also rapidly fluctuating. The diatoms disappeared towards 16 600 B.P. and saline conditions persisted from 16 000 to 12 500 B.P.

Clear differences between the evolution of the lakes appear during the period 21 000-12 500 B.P. Before 16 000-14 000 B.P., the humid phase registered in Lake Mobutu and in Lake Manyara was important enough to involve a lake water dilution at least as marked as during the preceding high lake-level stage (28 000-25 000 B.P.). On the contrary, there was too little water to sustain favourable lacustrine conditions in the Abhé and Ziway-Shala basins. In Ethiopia, there is a definite trend towards increasing aridity between 20 000 and 16 000 B.P. This trend culminated towards 17 000-16 000 B.P. when the Ethiopian lakes where close to, or below, their present levels. The pronounced divergence between the Ethiopian and the equatorial lakes imply substantial contrasts in the climates of these two regions between 20 000 and 14 000 B.P.

Nevertheless, very low lake-levels seem very general towards the close of the Late Pleistocene. From 14 000 B.P. until 12 500 B.P., extremely arid conditions characterized at least that part of East Africa under discussion here.

4. RE-ESTABLISHMENT OF WETTER CONDITIONS AND HOLOCENE LIMNOLOGICAL CHANGES

4.1 *12 500 - 10 000 B.P.*

Following the terminal Pleistocene arid phase, wetter conditions became re-established throughout the Ethiopian and equatorial lake districts between 12 500 and 10 000 B.P.

4.1.1 *Lake-level oscillations.* In the Afar, there is no evidence that Lake Abhé transgressed before 10 000 B.P. Nevertheless, the lowest neighbouring basins began to fill earlier, around 12 000-11 000 B.P. (Gasse 1975). Scattered data from Southern Afar (Williams et al. 1977) also confirm the return of lacustrine conditions by 11 500-11 000 B.P. In the Ziway-Shala basin, a brief but well-defined highstand has been dated to between 11 870 and 10 450 B.P. (Street 1978). After 12 500 B.P., the level of Lake Victoria rose sufficiently for overflow to begin, but a short subsequent episode of closure occurred around 10 000 B.P. (Kendall 1969). The 12 500-10 000 B.P. highstand is also evident in Lake Kivu (Hecky 1978).

Lake Mobutu regained its overflow level by 12 480 B.P. (Harvey 1976), and the waters in Lake Manyara became rapidly diluted between 12 500 and 12 400 B.P. (Holdship 1976). No changes are evident in these lakes around 10 000 B.P., but it is possible that the sample spacing is not sufficiently close to enable a brief lowstand to be detected.

4.1.2 *Diatom assemblages.* Species indicating alkaline and saline water became noticeable during the short-lived transgressive stage. Such transitional flora were rare during the Late Pleistocene despite the considerable dilution of the water during the highstands.

Cyclotella meneghiniana and *Thalassiosira rudolfi* were widespread. They

343

were often accompanied by other species characteristic of saline waters (such as *Rhopalodia gibberula, Anomoeoneis sphaerophora, Thalassiosira faurii* . . .). However, the synchronous or immediately successive spread of *Melosira agassizii* reflects rapidly decreasing salinity throughout the region.

The transgressive stage was followed by a dilute episode marked by the expansion of *Melosira* or *Stephanodiscus astraea. M.granulata* was dominant in the Abiyata core (Gasse & Descourtieux 1978). *S.astraea* was the most numerous species in Lake Mobutu (Harvey 1976) and Lake Kivu (Hecky 1978), but *M.agassizii* and *M.nyassensis* v. *victoriae* were of secondary importance at times in the former lake. *M.granulata* v. *angustissima,* in association with *Synedra acus* and *Nitzschia fonticola,* multiplied in Lake Manyara (Holdship 1976).

Neither cold-water nor oligotrophic species characterised this interval, which experienced tropical eutrophic environments. The warming related to African deglaciation undoubtedly occurred earlier than the re-establishment of wetter conditions. These data based on diatom flora are consistent with the pollen evidence. In equatorial Africa, the last ice age was coming to a close by 14 500 B.P. (Livingstone 1962), while the gradual spread of forests took place from 12 500 B.P. onwards (Hamilton 1972). The analysis of the Mt Badda core confirms that the ice had completely disappeared by 11 500 B.P. At this time, a small peat-bog with oligotrophic-acidic and subaerial community occupied the core site (Gasse 1978), although pollen analysis still records drier conditions than in equatorial Africa (Hamilton 1977).

4.2 *10 000 - 7 500 B.P.*

Between 10 000 and 9 000 B.P., the Ethiopian lakes underwent a sudden major transgression, and the Early Holocene lacustrine phase culminated around 9 500-8 400 B.P. The level of Lake Abhé was very stable, while diatom evidence indicates several minor fluctuations in the Ziway-Shala basin (Gasse & Descourtieux 1979). In the Mt Badda core, a marked alkaline peak dated at 10 000 B.P. could be interpreted as a climatic change reflected in an increase in local runoff (Gasse 1978). An abrupt rise in level is also recorded in most of the southern lakes (Turkana, Victoria, Kivu, Nakuru, Magadi) (Butzer et al. 1972). However, in lakes Mobutu and Manyara, dilute conditions seems to be continuous from 12 000-11 000 B.P. onwards. The marked peak in water-level at 9 500-9 000 B.P., known throughout the whole of tropical Africa (Rognon & Williams 1977), indicates the switchover to full interglacial conditions (Street & Grove 1976).

Pronounced similarities between the diatom assemblages are apparent. Around 9 000 B.P., *Stephanodiscus astraea* (or its varieties of similar ecological requirements) played a major role in the phytoplankton, although differences in the diatom communities reflect the influence of local conditions. In Lake Abhé, there was an almost monospecific population of *S.astraea* and its depositional rate has been evaluated at 40×10^6 frustules $cm^{-2}.y^{-1}$ (Gasse 1977b). Even Lake Afrera, a spring-fed saline lake, recorded a peak of *S.astraea* by 9 000

B.P. (Gasse 1975). *S.astraea* v. *minutula* again dominated in Lake Kivu (Hecky 1978). A *S.astraea/Melosira* assemblage developed in lakes Ziway-Shala, Mobutu, Manyara and Naivasha (Figure 3).

S.astraea is believed to require an oligohaline environment of moderate alkalinity (2-50 meq/l) (Richardson et al. 1978), and Kilham (1971) asserted that *S.astraea* dominates in lakes where the dissolved silica content falls below 1 mg/l. Therefore, extensive lakes with dilute water and low silica content were common in East Africa around 9 500-9 000 B.P. At this time, the ecological diversity which characterises the present lakes was undoubtedly attenuated. This lack of diversity must be attributed to major climatic factors acting on a very large scale.

In the Ethiopian lakes at least, alkalinity did not drop to the very low Late Pleistocene values. This may be the result of a higher evaporation rate associated with higher temperatures. Dissolved silica content fell to low levels throughout the region, either because of decreased silica input from the catchments or because of silica depletion caused by high diatom productivity of this interglacial episode.

Lakes Ziway-Shala, Abhé and Turkana, mainly fed by Ethiopian rivers, underwent a period of low levels between 8 400-6 500 B.P., suggesting that conditions in the Ethiopian plateaux became drier about this time. A new alkaline peak dated at 8 250 B.P. in the Mt Badda core might also indicate climatic change (Gasse 1978). A drop in water-level is registered in Lake Nakuru around 8 000 B.P. This temporary but well-defined regression has not been demonstrated in the other lakes. However, dilute conditions were coming to a close by 8 000 B.P. in Lake Manyara (Holdship 1976). In Lake Mobutu, an abrupt decrease in *Melosira* occurred at core-depth 160-150 cm (Harvey 1976), about 7 000 B.P. In the Naivasha core, the curve for feldspar shows a single peak at 19 m, suggesting possible inwash from a sudden flood. Microfossil peculiarities (abrupt increase in littoral diatoms) were also discovered at this level, which can be roughly dated to about 6 600-6 700 B.P. by interpolation (Richardson & Richardson 1972).

4.3 *7 500 - 3 500 B.P.*

Lakes fed from the Ethiopian highlands rapidly rose again (Figure 2). Lake Abhé and Lake Ziway-Shala were high from 7 200-6 500 B.P. to 4 000 B.P. (phases Abhé V and Ziway-Shala VI), although a minor regression is registered towards 5 700-5 800 B.P. Lake-levels fell rapidly from about 4 000 B.P. until the lakes reached their present levels or nearly so. Successions of diatom assemblages resemble those of the preceding highstand. Lake Turkana transgressed shortly before 6 600 B.P. A high level was maintained until after 4 400 B.P. and was followed by a rapid regression. Marked ecological changes are observed in the Mt Badda core around 6 200 and 4 350 B.P., and the biotope was apparently dry by 3 150 B.P. (Gasse 1978).

The southern lakes registered ecological fluctuations from 5 800-5 000 B.P.

onwards. In Lake Mobutu, *Melosira granulata* had completely disappeared by 5 000 B.P., and an almost monospecific community of *Stephanodiscus astraea* and its variety *minutula* developed (Harvey 1976). The flora was similar to that of lakes Abhé IV and Abhé V. Establishment of the low silica content characterising the modern lake apparently took place at this time. In Lake Manyara, fluctuating proportions of *Nitzschia fonticola, Thalassiosira rudolfi, Navicula elkab, Nitzschia* spp. and *Stephanodiscus astraea* v. *minutula* indicate lower salinity than at present, with occasional freshwater phases, between 8 000 and 5 200 B.P. The following disappearance of diatoms would mark the progressive onset of saline-alkaline conditions which culminated from 4 200 to 3 200 B.P. (Holdship 1976). In the Naivasha basin, the large lake existing from prior to 9 200 B.P. persisted until about 5 700 B.P. with an almost constant planktonic flora. Lake desiccation involved changes in the diatom community between 5 700 and 4 000-3 000 B.P. *Synedra berolinensis* assumed importance around 5 500 B.P. A *Melosira/Synedra* assemblage came progressively into being and diatom community bears resemblance to those presently found in Lake George. This shrinking phase appears to be associated with a trend to more dilute conditions. This apparent paradox indicates the closure of the connection between the main lake and the small crater lake from where the core was obtained. The following dry phase indicates a climate drier than today around 4 000-3 000 B.P. (Richardson & Richardson 1972).

The Middle Holocene was a period of expended lakes from about 7 000 to 4 000 B.P. However, lake-levels began to fluctuate and/or lacustrine conditions to deteriorate after 5 800-5 000 B.P. A widespread regression prevailed from 4 500 to 3 500 B.P. (except in Lake Mobutu). Most of the lakes recorded low levels and several of them dropped below their present level by 3 500 B.P.

4.4 *3 500 - 0 B.P.*

Paradoxically, detailed inter-regional correlations remain difficult for this period, because conditions fluctuated widely and their timescale is of the same order as error of the ^{14}C dates.

The Ethiopian lakes rose again between 2 700 and 1 000 B.P. (Figure 2), and numerous strandlines show strongly fluctuating water-levels during the last centuries. In the Mt Badda core, diatoms record a curious 400-450 year cycle in the Late Holocene evolution, which could reflect changes in local runoff. However, the climatic origin of these fluctuations is not demonstrated so far (Gasse 1978). Lake Turkana registered a final transgression a little before 3 000 B.P. and remained relatively low thereafter. A small lake developed anew in the Naivasha basin.

In Lake Abhé, favourable life conditions were never to be restored and the sedimentation remained clayey or silty. During the last few centuries, the organisms have tended to die out as a result of an extensive rise in pH and salinity. In the uppermost layers of the Mobutu core, *Stephanodiscus astraea* markedly decreased. *Nitzschia bacata* and *N.fonticola* multiplied and *Stepha-*

346

nodiscus hantzschii was becoming numerically abundant in the last 30 cm. The development of *N. bacata,* which would be numerous in the modern lake (Talling 1963), could mark the last transitional phase towards the present conditions.

The diatom study of a 2 m core from Lake George records the ecological history of the last 3 600 years (Haworth 1977). Three diatom zones can be discerned. The earlier change from a *Melosira/Fragilaria* assemblage to a *Nitzschia/Fragilaria* assemblage took place around 3 000 B.P. This might well be a response to a gradual enrichment of the lake combined with a possible increase in rainfall and perhaps a rise in lake-level. The upper zone is characterized by the modern assemblage *(Nitzschia/Synedra berolinensis)* and would indicate that the present ecosystem has remained stable for about 500 years.

No diatoms are known in Lake Manyara where present saline conditions prevailed since 3 200 B.P. (Holdship 1976).

The history of Lake Naivasha inferred from diatom studies reflects the existence of a small lake fluctuating near and below the modern level in the past 3 000 years. The *Melosira/Synedra* community is indicative of periods of low chemical concentration and of high water-level. The *Thalassiosira/ Cyclotella/Navicula elkab,* and sparce *Nitzschia* dominated communities marked times of lower lake-level. The modern lake seems to correspond to some of the highest levels recorded during this interval (Richardson & Richardson 1972).

The Late Holocene is a rather dry episode punctuated by short humid phases. The lakes experienced numerous but slight environmental changes. Present ecological conditions finally took place during the last centuries.

5. CONCLUSIONS

Recent investigations conducted by different workers have recovered a wealth of information on the changes in water-levels and diatom assemblages of various East African lakes.

Lake-levels fluctuations reflect changes in the P/E ratio. Some decreases in evaporation, considered a result of lower temperatures, seem discernible in the available diatom record. The Late Pleistocene highstand dated from 26 500-25 000 B.P. would be caused by both increased rainfall and decreased evaporation. Temperatures close to that of today were re-established by 12 000 B.P. and the following lake-level fluctuations appear mainly due to changes in precipitation.

The marked ecological diversity which characterises the modern lakes was firmly attenuated during various high lake-level stages. For instance, around 9 500-9 000 B.P., the occurrence of *Stephanodiscus astraea* indicates a general tendency towards dilute conditions, with medium low to medium alkalinity and low silica content. Diatom assemblages with different physical and chemical requirements developed in the Late Pleistocene lakes. The successions

347

of diatom communities remain difficult to interpret because they depend on numerous interrelated factors. They registered the regional changes in the P/E ratio involving changes in the water budget of the lakes. They also depended on qualitative and quantitative fluctuations in mineral-organic inputs from the catchments. Another factor is the trophic level and the productivity in the lakes. Why was the silica content lower in the Holocene lakes than in the Late Pleistocene lakes while the alkalinity was higher? Although we are not able to explain such chemical mechanisms, the similarities which occurred at times between the different lakes must be attributed to major environmental factors of regional influence.

Knowledge of the history of these East African lakes is of major importance for an understanding of depositional events along the Nile Valley. The Blue Nile flows from the Ethiopian plateaux; Lakes Victoria, Mobutu, George and Edward belong to the White Nile system. The palaeohydrological influence of the East African lake basins upon the late Quaternary evolution of the Nile is considered in a separate paper (Adamson, Gasse, Shakelton, Street & Williams, in preparation).

REFERENCES

Abbot, W.H. & S.van Landingham 1973. Micropaleontology and paleoecology of miocene non-marine diatoms from the Harper District, Malheur County, Oregon. *Nova Hedwigia* 23(4):847-906.
Butzer, K.W. 1971. Recent history of an Ethiopian Delta: The Omo River and the level of Lake Rudolf. *Univ. Chicago, Dept. Geogr., Res. Paper, 136.*
Butzer, K.W. 1976. The Mursi, Nkalabong and Kibish formations, Lower Omo Basin, Ethiopia. In: Y.Coppens, F.C.Howell, G.L.Isaac & R.E.F.Leakey (eds.), *Earliest Man and environments in the Lake Rudolf Basin.* Univ. Chicago Press, pp.12-23.
Butzer, K.W., G.L.Isaac, J.L.Richardson & C.Washbourn-Kamau 1972. Radiocarbon dating of East African lake levels. *Science 175:* 1069-1076.
Cholnoky, B.J. 1968. Die Okologie der Diatomeen in Binnengewassern. Cramer, Lehre.
Foged, N. 1964. Freshwater diatoms from Spitzbergen. *Tromsö Mus. Skrif. 11.*
Gasse, F. 1975. L'évolution des lacs de l'Afar Central (Ethiopie et TFAI) du plio-pleistocène à l'actuel. D.Sc. Thesis, Univ. Paris VI.
Gasse, F. 1977a. Evolution of Lake Abhé (Ethiopia and TFAI), from 70 000 B.P. *Nature* 265(5589):42-45.
Gasse, F. 1977b. Les groupements de diatomées planctoniques: Base de la classification des lacs quaternaires de l'Afar Central. In: *Recherches francaises sur le Quaternaire hors de France.* Comité National Français de l'INQUA, Paris, pp.202-234.
Gasse, F. 1978. Les diatomées holocènes d'une tourbière (4 040 m) d'une montagne ethiopienne: le Mont Badda. *Rev. Algol., N.S. 13*(2):105-149.
Gasse, F. & S.Delibrias 1976. Les lacs de l'Afar Central (Ethiopie et TFAI) au Pleistocène supérieur. In: S.Horie (ed.), *Palaeolimnology of Lake Biwa and the Japanese Pleistocene, 155*(4):529-575.
Gasse, F. & C.Descourtieux 1979. Diatomées et évolution de trois milieux éthiopiens d'altitude différente, au cours du Quaternaire supérieur. *Palaeecology of Africa 11.* p. 117-134.

Gasse, F., J.Ch.Fontes & P.Rognon 1974. Variations hydrologiques et extension des lacs holocènes du désert danakil. *Palaeogeogr., Palaeoclimatol. Palaeoecol. 15:*109-148.

Gasse, F., P.Rognon & F.A.Street 1979. Quaternary history of the Afar and Ethiopian Rift lake basins. In: M.A.J.Williams & H.Faure (eds.), *The Sahara and the Nile.* A.A. Balkema, Rotterdam (in press).

Gasse, F. & F.A.Street 1978. Late Quaternary lake-level fluctuations and environments of the northern Rift Valley and Afar region (Ethiopia and Djibouti). *Palaeogeogr., Palaeoclimatol., Palaeoecol. 24:*279-325.

Grove, A.T., F.A.Street & A.S.Goudie 1975. Former lake-levels and climatic change in the Rift Valley of southern Ethiopia. *Geogr. J. 141:*177-202.

Hamilton, A.G. 1972. The interpretation of pollen diagrams from highland Uganda. *Palaeoecology of Africa 7:*45-149.

Hamilton, A.G. 1977. An Upper Pleistocene pollen diagram from highland Ethiopia. In: *Xth INQUA Congr., Birmingham, Abstract,* p.193.

Harvey, T. 1976. The paleolimnology of Lake Mobutu Sese Seko, Uganda-Zaire: The last 28 000 years. Thesis Ph.D., Duke Univ.

Hastenrath, S. 1977. Pleistocene mountain glaciation in Ethiopia. *J.Glaciol. 18:*309-313.

Haworth, E.Y. 1977. The sediments of Lake George (Uganda) V. The diatom assemblages in relation to the ecological history. *Arch. Hydrobiol. 80*(2):200-215.

Hay, R.L. 1976. *Geology of the Olduvai Gorge. A study of sedimentation in a semiarid basin.* Univ. California Press, Berkeley.

Hecky, R.E. 1978. The KivuTanganyika basin: The last 14 000 years. *Pol. Arch. Hydrobiol. 25*(1/2):159-165.

Hecky, R.E. & E.T.Degens 1973. Late Pleistocene-Holocene chemical stratigraphy and paleolimnology of the Rift Valley Lakes of Central Africa. Woods Hole Oceanogr. Inst. Tech. Rep., WHOI, pp.72-78.

Hecky, R.E. & P.Kilham 1973. Diatoms in alkaline saline lakes; Ecology and geochemical implications. *Limnol. Oceanogr. 18*(1):53-91.

Holdship, S.A. 1976. The paleolimnology of Lake Manyara, Tanzania: A diatom analysis of a 56 meter sediment core. Thesis Ph.D., Duke Univ.

Hustedt, F. 1957. Die Diatomeenflora des Flussystems des Weser in Gebiet der Hansestadt Bremen. *Abh. Nat. Ver. Bremen 34:*18-440.

Kendall, R.L. 1969. An ecological history of the Lake Victoria Basin. *Ecol. Monogr. 39:* 121-176.

Kilham, P. 1971. A hypothesis concerning silica and the freshwater planktonic diatoms. *Limnol. Oceanogr. 16*(1):10-18.

Livingstone, D.A. 1962. Age of deglaciation in the Ruwenzori range, Uganda. *Nature 194:* 859-860.

Livingstone, D.A. 1976. The Nile. Paleolimnology of Headwaters. In: J.Rzoska (ed.), *The Nile, biology of an ancient River.* Junk, The Hague, pp.21-29.

Livingstone, D.A. 1979. Environmental changes in the Nile Headwaters. In: M.A.J.Williams & H.Faure (eds.), *The Sahara and the Nile.* A.A.Balkema, Rotterdam.

Messerli, B., H.Hurni, H.Kienholz & M.Winiger 1977. Balé Mountains: Largest Pleistocene mountain glacier system in Ethiopia. In: *Xth INQUA Congr., Birmingham, Abstracts.* p.300ff.

Potter, E.C. 1976. Pleistocene glaciation in Ethiopia: New evidence. *J. Glaciol. 17:*148-150.

Richardson, J.L. 1968. Diatoms and lake Typology in East and Central Africa. *Int. Rev. Hydrobiol. 53*(2):299-338.

349

Richardson, J.L. 1969. Characteristic planktonic diatoms of the lakes of tropical Africa. *Int. Rev. Hydrobiol. 54:*175-176.

Richardson, J.L. & A.E.Richardson 1972. History of an African Rift Lake and its Climatic Implications. *Ecol. Monogr. 42:*499-534.

Richardson, J.L., T.Harvey & S.A.Holdship 1978. Diatoms in the history of shallow East African Lakes. *Pol. Arch. Hydrobiol. 25*(1/2):341-353.

Rognon, P. & M.A.J.Williams 1977. Late Quaternary climatic changes in Australia and N. Africa: A preliminary interpretation. *Palaeogeogr., Palaeoclimatol. palaeoecol. 21:*285-327.

Street, F.A. 1977. Late Quaternary lake-level fluctuations in the Ziway-Shala Basin, Southern Ethiopia: Palaeohydrologic interpretations. In: *Xth INQUA Congr., Birmingham, Abstracts.* p.439f.

Street, F.A. 1978. Chronology of Late Pleistocene and Holocene lake-level fluctuations, Ziway-Shala Basin, Ethiopia. *Proc. 8th Panafr., Congr. Prehist. Quat. Studs. Nairobi* (in press).

Street, F.A. & A.T.Grove 1976. Environmental and climatic implications of late Quaternary lake-level fluctuations in Africa. *Nature 261*(5559):385-390.

Talling, J.F. 1963. Origin of stratification in an African Rift Lake. *Limnol. Oceanogr. 8:* 68-78.

Talling, J.F. & I.B.Talling 1965. The chemical composition of African lake waters. *Int. Rev. Hydrobiol. 50*(3):421-463.

Truckle, P.M. 1976. Geology and late cainozoic lake sedimentology of the Suguta Trough, Kenya. *Nature 263*(5576):380-383.

Williams, M.A.J., P.M.Bishop, F.M.Dakin & R.Gillespie. Late Quaternary lake levels in southern Afar and the adjacent Ethiopian Rift. *Nature 267:*690-693.

350

THE PALEOCLIMATE OF THE CENTRAL
SAHARA, LIBYA AND THE LIBYAN DESERT

H.-J. PACHUR / G. BRAUN
Translated by A.Beck
Institut für Physische Geographie, Freie Universität Berlin

This study describes the results of investigations made in the lowlying areas, between 500 m and about 100 m above sea level, in the Tibesti Mountains (Hagedorn & Pachur 1971). The main study area was just below the Tropic of Cancer between the western border of Libya (northern edge of the Murzuk basin) and the Libyan Desert in Egypt (Djebel Tartur, west of the Kharga oasis, the Abu Ballas escarpment below approximately 24°N, 27°E and Gilf Kebir).

The aim of the investigations was to prove the existence of late Quaternary wet phases outside the central Saharan mountains. This meant refuting the hypothesis of the existence of a 'Kernwüste' in the Serir Tibesti area (Meckelein 1959) and doing justice to the fact that the mountains provide particularly favourable climate provinces and that data found there which indicate periods of increased precipitation should be compared with data from lowlying areas before their general validity is assumed.

The literature relevant to the Libyan part of this question is discussed in Butzer (1967), Pachur (1975) and Gabriel (1978). Older studies by Bagnold et al. (1939) and Caton-Thompson (1952) are available on the western Libyan Desert, and Wendorf et al. (1976) recently published an important study based on archaeological investigations on the eastern edge of the Western Desert, in which climatic history from the Upper Acheulian (Riss) to the Holocene epoch is outlined.

I shall begin with results from the eastern central Sahara east of Gilf Kebir and west of the Kharga depression. The area is characterised by west-east striking escarpments with Jura to Upper Cretaceous strata (Klitzsch 1978). In front of the cuestas is a series of fossil depressions with a relative depth of more than 12 m in places. Such hollows were formed recently by the wind. They are a feature of an extremely arid climate and can thus be used as a climatic indicator.

The escarpment known as Abu Ballas consists of a sequence of fine sand and mudstone. Its relative height is over 110 m. The foreland of this escarpment contains deflation hollows, whose longitudinal axes are at right angles to the morphological strike. They contain a sequence of fine-clastic sedi-

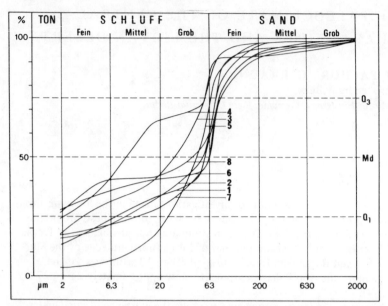

Figure 1. Typical cumulative curve of grain-size distribution of stillwater silt and eolian accumulation. Cross-section E of Abu Ballas, 24°14′N, 27°25′E. Sample interval 40 cm from base to top.

ments, which are currently being eroded by the wind. The cumulative curve of particle size is bimodal, ranging from fine sand to silt and clay (Figure 1). In the standard profile, limestone content does not exceed 3 per cent. Sand content consists of 90 per cent quartz, well-rounded grains with a frosted surface. Their particle size confirmed them to be flowing sand. The material is layered, the strata vary between millimetre and centimetre thickness. Some profile sections contain root horizons – probably phragmites. The root horizons are restricted to sand layers. Some layers are loess-like sediments according to Smalley & Vita Finzi (1968). The pollen content is too low for analysis. The salt content of some layers exceeds 61,88 mmS; therefore, sebkha conditions are not to be counted on. The sediment structure would also be incompatible. The lack of high-magnesium calcite* suggests a similar interpretation. In contrast, recent calcareous muds from salt lakes in Libya show high-magnesium calcite. Drift sand deeper than 1 m often forms the base of the fine-grained clastic sediments. The fact that these occur in deflation hollows indicates considerable eolian activity before sedimentation took place. According to our present level of knowledge, the structure and texture of the sediments may be interpreted as the result of interacting eolian pro-

* I am grateful to Mr U.Förstner and Mr P.Stoffers, Heidelberg, for dating the sediments.

cesses and flooding. Swamp-like environments, probably surrounded by reeds, developed; the swamps dried up intermittently. Organic traces are limited to numerous artefacts and some bone remains of gazelles, hares and bovides (smaller than cattle).

A climatic interpretation is possible owing to two additional factors:

1. The thickness of the deposits indicates long-term formation. Maximum 6 m are exposed. With hammer seismics, we determined thicknesses between 12 and 19 m, i.e. a thickness of more than 24 m.

2. Stillwater sediments were also found outside the escarpment area in flat depressions. This means they are not specific groundwater outlets of a fossil groundwater level, but are due to autochthonous rainfalls and a groundwater level near the surface with a limited catchment area. This is even more true of those stillwater sediments proved to exist in the middle of the dune chain of the Great Sand Sea. Evidence of this is provided by artefact finds in the lower third of the longitudinal dunes which possibly used to have a vegetational cover.

An approximate age for the stillwater accumulations was achieved by means of radiometric dating. This was undertaken by Prof M.A.Geyh (Hanover), to whom my thanks are due for his interest and for helpful discussions.

Near the Abu Ballas escarpment and some depressions, charcoal and wood remains from fire-places were dated. These are baked sediments and therefore morphologically more resistant, so that they stick up above the present erosion surface. It should thus be possible to deduce an age ante quem. Since these finds were made at different deflation levels, conclusions may be drawn as to the length of time the sedimentation conditions lasted. The ^{14}C ages lie between 10 000 and 7 100 B.P. and will be referred to again later.

The second data group comes from the Gilf Kebir plateau and is suitable for the problem in question for several reasons. It is situated almost at the latitude of the tropic, well away from coastal influences. Furthermore, it is high enough (maximum 1 000 m above sea level) to intensify the climatic phases to be traced and to make them morphologically identifiable and yet, on the other hand, low enough not to develop its own climatic regime (Pachur, Gabriel, in preparation).

In the backwater part of Wadi Wassa — blocked by fossil dune sand — stillwater sediments were found, exposed over 4 m. They had a bimodal grain size distribution. Radiometric dates between 8 695 and 8 460 were obtained directly from the sediment by means of bones, artefacts, pottery, charcoal and root horizons. The occurrence of pottery* at such an early period is new for this area and will be the subject of further investigations. Stone places on the sediment surface, together with tree roots, gave ages around 4 360 B.P. In addition, shallow lake formations occurred from time to time in northward draining valleys around 7 170 B.P.

In the mountain forelands, over 40 cm long stones with a groove round

* Thermoluminescence determination in preparation.

them to hold a rope were found on the surface of heavily rubificated accumulations. Considered together with rock paintings in the surrounding area (Djebel Auenat) these could be indirect proof of cattle rearing. The locality indicates an age of about 8 000 B.P.

To sum up:
Between 10 000 and at least 7 100 B.P.*, the west part of the Libyan desert had an environment characterized by widespread, shallow, swamp-like waterplaces. The vegetation was light, providing food for animals adapted to dry conditions. Cattle rearing was probably possible in areas with additional water. There is morphological proof of wind influence in the accumulations of drift sand at aerodynamically favourable spots, causing temporary dams in the upper reaches of wadis. Sedimentological evidence is provided by the proportion of drift sand in the stillwater sediments and also in the silts, which were probably wind-transported.

Let us compare these results with those from three localities in the western central Sahara: Serir Tibesti, Serir Calanscio and the northern edge of the Murzuk basin.

1. In Serir Tibesti a series of limnic sediments was found at about 23°30'N. They were classified as freshwater on the basis of strata-forming diatomites and molluscs (Pachur 1974). Limestone content is over 40 per cent. They are calcites with no magnesium content.** In contrast, recent calcareous muds from salt lakes in Libya show high Mg calcite. The thickness of the limnic sediments exceeds 5 m. The molluscs and calcareous muds were dated at between 8 880 and 7 500 B.P. Under the limnic sediments are fine-grained fluvial sands, which in turn filled up a probably windéroded hollow. Additional water from the Tibesti mountains influenced the balance of this lake.

At the top of the deposit, the calcareous muds are succeeded by silt-rich stratified sediments showing the end of the limnic phase and pass through a swamp stage comparable with the Western Desert deposits. The next overlying strata are fluvial with salt-crusted fine sands on which tamarisks grow.

Their base was dated at about 2 000 B.P., thus giving a date post quem for the fluvial phase. On the northern margin of Serir Tibesti beyond the influence of the Tibesti, an additional deposit containing lacustrime limestone with radiocarbon dates between 6 000 and 5 000 B.P. was found.

Outside the lake basin three fluvial accumulations were distinguished in the Serir Tibesti area. The youngest is the above-mentioned stratum directly overlying the lake sediments (Pachur 1974).

A fossil river crosses the southern part of the Serir Calanscio in a northeasterly direction. Fluvial sediments — channel bars, point bar deposits and

* Geyh & Jäkel (1974) begin of an arid phase in the western Sahara, based on statistical assay.
** Meyers, O.H., 1939, S.Bagnold et al.)

354

suspension load deposits – are developed. Populations of freshwater gastropods were proved to have existed in several localities in the wadi accumulations, confirming that we are not dealing with a chance occurrence due to unusually high precipitation. Pollen from Mediterranean flora including ferns (analysed by Schulz) gives the same interpretation. An elephant skeleton found in the most recent deposits of this wadi was ^{14}C dated at 3 420 years B.P.

2. The morphology and petrography of the fluvial deposits suggest that they were transported over a distance of about 800 km at a gradient of about 1 ‰ from the northern edge of the Tibesti (Enneri, Bardagué playa) into the Serir Calanscio. We used as tracers acid-volcanic pebbles (ignimbrite, trachyte, rhyolite). According to Vincent (1970) 'rhyolitic volcanic rocks are present only in the interior of Tibesti'.

The endorheic state of the central Sahara was interrupted and there was a fluvial link between the Tibesti and the Mediterranean; general considerations indicate that this took place within the radiometrically defined period around 8 000 B.P. This fact only makes sense if we postulate autochthonous rainfalls in the lowlying areas.

3. On the northern edge of the Murzuk basin, soils with hydromorphic characteristics were evaluated as climatic evidence. They are connected to shallow lake basins, from which a sapropel horizon and a calcareous crust ($>$ 90 cm) are being ^{14}C dated (in preparation).

We must also take into consideration the numerous finds of large mammals (giraffe, bovides, elephant) in the Serir Tibesti near the mountains (Gabriel 1972) and also the artefacts found all over the entire Serir Tibesti and Serir Calanscio (Pachur 1974). These can only be explained by autochthonous precipitation.

The finds introduced here lead to a very conservative estimate of precipitation levels about 8 000 B.P.: $<$ 300 mm annual rainfall in the western and $<$ 200 mm in the eastern central Sahara..

At present this estimate is being simulated in a lysimeter model based on a numerical solution of the Fokker-Planck-equation, describing the one-dimensional water diffusion processes in soils.

From Sonntag & Klitzsch et al. (1977), we know the groundwater age, e.g. from the surroundings of the Murzuk basin, as one boundary condition. The other is the ^{14}C age of the surface finds. If the estimated precipitation levels are realistic they ought to lead to groundwater recharge.

The lysimeter model was developed by G.Braun (cf. 1978) and calibrated on the basis of long-term meteorological-hydrological time series. The soil consists of medium-grained sand without impermeable strata. For each layer a function exists of matrix suction versus water content ($cm^{+3} cm^{-3}$) and hydraulic conductivity (cm/d) versus water pressure (cm).

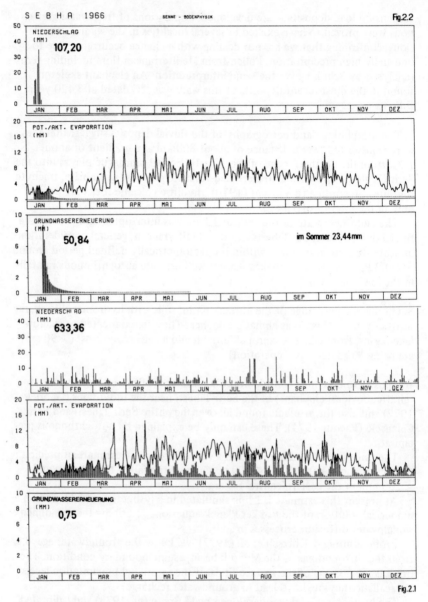

Figure 2. Computer outprint: simulation of rainfall. Soil physics parameter of lysimeter monolith, Senne, W.Germany. Climatic data from Sebha, Libya.

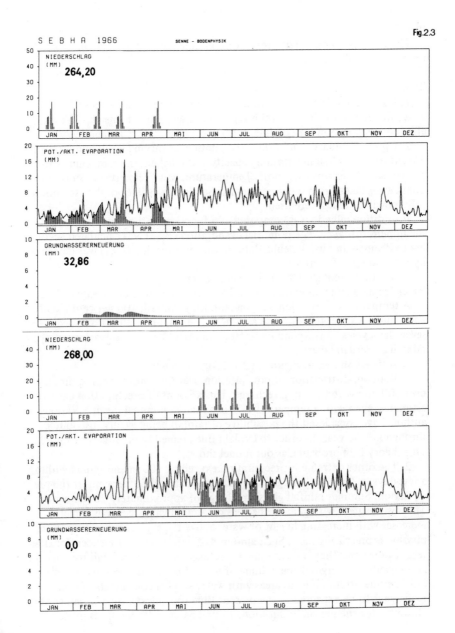

Fig.2.3

S E B H A   1966   SENNE - BODENPHYSIK

NIEDERSCHLAG (MM) 264,20

POT./AKT. EVAPORATION (MM)

GRUNDWASSERERNEUERUNG (MM) 32,86

NIEDERSCHLAG (MM) 268,00

POT./AKT. EVAPORATION (MM)

GRUNDWASSERERNEUERUNG (MM) 0,0

Initial condition:

The initial water-content profile was obtained after several preliminary runs, i.e. after several years.

Boundary conditions:

1. Lower boundary condition: a level is considered at a depth of 3.5 m; water flowing through this level is regarded as seepage. At this depth, the effect of evaporation is negligible, (cf. Dincer et al. 1974, Thoma 1979).

2. Upper boundary condition: precipitation of varying levels and intensity was infiltrated. When infiltration capacity was reached, the remaining water was considered as surface runoff. Temperature, wind velocity and sunshine hours were taken from measurements at Sebha; for radiation, the astronomically possible values were used.

In addition, albedo was computed as a function of the water content in the uppermost layer. The potential evaporation (E_{pot}) was calculated according to Penman and using Sebha data. Actual evaporation (E_{act}) was determined by the following function:

$$E_{act} = E_{pot} . \exp (h/3\,000), \quad h = \text{suction (cm)}$$

We chose data from an extremely arid area in order to obtain a safe estimate.

A temperature correction was made: during rainfall, 100 per cent cloud cover and $14°C$ depression were assumed. The total rainfall was divided into seven-day-cycles, corresponding to Heckendorf's (1977) observations of cyclone rainfall at Bardai/Tibesti.

Results are shown in Figure 2 (2.1-2.3):

1. Random distribution of precipitation over the whole year requires amounts over 600 mm with 0.75 mm recharge. This amount of precipitation exceeds all estimates, even with a temperature $7°$ less than the annual mean.* Various authors have considered the possibility of rainfall being evenly distributed throughout the year, in order to explain fine-grained sediments for example. This theory is improbable, as our model shows.

2. The other extreme is precipitation exceeding 100 m and falling within a few days in January. This gives a recharge of almost 50 per cent of the precipitation. Summer rainfall gives a recharge of approximately 20 per cent. This result agrees with general theories on groundwater recharge, which increases with increasing height of water above the surface when evapotranspiration is practically zero. Such amounts of precipitation are exceptional in recent times, but they indicate the importance of current rainfall for current groundwater recharge, if the volume of water from surface runoff is taken into consideration. Thus, in areas with water surplus considerable groundwater recharge is to be expected at the present time too, given a suitable substratum. However, for the period in question, around 8 000 B.P., the finds made

---

* Annual mean temperature (Sebha 1966): $\dfrac{1}{365} \cdot \displaystyle\sum_{i=1}^{365} \dfrac{T\max_i + T\min_i}{2} = 22.3°C$

throughout the study area show more vegetation, steppe fauna, human activity and paleopedology and indicate that a model assuming aridity throughout the year with few heavy rainfalls is less likely than the following assumption:

3. Total precipitation is higher and distributed over the six winter months with a groundwater recharge of about 12 per cent. For this at least 264 mm/a are necessary: just under the estimates based on fauna requirements and a comparison between the reconstructed forms about 7 500 B.P. and the amount of precipitation of comparable recent environments (Pachur 1974). With summer rainfall no recharge takes place, so we prefer to postulate winter rainfall for the period about 8 000 B.P.

However, it must be emphasized that this model is only valid for substrata with good water conductivity, e.g. fluvial sands without silt-bearing strata or dune sands – according to the proverb of the people of central Asia: 'Life is in sand'. Thus wide areas of the central Sahara can only be regarded with reservation as possible catchment areas for groundwater recharge. This is further evidence that our estimate is on the safe side. One aspect of this view is that possibly groundwater recharge does not take place in the middle of a humid phase but at the start of it, when vegetation does not hinder surface runoff and increase evaporation.

These results fit in with finds to date in the central Sahara area: in the early Holocene only a semi-arid climatic phase with marked dry seasons existed. Over several thousand years, a groundwater level near the surface fed lowlying swamps with fresh to slightly brackish water in Egypt and freshwater lakes in Libya. Evapotranspiration losses from a sparse vegetation (becoming more dense only along the channels and wet places) were made up by additional water in higher relief areas.

The following graph summarizes chronological and spatial aspects; Figure 3, cf. schedule of ^{14}C dates.

1. There are no data from the lowlands for the period between c. 11 000 and c. 25 000 B.P. Only older dates occur: freshwater diatomites from the Libyan Desert at 26 100, tufaceous limestone with freshwater snails about 28 800 and lake sediments from Harudj at 33 660. Wendorf et al. (1976) found similar magnitudes at the eastern edge of the Libyan Desert.

2. The following period is considered arid, also for stratigraphic reasons, because the limnic or telmatic sediments lie in hollows which were almost certainly cleaned out by the wind, or on top of drift sand. Fluvial accumulations also covered eolian sand in the main wadi of Gilf Kebir.

3. In the highlands, the early existence of a per definitionem humid climate is indicated by the presence of caldera lakes (Faure 1966). Limnic accumulation beings in the lowland around 10 000 B.P. and is well-documented around 8 000 B.P. in the western part of the central Sahara (compare Geyh & Jäkel (1974)).

A comparison of all the dates up to 5 000 B.P. (end of limnic sedimentation) shows a remarkable difference between west and east: coeval sediments were

359

Schedule of ^{14}C dates; see Figure 3

^{14}C-Labor	Age	Material	Position	Sampling by
Hv 2260	2690 ±435	Kollogen	N: 21° 30′ E: 17° 11′	Gabriel
Hv 2921	7380 ±110	CaCO$_3$	N: 21° 12′ E: 17° 25′	Molle
H 2938-2432	8295 ±190	CaCO$_3$	N: 21° 19′ E: 16° 04′	Hagedorn
Hv 2875	7570 ±115	o.CaCO$_3$	N: 23° 31′ E: 17° 28′	Pachur
Hv 2874	1435 ±50	org.C	N: 23° 30′ E: 17° 28′	Pachur
Gif 378	12400 ±400	CaCO$_3$	N: 20° 58′ E: 15° 31′	Faure
Gif 379	14790 ±400	CaCO$_3$	N: 20° 58′ E: 15° 31′	Faure
Gif 380	14970 ±400	CaCO$_3$	Trou Natron	Faure
B	8530 ±100	org.C	N: 21° 22′ E: 18° 32′	Messerli
Hv 2772	8055 ±90	org.C	N: 20° 37′ E: 17° 22′	Gavrilovic
Hv 3355	13760 ±185	o.CaCO$_3$	N: 17° 20′ E: 21° 16′	Molle
Hv 2775	7455 ±180	Apatit	N-Tibesti, Bardagué	Gabriel
Hv 2773	6435 ±225	Kollagen	N: 22° 40′ E: 16° 40′	Gabriel
Hv 3766	5125 ±185	Kollagen	N: 22° 40′ E: 18° 30′	Pachur
Hv 3761	1625 ±145	org.C	N: 25° 00′ E: 18° 20′	Pachur
Hv 3762	2300 ±145	org.C	Hv 3761	Pachur
Hv 3763	5110 ±295	o.CaCO$_3$	N: 25° 40′ E: 19° 10′	Pachur
Hv 3769	5295 ±145	CaCO$_3$	Hv 3763	Pachur
Hv 4206	8115 ±475	Hk	N: 21° 20′ E: 17° 03′	Jäkel
Hv 4801	6100 ±110	Hk	N: 25° 10′ E: 17° 35′	Gabriel
Hv 4113	5510 ±370	Hk	N: 26° 15′ E: 19° 20′	Gabriel
Hv 4117	5410 ±250	Hk	N: 27° 28′ E: 21° 15′	Gabriel
Hv 2937-2431	9340 ±85	Hk	N: 21° 38′ E: 16° 54′	Jäkel
Hv 5725	3420 ±230	Apatit	N: 27° 28′ E: 21° 15′	Gabriel/Pachur
Hv 8690	7405 ±275	Hk	N: 24° 14′ E: 27° 25′	Pachur
Hv 8691	8195 ±115	Hk	N: 24° 14′ E: 27° 25′	Pachur
Hv 8692	7170 ±210	Hk	N: 24° 14′ E: 27° 25′	Pachur
Hv 8693	9260 ±370	Hk	N: 24° 14′ E: 27° 25′	Pachur
Hv 8694	7790 ±340	Hk	N: 24° 14′ E: 27° 25′	Pachur
Hv 8311	8460 ±520	Hk	N: 23° 11′ E: 26° 02′	Pachur
Hv 8312	8695 ±570	CaCO$_3$	N: 23° 11′ E: 26° 02′	Pachur
Hv 8313	4365 ±95	Hk	N: 23° 11′ E: 26° 02′	Pachur/Gabriel
Hv 8315	28800 ±460	CaCO$_3$	N: 25° 14′ E: 30° 03′	Pachur/Gabriel
Hv 8316	4210 ±60	Hk	N: 25° 14′ E: 30° 04′	Pachur/Gabriel
Hv 8317	7755 ±80	Hk	N: 25° 30′ E: 29° 00′	Pachur/Gabriel
Hv 8318	10100 ±255	Hk	N: 24° 12′ E: 27° 25′	Pachur
Hv 8319	26100 ±340	Hk	N: 28° 00′ E: 27° 00′	Pachur
Hv 8240	6410 ±50	Hk	N: 25° 14′ E: 30° 03′	Pachur/Gabriel
Hv 8241	6405 ±325	Hk	N: 24° 12′ E: 27° 25′	Pachur/Gabriel
Hv 8242	3830 ±365	Hk	N: 23° 10′ E: 26° 05′	Pachur/Gabriel
Hv 5624	16120 ±215	CaCO$_3$	Bardai, Tibesti	Jäkel
Hv 3714	7825 ±85	CaCO$_3$	N: 21° 20′ E: 17° 03′	Jäkel
Hv 7370	33650 ±1750	CaCO$_3$	N: 28° 07′ E: 18° 03′	Pachur
Hv 7371	8545 ±130	o.CaCO$_3$	N: 23° 31′ E: 17° 28′	Pachur
Hv 7372	8880 ±130	o.CaCO$_3$	N: 23° 31′ E: 17° 28′	Pachur
Hv 7374	6535 ±175	org.C	N: 25° 40′ E: 19° 10′	Pachur
Hv 7373	1930 ±85	org.C	N: 23° 31′ E: 17° 28′	Pachur

B = Bern   Gif = Gif-sur-Yvette   Hv = Hannover, M.A. Geyh
o.CaCO$_3$= mollusc   Hk = char-coal   org.C = organic C (wood, limnic mud)

360

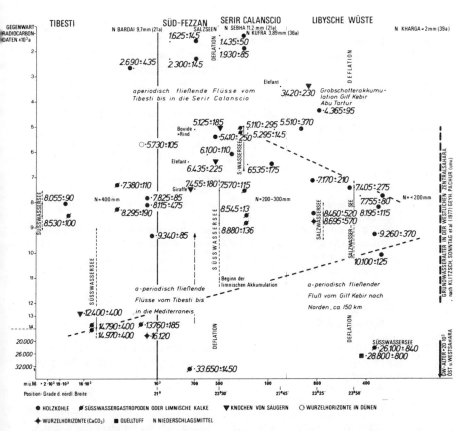

Figure 3. Radiocarbon dates of surface samples. Left column: dates from Bardai research group (summarized by Jäkel 1978). Other dates from lowland S.Libya and Western Desert, Egypt. Abscissa: geographical position, magnitude N and m above sea level.

formed in Libya in biologically highly active freshwater lakes; in Egypt, on the other hand, they were formed in flat, swampy environments with high salt content and active sand transport.

We can confirm the existence of a chronologically and spatially extensive semiarid phase in the lowlands and a spatial differentiation of the early Holocene semiarid phase according to decreasing humidity. If we compare these geomorphological finds with the totally independently obtained groundwater dates*, we have other good reasons to postulate the existence of a phase of

---

* Sonntag et al. (1978): Two groups of groundwater dates:
$> 20.10^3$ B.P. $6.10^{-3} - 11.10^{-3}$ B.P.

361

autochthonous rainfall in the period between 12 000 and 5 000 B.P. causing decreasing humidity from west to east.

4. The period after 5 000 B.P. shows a trend to decreasing precipitation, which was, however, interrupted about 3 000 B.P. The rainfall supply was, however, insufficient to cause lake formation. Heavier discharge of the rivers in Tibesti, Gilf Kebir and Djebel Tartur (Pachur, Gabriel) led to the formation of a coarse-material accumulation (the lower terrace whose existence can be proved throughout the entire Sahara). In Serir Tibesti, this accumulation represents the top layer of the lake sediments.

REFERENCES

Bagnold, R.A., R.F.Peel, O.H.Myers & H.A.Winkler 1939. Expedition to the Gilf Kebir and Kwenat. *Geogr. J. 93:*281-313.

Braun, G. 1978. A water balance model for weighable lysimeters. Proceedings JFIP. In: *JFIP Modeling, identification and control in environmental systems.* North-Holland Publishing Company, Amsterdam, pp.283-289.

Braun, G. & H.J.Pachur. Niederschlagshöhen im Frühholożän der Zentralsahara (in preparation).

Butzer, K.W. 1974. Quartäre Vorzeitklimate der Sahara. In: H.Schiffers (ed.), *Die Sahara und ihre Randgebiete.* pp.349-387.

Caton-Thompson, G., E.W.Gardner et al. 1952. *The Kharga Oasis in prehistory.* London, Athlone, p.213.

Dincer, T., Al-Mugrin & U.Zimmermann 1974. Study of the infiltration and recharge through the sand dunes in arid zones with special reference to the stable isotopes and thermonuclear tritium. *J. Hydrol. 23:*79-109.

Edmunds, W.M. & E.P.Wright 1979. Groundwater recharge and palaeoclimate in the Sirte and Kufra basins, Libya. *J. Hydrol. 40:*215-241.

Faure, H. 1965. Lacs quaternaires du Sahara. *Mitt. Int. Verein. Limnol. 17:*131-146.

Gabriel, B. 1977. Zum ökologischen Wandel im Neolithikum der östlichen Zentralsahara. *Berliner Geogr. Abh. 27.*

Geyh, M.A. & D.Jäkel 1974. Spätpleistozäne und holozäne Klimageschichte der Sahara auf Grund zugäng. Licher ^{14}C-Daten. *Z. Geomorph. N.F.18,* p. 82-98.

Geyh, M.A. & K.P.Obenauf 1974. Zur Frage der Neubildung von Grundwasser unter ariden Bedingungen. In: Berlin FU, *Pressedienst Wissenschaft 5:*70-91

Hagedorn, H. & H.J.Pachur 1971. Observations of climatic gemorphology and Quaternary: Evolution of landforms in South Central Libya. Symp. on the geology of Libya. Faculty of Science, Univ. Libya, pp.387-400.

Heckendorff, W.D. 1977. Untersuchungen zum Klima des Tibestigebirges. *Ber. Inst. Meteorol. und Klimatologie TU Hannover, 17:*347.

Jäkel, D. 1978. Eine Klimakurve für die Zentralsahara. In: *Sahara. 10 000 Jahre zwischen*

---

using samples from the Murzuk basin and the Western Desert, whereby younger groundwater in Egypt does not seem to be represented.

Edmunds, W.M. & Wright, E.P. (1979): Dates of groundwater from Sirte and N.Kufra:
13.10³ – 34.10³ B.P.   5.10³ – 7.10³ B.P.
and water possibly linked to the northern Tibesti foreland: 7 800 B.P.

Weide und Wüste. Handbuch, Ausstellung Rautenstrauch. Joest Museum für Völkerkunde, Institut für Urund Frühgeschichte der Universität zu Köln, Museum A. Koenig, Bonn, pp.382-396.

Klitzsch, E. 1978. Geologische Bearbeitung Südwest-Agyptens. *Geol. Rundschau 67*(2): 509-520.

Klitzsch, E. & C.Sonntag et al. 1976. Grundwasser der Zentralsahara: Fossile Vorräte. *Geol. Rundschau 65*(1):264-287.

Meckelein, W. 1959. Forschungen in der zentralen Sahara. *Klimageomorphologie.*

Pachur, H.J. 1974. Geomorphologische Untersuchungen im Raum der Serir Tibesti (Zentralsahara). *Berliner Geogr. Abh. 17.*

Pachur, H.J. 1975. Zur spàtpleistozànen und holozànen Formung auf der Nordabdachung des Tibestigebirges. *Die Erde 106:*21-46.

Pachur, H.J. 1978. Late quaternary of southern Libya. *2nd symposium geology of Libya* (in press).

Pachur, H.J. & B.Gabriel. Geomorphologische Untersuchungen in der Western desert, Agypten (in preparation).

Schulz, E. 1974. Pollenanalytische Untersuchungen quartärer Sedimente des Nord-West Tibesti. In: Berlin, FU, *Pressedienst Wissenschaft 5*(74):59-69.

Smalley, I.J. & C.Vita-Finzi 1968. The formation of fine particles in sandy deserts and the nature of desert loess. *J. Sed. Petrol. 38*(3):766-774.

Sonntag, C. & E.Klitzsch et al. 1978. Paläoklimatische Information im Isotopengehalt C-14 datierter Saharawässer: Kontinentaleffekt in D und O-18. *Geol. Rundschau 67* (2):413-424.

Thoma, G. 1979. Isotopenhydrologie in der ungesättigten Bodenzone: Eine neues in situ Verfahren zur Gewinnung von Bodenfeuchteproben aus grosen Tiefen. Anwendung auf der Europa Düne bei Archacon in Frankreich. Dissertation, J. Umweltphysik Ruprecht-Karl-Universität Heidelberg.

Vincent, P.M. 1970. The evolution of the Tibesti Volcanic Province, eastern Sahara. In: T.N.Clifford & G.Gass (eds.), *African magmatism and tectonics.* pp.301-319.

Wendorf, F. et al. 1976. The prehistory of the Egyptian sahara. *Science 93*(4248):103-114.

HOLOCENE BIOGEOGRAPHICAL VARIATIONS ALONG THE NORTHWESTERN AFRICAN COAST (28°-19°N). PALEOCLIMATIC IMPLICATIONS

N. PETIT-MAIRE
Laboratoire de Géologie du Quaternaire, C.N.R.S., Marseille-Luminy.

The Saharan part of the east Atlantic coast extends from Wadi Draa (28°30'N) to Nouakchott (17°N), along some 2 000 km. Except for its northern and southern margins, this desert area was practically a *terra incognita* as far as Quaternary palaeogeography was concerned. Between 1970 and 1974, a stretch of land not exceeding 50 km inland was observed and biogeographical data which could help to reconstitute the recent past environments were collected: human and large vertebrate bony remains, marine molluscs from *in situ* former sea beaches or from middens, and archaeological documents related to the living conditions of former populations (lithics, ceramics, bone artefacts, etc. . . .). Most of the sampled material was ^{14}C datable and a quite tight chronological frame could be provided (G.Delibrias, L.Ortlieb, N.Petit-Maire 1976; G.Delibrias & J.Evin 1979). No date resulted older than 10 500 B.P., which is quite logical considering the Upper Pleistocene marine and continental conditions: during the glacial maximum the water temperature lowered considerably, the Canary current and upwelling processes were stronger and caused intense fog and extreme drought in this whole area, thus totally unfavourable to vegetal and animal life (McIntyre et al. 1976).

This short survey of recent littoral palaeobiology thus only covers the Holocene.

1. PRESENT CONDITIONS

From the Draa down to the 23°N, the shore is topped, at 1 m to 80 m, by the Moghrebian (plio-pleistocene) platform, which, in places, may be eroded, buried under littoral sand formations, fractured, interrupted by low sandy beaches or large sebkhas, or, in one continuous area between 25°43'N and 24°42'N, vertical, uninterrupted, and 40 to 80 m high. Between 23° and 21°N (Amtal sebkha – Cape Blanc), the eroded Aguerguer gresocalcareous formation lines the coast (L.Ortlieb 1975). It is often sand-covered or edged by small depressions or sebkhas: these are separated from the sea by sand bars; their surface varies from a few square kilometres to more than 150 km². Between 21° and 19°N (Cape Timiris), the ria coast is marked by rocky capes, active dunes reach the shore line. South of 19°N, the sandy low coast is lined by deep sandy bays limited by rocky points (L.Hebrard 1973).

Although in a lesser way than during the glacial episodes, the climate of this part of the Atlantic coast is still strongly influenced by the cold Canary Islands current (always less than 20°C) and other processes related to the strong northeastern trade winds. This dynamic and thermic barrier results, in the stretch of land observed in this research, in an hyperarid foggy and windy desert, especially from Cape Bojador to Cape Blanc (26-21°N) where the current year round is parallel to the shore (Personal communication, J.Dubief 1979, after the 'Pilot Chart of the North Atlantic Ocean').

North to south, the annual rainfall means are (J.Dubief 1963):

Cape Juby	43.0 mm
Aiun	51.6 mm
Dakhla (Rio de Oro Bay)	37.6 mm
Nouadhibou (Cape Blanc)	28.7 mm
(Nouakchott	140.5 mm)

Surface air temperatures are quite similar all along this coast (J.Dubief 1959):

	Annual means		Absolute annual	
	minima	maxima	minima	maxima
Juby	16°1	21°6	5°0	19°6
Dakhla	16°5	24°5	8°8	41°8
Nouadhibou	16°4	27°6	6°6	45°4
Nouakchott	19°0	32°2	4°5	47°0

Vegetation is, of course, very scarce, in particular between 21°N and 26°N. It includes Chenopodiaceae (*Traganum, Suaeda, Salsola, Atriplex, Cornulaca, Arthiocnemum*), Asclepiadaceae (*Leptodenia*), Graminae (*Panicum, Aristida, Sporobolus, Aeluropus, Spartina*), Zygophyllaceae (*Zygophyllum waterlotii*), Caryophyllaceae (*Polycarpaea*), together with other less significant families (Quezel 1965, at Cape Blanc). A relict mangrove patch (*Avicennia* sp.) is still found in Cape Timiris area.

Present marine mollusc repartition is given in Figure 1: no warm water littoral species now live north of Cape Blanc, except *Anadara senilis, Venericardia ajar* and *Donax rugosus* (Th.Monod 1948, G.Nickles 1950, 1955, J.C. Rosso & N.Petit-Maire 1979). In fact, most West African species now living as far south as Guinea or Angola do not get beyond the 19°N (J.C.Rosso & J.Monteillet 1978), on account of the low water temperature above this latitude (15° to 18°C at Cape Blanc in the winter time).

As for continental large mammals, the only one now present in the studied area is *Gazella dorcas (Oryx algazel* lives further inland in the Rio de Oro area). *Lepus capensis, Hyaena hyaena* and *Canis aureus* were rarely met.

Man, as small nomadic groups, avoids the cold and foggy, or windy and dusty coastal seasons. Only by 19°N does one small village of fishermen subsist upon scarce water spots and imported water.

366

On the whole, the Occidental Sahara continental border nowadays shows a striking biological nakedness.

2. THE PRE-NOUAKCHOTTIAN PERIOD (10 500-6 000 B.P.)

By 10 000 B.P., sea level was still some 35 m lower than nowadays and thus we have no information about life — especially human settlements — which could have taken place upon the continental shelf area now under water. Iso-baths indicate a shelf extension (200 m water depth) varying between 50 km off-shore to the north to 100 km to the south (Banc d'Arguin).

Upon the continental littoral stretch topping the dead cliff between 28° and 26°N, sparse human remains were found, associated with epipaleolithic tools without any ceramics. For one of these sites, located at 28°N (M.Charon, L.Ortlieb, N.Petit-Maire 1973) radiocarbon dates range from 10 500 to 6 100 B.P.

South of Cape Bojador, no trace of this culture is found; all dates for human settlements are later and associated with Neolithic artefacts.

The only large mammal collected in association with this period is *Lycaon pictus* (det. J.Bouchud), the African wolf, normally living in packs in savannas and mountains, feeding on antelopes or gazelles (J.Dorst & P.Dandelot 1972). However, it may live in sahelian and even saharan biotopes, subsisting on rodents, birds and domestic animals; it was seen, in 1930, even in the Tanez-rouft (J.Dekeyser 1956). Thus, its presence at Izriten, near Cape Juby, during the early Holocene, is not meaningful.

Except for *Amygdala decussata (= Venerupis decussatus)* integrated to the Izriten epipaleolithic context and ^{14}C dated 7 860 ± 170 and 8 100 ± 110 B.P. (D.Grébenart 1975) and which does not give any meaningful ecological indication, all the other marine molluscs determined from the middens along this northern part of the coast have not been dated and can belong as well to the early as to the late Holocene period (Neolithic sites being superimposed and mixed by deflation with Epipaleolithic ones). This is the case for one *Cymbium tritonis* (det. Ph.Brebion) collected at Izriten, which probably belongs to the Nouakchottian tropical species migration noted in three other places between 27° and 28°N.

Thus, biogeography indicates, in the early Holocene, human life down to Cape Bojador, while no trace of it could be found more to the South, probably still under the influence of strong upwelling and consequent climate in the coastal area.

3. THE NOUAKCHOTTIAN TRANSGRESSION PERIOD (6 000-4 000 B.P.)

The Holocene transgression is well-known along this coast (= Nouakchottian, P.Elouard 1968) by the important shell deposits (mainly *Anadara senilis*,

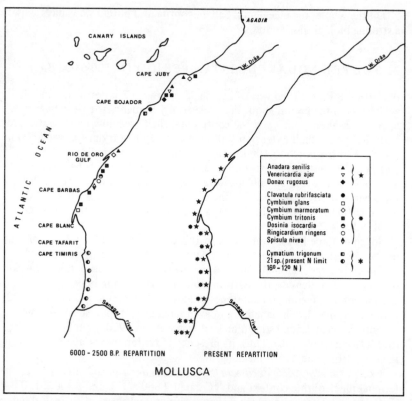

Figure 1. Comparative repartition of marine Molluscs in the last 6 000 years (graph prepared by J.C.Rosso.

Cerastoderma edule, Dosinia isocardia, P.Elouard 1968) found in the littoral sebkhas south of the 24° parallel, between −1 m and +2 m relative to MSL (L.Hebrard 1973, L.Ortlieb 1975): while, to the north, the shoreline was little or not modified by this sea level rise, on account of its morphology — it was, to the south, deeply festooned by shallow gulfs penetrating as far as 100 km inland (Ndramcha sebkha, L.Hebrard 1968, 1973). The transgression is dated 6 800 to 4 200 B.P., with its peak — at Nouakchott — dated 5 500 B.P. The shoreline morphology as well as the mollusc ecology point to a calm sea with little water movement (P.Elouard 1975).

We have good information about Mollusc populations along the Nouakchottian shoreline, both through *in situ* deposits and human settlements middens, found nearly all along the coast, except for that part of it between 26° and 23°30' where access to the sea was impeded by the cliff morphology (N.Petit-Maire 1979).

368

Since the region here studied covers the contact area between two inverte-brate faunal 'provinces' (Morocco-Lusitanian and Senegalian provinces), it indeed may provide valuable information about biotope dynamics: in fact, striking changes in nearly 30 species repartition are registered between 6 000 and 2 500 B.P., the frequency of occurrence and the density of shells indicat-ing a progressive shift in ecological conditions with an optimum towards the end of the transgression, slowly degrading after 3 500 B.P.

Figure 1 shows, relative to present repartition (M.Nickles 1950, 1955), the Nouakchottian repartition for some significant species. Since about 3 000 years:

– *Cymbium tritonis* and *C.marmoratum* have shifted 6° to 7° south (J.C. Rosso & N.Petit-Maire 1979).

– *Anadara senilis, Venericardia ajar* and *Donax rugosus* have shifted 3° to 4° south (J.C.Rosso & N.Petit-Maire 1979).

– *Spisula nivea, Ringicardium ringens* and *Cymbium glans* shifted 1° to 2° south (J.C.Rosso & N.Petit-Maire 1979) as well as *Dosinia isocardia* (Lecoin-tre 1964).

– 21 species, not listed on Figure 1, have shifted 3° to 7° south (Cape Timiris down to 12-16°N):

Neritina glabrata	*Conus genuanus*
Turritella ungulina	*Conus grayi (?)*
Pachymelania fusca	*Odostomia lamothei (?)*
Pachymelania aurita	*Crassostrea gazar*
Tympanotonus fuscatus	*Brachidontes atropurpureus*
Aclis beddomei	*Venericardia lacunosa*
Hexaplex angularis lyratus	*Congeria africana*
Thais forbesi	*Iphigenia laevigata*
Marginella aurantia (?)	*Corbula trigona*
Surcula coerulea	*Dentalium katchekense*

(J.C.Rosso, P.Elouard, J.Monteillet 1977, J.C.Rosso, N.Petit-Maire 1979).

All the species here listed are of tropical ecology and nowadays live at least down to the Angola coast. Their presence, during the Nouakchottian, up to Cape Juby or Rio de Oro Bay undoubtedly infers warmer waters than now-adays. The change of littoral profile to large shallow gulfs – in the south – also contributed to the extraordinary proliferation of some species (*Ana-dara senilis* for example), but this is not the case to the north where coastal morphology hardly changed during the transgression.

The question is whether one deals with a migration from north to south resulting in the present repartition, from positions already held before Nou-akchottian times (J.C.Rosso & N.Petit-Maire 1979) or with a south to north migration begun by 6 000 B.P. from more southerly positions. The slightly positive oxygen isotopic values, measured upon Ouljian shells carbonates, imply either a concentration of salt or a relatively low water temperature (L. Ortlieb 1975). This may mean that the cold current and/or the upwelling were

369

active even during this warm period (which was not the case during the Moghrebian, Plio-Pleistocene transgression), dated 75 000 to 95 000 B.P. in Morocco (C.Stearns & D.Thurber 1965). During the Upper Pleistocene glacial maximum, the polar front shifted south to about 42°N and intensification of cold Canary Current and of upwelling resulted in an abrupt fall of water temperatures (McIntyre et al. 1976): these can be estimated to 10°C in the winter and 15°C in the summer, off Cape Blanc at 18 000 B.P. (U. Pflaumann 1979). The present seasonal temperatures now vary respectively from 15° to 27°C, between 21° and 19°N (L.Hebrard 1973).

Under such conditions, it seems quite impossible for warm water molluscs to have subsisted as far North, from the Moghrebian to the Holocene, and then been able to undergo a 'renewal' in the mid-Holocene climatic optimum. They must, then, have migrated from south to north: since larvae are unable to ascend strong currents or stand cold waters, one must assume a weakening (or even a cessation) of the Current and of the upwelling during that period. These data correlate with the research by M.Sarnthein & Koopmann (1979), L.Diester-Haass (1979) and M.Rossignol-Strick & Duzer (1979), pointing to a marked weakening of the upwelling at the same period, due to 'a breakdown of the anticyclonic cell over the East-Atlantic margin and a generally very sluggish harmattan' (M.Sarnthein & Koopmann, id.). The present time situation resembles 'more that of the glacial regime than that of the mid-Holocene' (M.Sarnthein & Koopmann, id.), and explains the southward migration of the same species after 2 000 B.P.

These important changes undergone, during the Nouakchottian, by the Invertebrate marine fauna are not accompanied by such large differences in the continental coastal band.

North of Cape Bojador, the human epipaleolithic occupation disappears. Some of the anthropological features of these populations ('Cro-Magnoïd' or late Mechta-el-Arbi type) may perhaps be recognized in later groups, but South of the 23°N only (Mahariat), at later dates (N.Petit-Maire et al. 1979). But the oldest Canarians (Guanche from Guyadeque, oral communication, D. Ferembach) also have similar Cro-Magnoïd traits (although missing the corresponding lithics). One does not know when these populations reached the Canary Islands, but it is now proven that intra-mediterranean navigation took place in prehistoric times as early as 8 000 B.P. (G.Camps 1979). Even nowadays, Canarian fishing boats may sometimes, especially in the summer time, find favourable wind conditions to sail back from Morocco coast to Fuerteventura island (J.Dubief, *in litt.* 1979). Thus, one must not wonder at the possibility of such a crossing, especially during the probable weakening of Canary Current implied by Mollusc fauna changes after 6 000 B.P.

All along this coast, except for the stretch extending South of Bojador, neolithic life begins by 6 000 B.P. No common anthropological or archeological traits are found in the populations north and south of the 23-25°N natural barrier, the inhospitable character of which having obviously prevented passage, at least along the littoral band we observed.

370

This Nouakchottian habitation is still quite scarce, as indicated by low frequency of sites and possibly by large size of *Anadara senilis* and other shells collected in 6 000-5 000 B.P. middens, which may suggest the possibility for selectivity in human consumption (or optimal ecological conditions for the species). No trace of settlements were found; this could mean that the populations did not live along the coast at that time, but only roamed from inland to collect sea-food. This hypothesis is strengthened by the observation during that period of high lake levels 50 to 100 km inland (G.Delibrias, L.Ortlieb, N.Petit-Maire 1976), which might have been more attractive to man and other mammals than the then driest coastline, where no trace of large mammalian life was found: the Neolithic trips to the coast must have been for fishing-gathering rather than hunting-gathering. Further research inland should test this hypothesis.

In brief, the Nouakchottian transgression brought great change in littoral marine biogeography, but, on the continent, important differences in human and mammalian life did not yet appear.

4. THE TAFOLIAN STAGE (4 000-2 000 B.P.)

By 4 200 B.P., the transgressive episode is over (although two minor oscillations may have occurred at 3 500 and 2 500 B.P., G.Einsele et al. 1974) and the shallow gulfs progressively turn into lagoons. By 2 000 B.P., these have completely evaporated into sebkhas: no marine shell younger than this date has ever been found either *in situ* or in middens (L.Ortlieb 1975, Delibrias, Ortlieb, Petit-Maire 1976).

Radiocarbon dates show that Nouakchottian proliferation of tropical molluscs south of 23°N lasted at least until 3 000 B.P. However, the shells *(A. senilis, C.edule)* are then smaller and thinner, on account of abnormal salinity. During that period, mangrove spreads as far north as Cape Tafarit, as proven by the presence of the *Avicennia* (L.Hebrard 1973).

On the continent, lake levels are then very low and will by 3 000 B.P. reach total evaporation (G.Delibrias, L.Ortlieb & N.Petit-Maire 1976). The minor marine oscillation at 3 500 B.P. may have resulted, along the coast, in a rise of phreatic level creating small water-pools (Einsele et al. 1974). These two processes might be related to the major biological changes observed in Atlantic Sahara during the 'Tafolian' (L.Hébrard, 1972).

Neolithic groups have, during that period, passed or lived all along the Saharan coast (Figure 2). Except for the inhospitable 23-25°N part, the large extent of sites, the archeological density and quality of artefacts, as well as the existence of large necropolis (26°, 22°, 21°, 20°N. N.Petit-Maire et al. 1973-1979), prove continuous or at least seasonal settlement.

However, a great difference in way of life characterizes the settlements between 21° and 24°N: there, no frequent adornments, no elaborate burials, very few ceramics, no grindmills, no differentiation in site topology. In

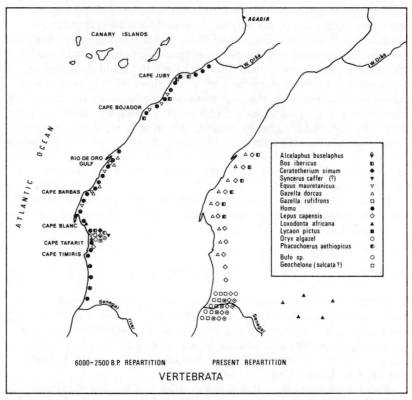

Figure 2. Comparative repartition of large mammalian fauna and human occupation, in the last 6 000 years.

contrast, south of Cape Blanc, at the same period and in the same anthropological context (common artefacts, rites and physical traits), shell pendeloques, rock and shell beads necklaces, large numbers of pottery vases with varied shapes and patterns (H.Camps-Fabrer 1979, E.de Longrée 1979), division of sites into precise topological areas (J.P.Carbonnel 1979) indicate aesthetic intentions inconsistent with a hard lifetime, busily filled with survival preoccupations.

Moreover, grindmills and mortars are very numerous south of Cape Blanc only (sometimes about 100 upon a single dune site). They imply a large quantity of Gramineae in the nearby environment, and corroborate the $^{13}C/^{12}C$ ratio results, measured in human remains from burials from this period at $20°$ and $20°30'N$, indicating an alimentation based on vegetals or on herbivorous animals (G.Delibrias & J.Evin 1979).

Palynological studies have as yet been made only between $20°$ and $21°N$,

372

no comparison being available, for the moment, with the northern strip. The vegetation associated with sites dated 4 500-2 500 is chiefly of Chenopodiaceae, Graminae and Cyperaceae. These latter, not found nowadays in the same area (P.Quezel 1965), probably indicate the existence of water-pools or marshy spots. The percentage of Graminae, 22 per cent, is very high if one takes into account the strong impact of Chenopodiaceae in littoral areas and the consequent flora inversion found more or less clearly in other known sea or lake coastlines (R.Bonnefille & A.Vincens 1977). A 22 per cent littoral Graminae rate may thus correspond to an 80 per cent rate inland (oral communication M.Rossignol-Strick 1979).

The large mammalian fauna associated with the neolithic sites, from north to south, show the same discrepancy between the two areas north and south of Cape Blanc (Figure 2).

In three places north of Bojador and two places between the Tropic and Cape Blanc, fragmentary bones of small Bovids (*Bos ibericus* ?) and Equids (*Equus mauretanicus* ?) have been found. These remains, although scarce, indicate in Neolithic times feeding and drinking possibilities inconsistent with the modern situation.

A great number of marine mammal bones (*Tursiops* sp., *Physeter* sp., *Monachus monachus*) are associated with the middens north of 26°N.

South of Cape Blanc, the Vertebrate repartition was completely different from nowadays, and implies continental migrations as important as the ones observed for marine molluscs during the Nouakchottian (Figure 2). Between 4 500 and 2 500 B.P., the following species were present by 20°30'N (determinations J.Bouchud & C.Guérin):

Loxodonta africana
Ceratotherium simum
Phacochoerus aethiopicus
Alcelaphus buselaphus
Gazella rufifrons
Bos ibericus
Equus mauretanicus
Bubalus caffer (?) (C.Guérin)
Conochaetes gnu (?) (J.Bouchud)

This association is clearly of steppe or savanna type.

The elephant still lived about 50 years ago up to 17°N, from Dakar to Tchad. Nowadays, a small relict group subsists in Assaba South Mauretania area. Under normal conditions, one individual daily drinks and bathes and it eats about 300 kg vegetal food. The white rhino was formerly abundant in most tropical African savannas; it now only exists in Ubangui and along the white Nile. It needs water and feeds on grass (Dekeyser 1956). The phacochoerus is a typically sahelo-sudanese sanimal and still lives in Mauretania, south of 17°N; it often drinks and bathes. Solitary individuals may be found in arid areas, but the case is abnormal, since water is 'for this species a first rate biological factor' (Dekeyser 1956). This wild Suid feeds on grass, wild

373

fruit and the roots of aquatic vegetals. Its bony remains are found in large quantities at Cape Tafarit latitude. The bubale antelope is an open bushland dweller and is never found in sub-saharan desert areas (J.Dorst & P.Dandelot 1976). It is herbivorous and, under normal conditions, regularly drinks; however, it may remain without any water for long periods (J.Dorst & P.Dandelot 1976). It lives nowadays south of 16°N, but a small size bubale possibly still subsists in Morocco and Niger (Dekeyser 1956). The red forehead gazella is now only known south of 17°N. In contrast to *G.dorcas,* it cannot live in the Sahara; however, it can stay long without drinking, provided succulent vegetals are available. It is often associated with *Acacia* and *Balanites,* both not existing beyond the 100 mm isohyet. The African buffalo and the gnu are also savanna animals, never living far from water spots. As for *Bos ibericus* and *Equus mauretanicus,* they are now extinct but indicate also a grassland environment.

Thus, all data about coastal continent biogeography during the Tafolian correlate and point to a 3° to 4° northern shift of the sahelian steppe, between 4 500 and 2 500 B.P.

5. THE POST-TAFOLIAN PERIOD (2 000-0 B.P.)

Since about 2 500 B.P., a strong regression in human and other continental life takes place along the littoral. Little by little, sites get poorer and, by 2 000 B.P., man and animals have migrated away. Only nomadic small groups will now cross this land. In fact, aridification processes, begun as early as 3 000 B.P. with dune formation, chased life away from the whole area (L. Hébrard 1973).

6. CONCLUSIONS

This multidisciplinary study of invertebrate, vertebrate and human repartition along the Atlantic Sahara coast in the last 10 000 years has important palaeoclimatic implications.

1. A weakening (or cessation) of upwelling and a consequent warming up of coastal water took place during the Nouakchottian (marine Molluscs data). Upon the continent, Vertebrate and human life were scarce along the coastal strip: high lake levels being known 50 to 100 km inland (21°to 20°N) during that period, these more favourable conditions may account for the relative low peopling of the coast.

2. After the Nouakchottian episode, vertebrate and human life develops significantly along the whole low shore strip. South of Cape Blanc, a steppe type environment is characterized by high increase in Gramineae and presence of Cyperaceae, associated with a large mammalian fauna now living south of 17°N and sedentary or at least seasonal human settlements.

374

However, the density of human and animal occupation — especially north of 21°N — must not be overestimated in terms of fresh water resources:
— most of the sites were occupied successively and not simultaneously (the large *Anadara senilis* middens at Cape Tafarit cover a 400 to 600 year span, G.Delibrias & J.Evin 1979).
— the majority of settlements were located near wells or water spots (now dry), showing that rainfall was not an important factor for survival. No evidence for run-off was found in morphology; no oysters were found in middens, except at sebkha Laasailia (26°53'N).
— large vertebrates (Elephant, Rhino, etc. . . . .) were a 'relict fauna' in northern and southern Sahara since the early Holocene, located in the most favourable areas. (They probably migrated to coastal phreatic outcrops after the Nouakchottian lakes evaporated, together with neolithic man).
— large mammals, including man, stand dry conditions much better than small ones, the ratio of water evaporation to body weight being much less.

In spite of these considerations, the association of steppe vegetation and animal life with important cultural neolithic traits, by 3 500 B.P., as far north as 20-21°N, implies climatic conditions very different from the hyperarid present ones and a minimum annual rainfall of about 150 mm (cf. 28 mm at present).

The ex-Spanish Sahara coast extended beyond the belt margin of the Sahelian steppe, but on account of the attenuated meteorological and oceanic processes formerly mentioned, was less arid than nowadays, to the north and the south in particular.

By 2 000 B.P., the steppic belt shifted back south, maybe quite rapidly: in Sahel areas with a low mean annual rainfall 'a shift of the climatic zones of only 1° in latitude within 100 years would mean a decrease in rainfall of some 40-50 per cent' (J.L.Cloudsley-Thompson 1977, p.41).

REFERENCES

Bayle des Hermens, R.de, E.de Longrée, A.Smith & D.Vialou 1979. Les sites nécropoles de Tintan et Chami: Contexte archéologique. In: N.Petit-Maire, et al., *Le Sahara atlantique à l'Holocène. Peuplement et écologie*. Mémoire CRAPE XXVIII, Alger. 340p., 76 pl.

Beucher, F. 1979. Les sites nécropoles de Tintan et Chami: Contexte palynologique. In: N.Petit-Maire, et al., *Le Sahara atlantique à l'Holocène. Peuplement et écologie*. Mémoire CRAPE XXVIII, Alger, 340p., 76 pl.

Bonnefille, R. & A.Vincens 1977. Représentation pollinique d'environnements arides à l'est du Lac Turkana (Kenya). *Suppl. Bull. AFEQ 1*(50):235-247.

Brebion, P. 1973. Nouvelles recherches sur les Gastéropodes pliocènes et quaternaires du Maroc atlantique. *C.r. Acad. Sci. Paris 277:*489-492.

Brebion, P. & L.Ortlieb 1976. Nouvelles recherches géologiques et malacologiques sur le Quaternaire de la province de Tarfaya (Maroc méridional). *Géobios 9*(5):529-550.

Camps, G. 1979. La navigation préhistorique en Méditerranée. *Conf. Univ. Marseille-Luminy.*

Camps-Fabrer, H. 1977. Les objets de parure et l'industrie de l'os des gisements de Laasailia à Amtal. In: N.Petit-Maire et al., *Le Sahara atlantique à l'Holocène. Peuplement et écologie*. Mémoire CRAPE XXVIII, Alger, 340p., 76 pl.

Charon, M., L.Ortlieb & N.Petit-Maire 1974. Occupation humaine holocène de la région du Cap Juby (Sud-Ouest marocain). *Bull. Mém. Soc. Anthrop. Paris*, t.X, sér.XIII, no. 4, pp.279-412.

Cloudsley-Thompson, J.L. 1977. *Man and the biology of arid zones*. Edward Arnold, London.

Dekeyser, P. 1956. *Les Mammifères de l'Afrique noire francaise*. IFAN, Dakar.

Delibrias, G., L.Ortlieb & N.Petit-Maire 1976. New ^{14}C data for the Atlantic Sahara (Holocene). *J. Hum. Evol. 5*:535-546.

Delibrias, G. & J.Evin 1979. Les sites nécropoles de Tintan et Chami: Datations par la méthode du Carbone 14. In: N.Petit-Maire et al., *Le Sahara atlantique à l'Holocène. Peuplement et écologie*. Mémoire CRAPE XXVIII, Alger, 340p., 76 pl.

Diester-Haas, L. & H.J.Schrader 1979. Neogene coastal upwelling of northwest and south-west Africa. *Marine Geol. 29*:39-54.

Dorst, J. & P.Dandelot 1976. *Guide des grands Mammifères d'Afrique*. Delachaux et Niestlé, Lausanne.

Elouard, P. 1968. Le Nouakchottien, étage du Quaternaire de Mauritanie. *Ann. Fac. Sci. Dakar, 22*:121-137.

Elouard, P., H.Faure & L.Hébrard 1969. Quaternaire du littoral mauritanien entre Nouakchott et Port-Etienne. *Bull. Liais. Ass. Sénégal. Et. Quatern., Dakar 23/24*:15-24.

Elouard, P. 1979. Les sites nécropoles de Tintan et Chami: Mollusques testacés. In: N. Petit-Maire et al., *Le Sahara atlantique à l'Holocène. Peuplement et écologie*. Mémoire CRAPE XXVIII, Alger, 340p., 76 pl.

Einsele, G., D.Herm & H.U.Schwarz 1974. Holocene eustatic (?) sea level fluctuations at the coast of Mauritania. *'Meteor' Forsch.-Ergebn. C 18*:43-62.

Einsele, G., D.Hermn & H.U.Schwarz 1977. Variations du niveau de la mer sur la plateforme continentale et la côte mauritanienne vers la fin de la glaciation de Würm et à l'Holocène. *Bull. Liais. Ass. Sénégal. Et. Quatern., Dakar 51*:35-48.

Fontes, J.C., P.Elouard & H.Faure 1972. Essai d'interprétation isotopique et paléoécologique du Quaternaire marin de Mauritanie. *Actes 6ᵉ Congr. panafr. Préhist. Et. Quatern. Dakar*:493-497.

Grébenart, D. 1975. Cénotaphe et inhumations épipaléolithiques de la nécropole d'Izriten (Sahara atlantique marocain). *Antiquités Afr.*

Hébrard, L. 1972. Un épisode quaternaire en Mauritanie (Afrique occidentale) à la fin du Nouakchottien: le Tafolien, 4 000-2 000 ans avant le présent. *Bull. Liais. Ass. Sénégal. Et. Quatern., Dakar 33/34*:5-16.

Hébrard, L. 1978. Contribution à l'étude géologique du Quaternaire du littoral mauritanien entre Nouakchott et Nouadhibou (18-21° latitude Nord). *Doc. Lab. Géol. Fac. Sci. Lyon 71*.

Lecointre, G. 1964. Les relations du Quaternaire marin de Mauritanie avec celui des régions avoisinantes. *Bull. Bur. Rech. géol. minières 2*:91-109.

McIntyre, A., N.G.Kipp et al. 1976. Glacial North Atlantic 18 000 years ago. *Geol. Soc. Amer. Mem. 145*:43-76.

Mauny, R. 1956. Préhistoire et zoologie: la grande 'faune éthiopienne' du Nord-Ouest africain du Paléolithique à nos jours. *Bull. Inst. fr. Afr. noire, Dakar*, série A: pp.246-279.

Monod, T. 1945. La structure du Sahara atlantique. *Trav. Inst. Rech. Sahar., Alger, 3*:27-55.

Monod, T. 1961. The late Tertiary and Pleistocene in the Sahara and adjacent southerly regions with implications for primate and human distributions. In: *African ecology and human evolution*. Wenner-Gren Found, p. 117-229.

Nickles, M. 1950. *Mollusques testacés de la Côte occidentale d'Afrique*. Lechevalier, Paris.

Nickles, M. 1955. *Scaphopodes et Lamellibranches récoltés dans l'Ouest africain*. Atlantide Report 3. Danish Science Press, Copenhagen.

Ortlieb, L. 1975. *Recherches sur les formations plio-quaternaires du littoral ouest-saharien (28°30' - 20°40' lat. N)*. Trav. et Doc. 48. ORSTOM, Bondy, France.

Petit-Maire, N. 1979. Prehistoric palaeoecology of the Sahara Atlantic coast in the last 10 000 years. A synthesis. *J. Arid Environments 2:* 85-88.

Petit-Maire, N. & J.P.Fléchier 1979. La sebkha Laasailia. La sebkha Amtal. La sebkha Mahariat. La sebkha Lemheiris. La sebkha Edjaila. In: N.Petit-Maire, et al., *Le Sahara atlantique à l'Holocène. Peuplement et écologie*. Mémoire CRAPE XXVIII, Alger, 340p., 76 pl.

Petit-Maire, N. & M.Charon 1979. Les sites nécropoles de Tintan et Chami: Les restes humains. In: N.Petit-Maire et al., *Le Sahara atlantique à l'Holocène. Peuplement et écologie*. Mémoire CRAPE XXVIII, Alger, 340p., 76 pl.

Petit-Maire, N. 1979. Les sites nécropoles de Tintan et Chami: Autres faunes. In: N.Petit-Maire et al., *Le Sahara atlantique à l'Holocène. Peuplement et écologie*. Mémoire CRAPE XXVIII, Alger, 340p., 76 pl.

Pflaumann, U. 1979. Variations of the surface-water temperature at the Eastern North Atlantic continental margin (Sediment surface samples, Holocene climatic optimum, and Late Glacial maximum). *Palaeoecology of Africa 12*.

Rosso, J.C. 1974. Contribution à l'étude paléontologique du Quaternaire sénégalo-mauritanien: Mollusques du Nouakchottien de Saint-Louis (Sénégal). I. Gastropodes et Scaphopodes. *Bull. Inst. fond. Afr. noire, A 36*(1):1-50.

Rosso, J.C. 1975. Contribution à l'étude paléontologique du Quaternaire sénégalo-mauritanien: Mollusques du Nouakchottien de Saint-Louis (Sénégal). II. Lamellibranches. *Bull. Inst. fond. Afr. noire, A 36*(4):789-841.

Rosso, J.C. 1976. Mollusques du Nouakchottien de Mauritanie atlantique. Catalogue systématique et critique. *Doc. Lab. Géol. Fac. Sci., Lyon*.

Rosso, J.C., P.Elouard & J.Monteillet 1978. Mollusques du Nouakchottien (Mauritanie et Sénégal septentrional): Inventaire systématique et esquisse paléoécologique. *Bull. Inst. fond. Afr. noire, A 39*(3):465-486.

Rosso, J.C. & N.Petit-Maire 1979. Amas coquilliers du littoral atlantique saharien. *Bull. Musée d'Anthrop. Préhist., Monaco 22:* 79-118.

Rossignol-Strick, M. & D.Duzer 1979. Late Quaternary pollen and dinoflagellate analysis of marine cores off West Africa. *'Meteor' Forsch.-Ergebn.* C30, 1-14.

Sarnthein, M. 1978. Sand deserts during glacial maximum and climatic optimum. *Nature 271:* 43-46.

Sarnthein, M. & B.Koopmann 1979. Late Quaternary deep-sea record on NW African dust supply and wind circulation. *Palaeoecology of Africa 12*.

Sarnthein, M. & L.Diester-Haass 1977. Eolian sand turbidities. *J. Sed. Petrol. 47:* 868-890.

Stearns, C.E. & D.L.Thurber 1965. Th^{230}/U^{234} dates of late pleistocene marine fossils from the Mediterranean and Moroccan littorals. *Quaternaria 7:* 29-42; In: *Progress in oceanography*. Pergamon Press, Oxford (1967).

Thompson, J. 1978. Ocean deserts and ocean oases. *Climatic Change 1:* 205-230.

GENERAL ASPECTS

COMPARISON OF LATE-QUATERNARY CLIMATIC EVOLUTIONS IN THE SAHARA AND THE NAMIB-KALAHARI REGION

E. M. VAN ZINDEREN BAKKER, Sr
Institute for Environmental Sciences, University of the Orange Free State, Bloemfontein

INTRODUCTION

The purpose of this paper is to make a comparison between late-Quaternary palaeoenvironments of the Sahara and the Namib-Kalahari region. This comparison will be limited to the periods of the last glacial maximum and the Holocene Climatic Optimum. This exercise will disclose remarkable similarities and correlations as well as differences which may be explained by taking the disparity of the geographic setting of the two regions into account.

The age of the two deserts has not been settled finally. According to Axelrod & Raven (1978) botanical evidence shows that the vast Sahara region dried out gradually and only reached arid conditions in Pliocene-Quaternary times. However, Diester-Haass (1977) concluded from studies of D.S.D.P. cores at site 369 (±27°N) of the coast of Rio de Oro that in early and middle Miocene times, the climate in the adjacent northwestern Sahara was arid and that the climate from the late Miocene to the Pleistocene was similar to that of the late Quaternary.

The age of the arid Namib can also not be fixed with certainty. The correlation between the Antarctic glacial history and the upwelling along the southwest African coast has been described by Van Zinderen Bakker (1975). For the timing of the desertification of the coastal region determinations of the palaeotemperature of deep water were used (Shackleton & Kennett 1974). The desert was consequently assumed to be of Oligocene age. In this assumption, the quantity of the cold upwelling water, needed to cause aridity along the Namibian coast, was not taken into account.

Comparative studies of fossil evidence by Axelrod & Raven (1978) and Tankard & Rogers (1978) came to the conclusion that the Namib desert is very youthful and is of Quaternary age. Definite proof for an intermediate solution between these extremes has recently been produced by Siesser (1978) who showed that the upwelling of South Atlantic Central Water, which existed spasmodically along the Namibia coast since the Middle or Late Oligocene, was strongly intensified from the early Late Miocene onward. The volume of cold upwelling water then increased as a consequence of the rapid accumulation of the East Antarctic ice sheet at about 10-14 m.y. ago (Shackleton &

Kennett 1975) with the consequence that existing arid regions gradually enlarged to the present day coastal desert. The intensification of the aridity dates back to ~ 10 m.y. ago (Siesser 1978; Coetzee 1978).

The northern and southern parts of Africa have a similar orientation with respect to latitude, a situation which is unique for a continent. They are hence influenced by identical climatic zones. In the extreme case of the last ice age, they were both exposed on their poleward side to nearby glacial conditions, while their opposite margins were influenced by tropical climates.

The most fruitful comparison can be made between the western Sahara and the Namib desert along a profile running from Algeria to Zaire and thence to Cape Town (Trewartha 1966, p.62). Along this transect a comparison will be made of the following regions which have a more or less similar latitudinal and climatic setting in the north and in the south:

1. The Maghreb and the southwestern Cape region.
2. The western Sahara on the one hand and southern Kalahari and Namib on the other hand.
3. The zone of the Sahel and the region of pans in northern Botswana (19-21°S).

PRESENT CLIMATIC SETTING

The oceanic environment

Important differences and similarities exist in the exposure to maritime influences along the western and temperate coasts of the two regions under discussion. The oceanic environments along the temperate coasts of northern and southern Africa differ considerably. The north African coast faces the temperate Mediterranean, while the south African coast is exposed to the vast Southern Ocean, which grades from a temperate sea southward into the ice-covered Antarctic. The longitudinal extent of the coastal regions also has a great influence on the synoptic situation.

A striking similarity, however, exists between the oceanic setting along the west coast of northern and southern Africa, where cold currents from higher altitudes touch the coast. As a consequence of the configuration of the coast line both these currents have a limited latitudinal extent. The Canaries current is only aligned with the coast from 30°N to Cape Blanco at 21°N, while the Benguela current touches the coast between ca. 30°S and Cape Frio at 18°S. These cool currents are accompanied by upwelling which brings cold water of polar origin to the surface with the consequence that the isotherms along these coasts bend sharply equatorward. This cold surface water causes condensation and fog and prevents moisture from penetrating the continent.

Solar radiation

The solar radiation received by the Sahara and the Namib does not differ very much. The incoming solar radiation depends on many factors of which cloud cover and atmospheric moisture are the most important. During the southern summer, the radiation in southern Africa reaches a maximum in the erg of the southern Namib, viz. 300.10^5 Jm^{-2} day^{-1} (Schulze & McGee 1978). The South Atlantic anticyclone is responsible here for a clear sky and dry air. Similar values are known for the Sahara where in summer a maximum of 308.10^5 Jm^{-2} day^{-1} occurs in the large depression of the Libyan-Egyptian desert (Rognon & Williams 1977).

The maximum temperatures for the hottest day of the Namib and Sahara differ very little and are respectively 45°C and 48°C (Schulze 1974, Rognon & Williams 1977). These facts raise the evaporation and aggravate the aridity in the deserts.

Atmospheric circulation and precipitation

Except for their tropical and temperate extremities, northern and southern Africa are dominated by subtropical anticyclonic belts. The descending air over the Sahara is in summer provided by the Tropical Easterly Jet which originates over Tibet and affects the vast landmass stretching from the Red Sea to the Atlantic (Flohn 1964, 1966). This jet stream increases subsidence and most probably prevents the extension of tropical summer rains north of about Latitude 16°N. The aridity is aggravated on the ocean margins by up-wellings in the cold Somali and Canaries currents off the east and the west coast respectively and by the dry northeastern monsoon from southwestern Asia which traverses the eastern Sahara in winter. Other existing minor causes of aridity are not relevant to our present discussion.

The northern margin of the Sahara receives winter rainfall from cyclones which gradually lose their influence as they travel from west to east. The southern tropical margin of the Sahara is in summer influenced by the 'monsoon' which brings moisture from the Atlantic as far north as the Sahel. This rainfall affects a much larger area south of the equator. The isohyets in this tropical region run along the parallels of latitude indicating a zonal control of precipitation.

Although the climatic mechanisms of northern and southern Africa are comparable in broad outline, they differ substantially in their geographical setting. As a consequence of the vast surrounding oceans, the climatic system of Southern Africa is in general less complex than that of the Northern Hemispheric part of the continent. The subtropical jetstream along the southern upper edge of the Hadley Cell runs apparently undisturbed from west to east and is situated over the subtropical anticyclonic belt. This belt is centered around 32°S and shows seasonal meridional shifts of about 4° latitude. This anticyclonic belt exerts an aridifying influence over most of the subcontinent, especially over the western half. The belt is divided into three parts: a high situated over the southern Atlantic near the southwest African coast, a cell aloft over the subcontinent which is weakened in summer and thirdly a high pressure system over the Indian Ocean. The strong South Atlantic anticyclone and the upwelling accompanying the Benguela current are responsible for the concentration of the driest climate along the west coast in the Namib desert. From here the climate ameliorates gradually toward the east. The aridifying influence reaches as far as the middle of the subcontinent where the 400 mm isohyet divides southern Africa into a dry western half and a wetter eastern half. This situation is totally different from northern Africa where the Sahara stretches in an east-west direction and has its largest meridional extent in the eastern part.

Depressions are formed regularly in the southern westerlies over the south Atlantic where the surface westwinds are in contact with warmer tropical air. These sub-tropical cyclones travel east- and south-eastwards. In winter, their trajectories cross the southwestern corner of the continent and bring rain. In summer, they are usually displaced further southwards.

The geographical pattern of the rainfall distribution in southern Africa differs from that in north Africa, where the summer and winter rainfall areas do not overlap but are separated by the hyper-aridity of the Sahara which covers about $15°$ of latitude (Rognon & Williams 1977, Figure 2). The Namib desert on the other hand receives minimal summer and winter rain, the areas of which overlap as in South Africa. The strictly winter rainfall zone of South Africa is confined to the extreme southwestern area, but from this centre, the cyclonic rain invades the continent from September to November. Some winter rainfall, which is partly of an orographic nature reaches the latitude of nearly $22°$S. These invasions of winter rainfall are accompanied by incursions of cold polar air from high altitudes, causing severe frost. Blocking of these cold air masses, as described for the Atlas (Rognon & Williams 1977) does not occur in southern Africa.

Along the tropical margin of southern Africa, north of $20°$S, the summer rainfall is connected with the southward passage of the ITCZ. During the southern solstice season, humid tropical air masses are present along this latitude and can invade southern Africa under favourable conditions where these incursions cause rainfall by convection and convergence (Schulze 1974, p.268). The advection of such tropical humid air occurs when an equatorial trough is formed over the western interior of southern Africa. A similar penetration of tropical air results from the establishment of a low cut off in the westerlies over South Africa. Equatorial and maritime humid air from the Indian Ocean can also invade the plateau if a high pressure cell settles offshore along the east coast and draws in maritime air with an anti-clockwise circulation. These different weather types have been described by Vowinckel (Trewartha 1966, Schulze 1972).

The summer rains advance southward from November onward and reach a maximum over eastern South Africa in January and over the western parts in March (Schulze 1974, Figure 141). In April, summer rainfall stops abruptly and in May is replaced by winter rainfall in the south.

384

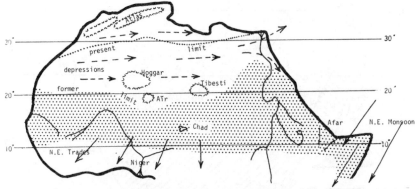

Figure 1. Rainfall pattern for Northern Africa c. 18 000 B.P. during northern winter. The present and former southern limits of regular winter rainfall are given. The stippled area represents the region which receives less than about 150 mm rainfall. (Butzer, 1958; Lauer & Frankenberg, 1979).

THE LATE PLEISTOCENE GLACIAL MAXIMUM (~ 18 000 B.P.)

The oceanic environment

This time span covers the coldest part of the Late Pleistocene in the Southern Ocean (Hays et al. 1976), tropical Africa (Coetzee 1967) and the Northern Hemisphere.

The oceans which were relatively warm before 20 000 B.P. now became cooler except for the equatorial region where the surface water temperature decreased only slightly (Gardner & Hays 1976, Bé et al. 1976). The Atlantic north of 42°N was transformed into a polar ocean (McIntyre et al. 1975), while the western Mediterranean was in summer 8°C colder than today (McIntyre et al. 1976). In winter, strong upwelling along the northwestern African coast caused a decrease in water temperature of 5-6°C (Gardner & Hays 1976; Diester-Haass, this volume).

Important changes also occurred in the Southern Ocean. Pack-ice even in summer reached 55°S. latitude which increased the albedo of the Earth. In the South Atlantic and the Indian Ocean, the Antarctic Polar Front shifted from 6-10° northward but it was rather stationary south of Africa (Hays et al. 1976). The Subtropical Convergence practically had the same position as at present, so that the subantarctic zone was latitudinally compressed (ibid.). The upwelling along the Namibian coast increased in winter and spread further north, bringing aridity to the Namib, Angola and Gabon (Van Zinderen Bakker 1967, de Ploey 1969, Andel & Calvert 1971, Bornhold 1973, Siesser 1978). The Benguela-South Equatorial Current strengthened especially in winter and was responsible for a cooling of the South Equatorial region of 8°C (Gardner & Hays 1976). All these phenomena point to a steeper thermal gradient and con-

385

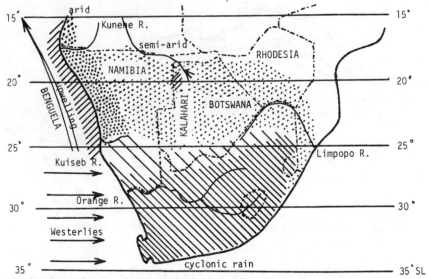

Figure 2. Rainfall pattern for Southern Africa c. 18 000 B.P. Southern winter.

sequently a strengthening of the atmospheric and oceanic circulation and more vigor of the trade winds (Bé et al. 1976, Prell et al. 1976, Gardner & Hays 1976) and not necessarily to an extensive shift of the climatic zones. These conditions with a stronger winter circulation occurred during most of the past 200 000 years (Gardner & Hays 1976), while present day conditions with a lower thermal gradient, weakened circulation and upwelling were only found about 125 000 B.P. (ibid.). The cooling of large regions of the oceans resulted in a diminished evaporation on the ocean surface and a generally drier climate with the consequence that the ice accretion climate of the polar ice sheets changed into an ice ablation phase. The production of cold Bottom Water and the intensity of upwelling increased.

Changes in the distribution of cyclonic rainfall

— The problem of the distribution of cyclonic winter rainfall during the last glacial maximum (18 000 B.P.) in northern and southern Africa has not been solved yet. The former assumption that the climatic belts moved equatorward during this time is at present replaced by the postulate that latitudinal movements of the pressure belts and the ITCZ were minimal, but that the circulation system was strengthened (Bé et al. 1976, Prell et al. 1976, Gardner & Hays 1976). The consequence would be that more depressions penetrated the northern and southern ends of Africa and brought here more rainfall especially in winter. These events would be accompanied by invasion of cold polar air.

— The available evidence for this wider spread of cyclonic rainfall is equivocal. In the Sahara according to Rognon & Williams (1977, p.314), winter rain could reach the Hoggar. This view of more rainfall in the northern Sahara during the last glacial maximum is also supported by studies on ocean sediments off the Northwest African coast (Chamley, this volume, Diester-Haass ibid). Rossignol (this volume) studying pollen occurring in ocean cores, however, concludes that from 19 000 to 12 500 B.P., the climate of the western Sahara was very arid. Brunnacker (this volume) also infers from studies on loess deposits that the glacial climate of the northern Sahara was arid except for the transitional periods at the beginning and the end of it.

– The situation was different in the region of the Atlas mountains where orographic obstructions caused heavy precipitation. Beucher (1971) studied the pollen flora deposited around lakes in the Northwest Sahara, south of the Atlas range, which were fed by this orographic rainfall. The Saourian and Soltanian terraces, respectively in Algeria and Morocco were laid down from > 40 000 to 12 500 B.P. This so-called pluvial was of a very complex nature as has been revealed by detailed stratigraphic studies at Cape Rhir (Sabelberg 1977).

– In southern Africa, the distribution of the cyclonic rainfall during the last glacial maximum is also not well-known. Here pollen evidence from the glacial maximum is so far only available from one core (Rietvlei 2) at the Southwestern Cape (Schalke 1973). The spectra can only be construed to indicate a cold, rainy climate with strong winds which are typical for the 'glacial' climate and led to dune formation. Forest did not occur in the region. The *Podocarpus* pollen percentage of ca. 2 per cent is far too low to indicate such forests (Coetzee 1967, pp.59, 60). The inference of an open vegetation with much grassland especially on wide emerged coastal plains, is supported by palaeontological evidence from caves occurring along the coast. Klein (1974) describes faunas of grazing ungulates from these sites.

– In the part of southern Africa south of about 24-25°S latitude more humid conditions prevailed during the last glacial maximum as can be inferred from information from the impressive Gaap sequence (Butzer et al. 1978), the Wolkberg Cave (Talma et al. 1974), the Alexandersfontein Basin (Butzer, Fock et al. 1973) and the Molopo Valley (Heine 1978b). The fact that an undisturbed pan surface in the coastal Namib south of the Kuiseb river mouth at 23°41'S has a Th/U age of 210 000 ± 10 000 years shows that winter rainfall of last glacial age could not have penetrated so far north (Selby 1977). Circumstantial evidence for heavier rainfall in the region of the Orange River mouth further south is described by Eriksson (1978) who found four superimposed lacustrine phases here.

– It is envisaged that a complex situation existed during the glacial maximum in the southern erg. In winter conditions were cold and windy and some rainfall of southern origin could penetrate here. During the warmer dry summer the strong southern winds could move the longitudinal dunes northward. At the northern edge of the erg sand blew into the Kuiseb river and was transported to the ocean, washed by the Benguela current onto the widely exposed shelf north of the river mouth and built up new longitudinal dunes along the present shore. A similar process occurred at the northern edge of the Namib where longitudinal dunes were formed along the southern Angola coast (Torquato 1972).

The meridional margins of the deserts

These regions received very sparse rainfall and arid conditions prevailed here. The lake basins of the Sahel were dry or showed low water levels, while dunes crossed the present southern boundary of the desert some hundreds of kilometres. The rainfall is estimated to have been 15-20 per cent of the present (Lauer & Frankenberg 1979). The aridity was associated with cold conditions (Servant-Vildary 1979).

In southern Africa, strong evidence exists for a very large lake occurring in northern Botswana, the southern counterpart of the Sahel, between 30 000 and 18 000 B.P. (Heine 1978b). During the Late Glacial rainfall of tropical or maritime origin could penetrate the southern Kalahari (Lancaster 1979, Cooke 1975, Grey & Cooke 1977).

THE HOLOCENE CLIMATIC OPTIMUM

A comparison of the palaeoclimatic conditions of the Sahara and the Namib deserts for this period poses problems as the Climatic Optimum of the Northern and Southern Hemispheres may not have been of the same age. In the Southern Ocean, this optimum is dated by Hays et al. (1976) at 9 400 ± 600 B.P., which is more than 3 000 years before the generally accepted age of this stage in the Northern Hemisphere. As it is not yet known with certainty whether this discrepancy also applies to southern Africa, the climatic trends of the earlier half of the Holocene of the Southern Hemisphere will be discussed and compared with the Hypsithermal of the Northern Hemisphere.

Another serious problem in making the envisaged comparison is that many dated events are known from the Sahara, while for the Namib-Kalahari region inferences have to be based mainly on information from the rest of southern Africa.

The Western Sahara

A wealth of dated information is available on the climatic conditions in the Sahara during the Climatic Optimum round 6 000-5 000 B.P. The temperature was in general 1-2°C higher than at present and the entire Sahara received more rainfall. A summary of these conditions is given by Lauer & Frankenberg (1979). Compared with the present, higher rainfall is recorded from the Antiatlas, southern Morocco, southern Tunisia, the Saoura, coastal Mauritania (400 per cent), inland Mauritania, the Sahel (200-400 per cent) and even from the present-day hyperarid Central and Western Sahara. The lakes in the Sahel

388

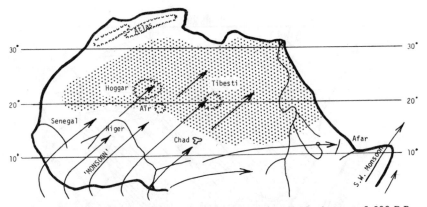

Figure 3. Rainfall pattern for Northern Africa during Climatic Optimum c. 5 500 B.P. during northern summer. Stippled area received less than 150 mm rain.

region had alternating, but generally high levels from 9 200-4 000 B.P. Lake Chad reached a maximum between 6 000 and 5 000 B.P. From 9 000-5 000 B.P. a fauna with tropical and more humid affinities lived in the northern Sahara and was after that time replaced by biota typical of drier open country. Southern savanna vegetation could under the more humid conditions invade the Sahara from the south and mediterranean vegetation migrated into the desert from the north.

In the Morocco mountains, the Cedrus forest was during this warm phase of the Atlanticum replaced by deciduous oak forest (Reille 1976, 1977). The more humid phase of early and middle Holocene age was marked by prehistoric occupation by Neolithic pastoralists. There are indications that this culture spread from the central Sahara, where it is known from 8 100-4 400 B.P. to the northern Sahara, where many sites date from 6 500-5 000 B.P.

The climatic factors responsible for these favourable conditions were: a weakening of the subtropical high pressure cells, an interplay of the influences of the Azores and the St Helena anticyclones, a weakening of the upwelling (Diester-Haass, this volume) and a higher evaporation from the ocean surface. The northward penetration of the monsoon in summer was probably possible as a consequence of the weakening of the tropical easterly jet stream which descends onto the Sahara in summer (Lauer & Frankenberg 1979). The consequence of these events was that southern summer rains and northern winter rains alternated and overlapped considerably in area. An important factor which may have contributed to the higher precipitation in the northern and southern Sahara may have been the numerous depressions described as easterly waves by Flohn (in: Lauer & Frankenberg 1979).

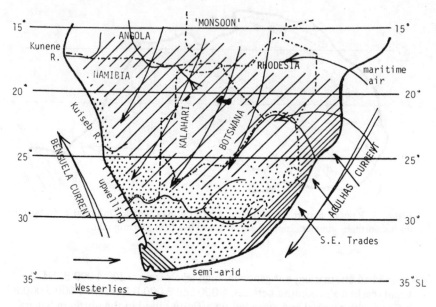

Figure 4. Rainfall pattern for Southern Africa. Climatic Optimum. Southern summer.

The Namib

As stated previously, practically only data from the Kalahari and adjacent parts of southern Africa are available for a brief general approximation of the Holocene palaeoclimates of the Namib.

During the Holocene, the general trend of the climate in South Africa points to more aridification. The northward invasion of the cyclonic winter rainfall of the last glacial maximum was during the sudden warming around 12 500 B.P. replaced by a southward penetration of summer rain of tropical origin which probably invaded the subcontinent in the western part as far as about 25°S Latitude. The consequence of this reversal in the rainfall pattern was that extensive lakes developedin the large depressions in the Kalahari. These lakes at 20-22°S latitude form the counterparts of the large lakes which originated in the Sahel at the same time (Heine 1978a).

As the temperature rose during the early Holocene, the upwelling weakened considerably and moved southward. Consequently, summer rainfall could at intervals penetrate the entire Namib desert. Even rivers originating in the escarpment bordering the southern erg could then bring water to pans in the dune field.

The southern coast of the continent was still under the influence of the westerlies, and received winter rainfall. As a consequence of the general warming and the rise in sea level the open vegetation partly consisting of grassland along the southern coast was between 10 000 and 8 000 B.P. replaced by bush, macchia and scrub.(Klein 1972, 1977, 1978, Moffett & Deacon 1977, Opperman 1978).

During the middle Holocene, the rainfall in southern Africa either diminished or the higher temperature raised the E/P ratio so that central South Africa became more arid. This drier climate even reached the south coast in the area of Groenvlei (Martin 1968). During this time, oscillations in humidity are known from the studies on fossil pollen by Schalke (1973).

In general, the climatic evolutions in the Sahara and the Namib-Kalahari region show interesting correlation during the early and middle Holocene. It appears, however, that aridity in South Africa spread before the final desertification of the Sahara set in. More dated information is needed on this discrepancy.

ACKNOWLEDGEMENTS

I wish to thank Dr J.A.Coetzee, Prof H.Flohn and Prof K.Heine for reading the manuscript and for fruitful discussion.

REFERENCES

Andel, Tj. H. & S.E.Calvert 1971. Evolution of sediment wedge, Walvis Shelf, Southwest Africa. *J. Geol. 79:*585-602.
Axelrod, D.I. & P.H.Raven 1978. Late Cretaceous and Tertiary vegetation history of Africa. In: M.J.A.Werger (ed.), *Biogeography and ecology of Southern Africa.* Junk, The Hague, I, pp.77-130.
Bé, A.W.H. et al. 1976. Late Quaternary climatic record in western equatorial Atlantic sediment. In: R.M.Cline & J.D.Hays (eds.), *Investigations of Late Quaternary paleooceanography and paleoclimatology.* Geol. Soc. Amer. Mem. 145:165-200.
Beucher, F. 1971. Etude palynologique de formations Néogènes et Quaternaires au Sahara nord-occidental. Thesis, Paris Univ.
Bornhold, B.D. 1973. Late Quaternary sedimentation in the eastern Angola Basin, Woods Hole Oceanogr. Inst. WHOI, 73-80. Unpublished manuscript.
Butzer, K.W., G.J.Fock, R.Stuckenrath & A.Zilch 1973. Paleohydrology of late Pleistocene lake Alexandersfontein, Kimberley, South Africa. *Nature 243:*328-330.
Butzer, K.W., N.A.Stuckenrath, A.J.Bruzewicz & D.M.Helgren 1978. Late Cenozoic paleoclimates of the Gaap Escarpment, Kalahari margin, South Africa. *Quatern. Res. 10*(3):310-339.
Chamley, H., L.Diester-Haass & H.Lange 1977. Terrigenous material in East Atlantic sediment cores as an indicator of NW African climates, *'Meteor' Forsch.-Ergebn. C 26:*44-59.
Coetzee, J.A. 1967. Pollen analytical studies in East and Southern Africa. *Palaeoecology of Africa 3:*149 pp.

Coetzee, J.A. 1978. Late Cainozoic palaeoenvironments of southern Africa, In: E.M.van Zinderen Bakker (ed.), *Antarctic glacial history and world palaeoenvironments.* A.A. Balkema, Rotterdam. pp.114-127.

Cooke, H.J. 1975. The palaeoclimatic significance of caves and adjacent landforms in western Ngamiland, Botswana. *Geogr. Journ. 141:*430-444.

Diester-Haass, L. 1976. Late Quaternary climatic variations in Northwest Africa deduced from East Atlantic sediment cores. *Quatern. Res. 6:*229-314.

Diester-Haass, L. 1977. Influence of carbonate dissolution, climate, sea-level changes and volcanism on neogene sediments off northwest Africa (Leg 41). In: Y.Lancelot, E. Seibold et al. (eds.), *Initial Reports DSDP, 41:*1033-1047.

Diester-Haass, L. & S.van der Spoel 1978. Late Pleistocene Pteropod rich sediment layers in the northeast Atlantic and Protoconch variation of *Clio pyramidata* Linné 1967.

Eriksson, P.G. 1978. An investigation of Quaternary aeolian-lacustrine sediments in Namaqualand. *Palaeoecology of Africa 10:*41-46.

Flohn, H. 1964. Investigations on the Tropical Easterly Jet. *Bonner Meteor. Abh. 4.*

Flohn, H. 1966. Warum ist die Sahara trocken? *Z. Meteorol. N.F., 17*(9-12):316-320.

Gardner, J.V. & J.D.Hays 1976. Responses of sea-surface temperature and circulation to global climatic change during the past 200 000 years in the eastern equatorial Atlantic Ocean. In: R.M.Cline & J.D.Hays (eds.), *Investigations of Late Quaternary paleooceanography and paleoclimatology.* Geol. Soc. Amer. Mem. 145. pp.221-246.

Grey, D.R.C. & H.J.Cooke 1977. Some problems in the Quaternary evolution of the landforms of northern Botswana. *Catena 4:*123-133.

Hays, J.D., J.A.Lozano, N.Shackleton & G.Irving 1976. Reconstruction of the Atlantic and Western Indian Ocean sectors of the 18 000 B.P. Antarctic Ocean. In: R.M.Cline & J.D.Hays (eds.), *Investigations of Late Quaternary paleooceanography and paleoclimatology.* Geol. Soc. Amer. Mem. 145. pp.337-372.

Heine, K. 1978a. Radiocarbon chronology of Late Quaternary lakes in the Kalahari, Southern Africa. *Catena 5:*145-149.

Heine, K. 1978b. Jungquartäre Pluviale und Interpluviale in der Kalahari (Südliches Afrika). *Palaeoecology of Africa 10:*31-39.

Klein, R. 1972. The late Quaternary mammalian fauna of Nelson Bay Cave (Cape Province, South Africa): Its implications for megafaunal extinctions and environmental and cultural change. *Quatern. Res. 2*(2):135-142.

Klein, R.G. 1974. Environmental and subsistence of prehistoric man in the Southern Cape Province, South Africa. *World Archaeol. 5*(3):249-284.

Klein, R.G. 1978. A preliminary report on the larger mammals from the Boomplaas stone age cave site, Cango Valley, Oudtshoorn district, South Africa. *S. Afr. Arch. Bull. 33:* 66-75.

Lancaster, I.N. 1979. Evidence for a widespread late Pleistocene humid period in the Kalahari. *Nature 279:*145-146.

Lauer, W. & P.Frankenberg 1979. Zur Klima- und Vegetationsgeschichte der westlichen Sahara. *Akad. Wiss. und Lit. Abh. Math. Naturwis. Klasse 1979 1:*1-61.

Martin, A.R.H. 1968. Pollen analysis of Groenvlei Lake sediments, Knysna (South Africa). *Rev. Palaeobot. Palynol. 7:*107-144.

McIntyre, A. et al. 1975. Thermal and oceanic structures of the Atlantic through a glacial – interglacial cycle. In: *Proc. WHO/IAMAP Symposium on Long-term Climatic Fluctuations.* Norwich, WMO, Geneva. pp.75-80.

McIntyre, A. et al. 1976. The surface of the Ice-age Earth. *Science 191*(4232):1131-1137.

Moffett, R.O. & H.J.Deacon 1977. The flora and vegetation in the surrounding Boomplaas

Cave, Cango Valley. *S. Afr. Arch. Bull. 32*(126):127-134.

Oppermann, H. 1978. Excavations in the Buffelskloof rock shelter near Calitzdorp, Southern Cape. *S. Afr. Arch. Bull. 33:*18-38.

Prell, W.L. & J.D.Hays 1976. Late Pleistocene fauna and temperature patterns of the Colombia Basin, Caribbean Sea. In: R.M.Cline & J.D.Hays (eds.), *Investigations of Late Quaternary paleooceanography and paleoclimatology.* Geol. Soc. Amer. Mem. 145. pp.201-220.

Reille, M. 1976. Analyse pollinique de sédiments post-glaciaires dans le Moyen Atlas et le Haut Atlas marocaines. Premiers résultats. *Ecologia Mediterranea 2:*153-170.

Reille, M. 1977. Contribution pollenanalytique à l'histoire holocène de la végétation des montagnes du Rif (Maroc septentrional). *Bull. AFEQ. suppl. 1*(50):54-76.

Ploey, J.de 1969. Report on the Quaternary of the Western Congo. *Palaeoecology of Africa 4:*65-68.

Rognon, P. & M.A.J.Williams 1977. Late Quaternary climatic changes in Australia and North Africa: A preliminary interpretation. *Palaeogeogr. Palaeoclim. Palaeoecol. 21:* 285-327.

Sabelberg, U. 1977. The stratigraphic report of Late Quaternary accumulation series in southwest Morocco and its consequences concerning the pluvial hypothesis. *Catena 4:*209-214.

Schalke, H.J.W.G. 1973. The Upper Quaternary of the Cape Flats area (Cape Province), South Africa. Thesis, Univ. Amsterdam.

Schulze, B.R. 1972. South Africa. In: J.F.Griffiths (ed.), Climates of Africa. *World Survey of Climatology 10:*501-586.

Schulze, B.R. 1974. *Climate of South Africa.* Part 8: General Survey. Weather Bureau, Pretoria.

Schulze, R.E. & O.S.McGee 1978. Climatic indices and classifications in relation to biogeography of Southern Africa. In: M.J.A.Werger (ed.), *Biogeography and ecology of southern Africa.* Junk, The Hague, pp.19-52.

Selby, M.J. 1977. Report in: Geomorphological Research in the Namib. *Namib Bull. Suppl. 2, Transvaal Mus. Bull. Oct.:*8.

Servant-Vildary, S. 1979. Paléolimnologie des lacs du bassin Tchadien au Quaternair récent. *Palaeoecology of Africa 11:*65-78.

Shackleton, N.J. & J.P.Kennett 1974. *Palaeotemperature history of the Cenozoic and the initiation of Antarctic glaciation: Oxygen and carbon isotope analyses in DSDP sites 279, 277 and 281.*

Shackleton, N.J. & J.P.Kennett 1975. Palaeotemperature history of the Cenozoic and the initiation of the Antarctic glaciation. *Initial Reports DSDP 29:*743-755.

Siesser, W.G. 1978. Aridification of the Namib desert: Evidence from oceanic cores. In: E.M.van Zinderen Bakker (ed.), *Antarctic glacial history and world palaeoenvironments.* A.A.Balkema, Rotterdam, pp.105-114.

Talma, A.S., J.C.Vogel & T.D.Partridge 1974. Isotopic contents of some Transvaal speleothems and their palaeoclimatic significance. *S. Afr. J. Sci. 70:*135-140.

Tankard, A.J. & J.Rogers 1978. Late Cenozoic palaeoenvironments on the west coast of Southern Africa. *J. Biogeogr. 5:*319-337.

Torquato, J.R. 1972. Origin and evolution of the Moçamedes Desert (Angola). In: T.F.J.Dessauvagie & A.J.Witemar (eds.), *African Geology. Proc. Conf. Afr. Geol. Univ. Ibadan (1970),* pp.449-460.

Trewartha, G.T. 1966. *The earth's problem climates.* Methuen, London.

Van Zinderen Bakker,E.M. 1967. Upper Pleistocene and Holocene stratigraphy and ecology on the basis of vegetation changes in sub-Saharan Africa. In: W.W.Bishop & J.D.Clark (eds.), *Background to evolution in Africa*. Univ. Chicago Press, pp.125-147.

Van Zinderen Bakker, E.M. 1975. The origin and palaeoenvironment of the Namib Desert biome. *J. Biogeogr. 2:*65-73.

Van Zinderen Bakker, E.M. 1976. The evolution of Late Quaternary palaeoclimates of Southern Africa. *Palaeoecology of Africa 9:*160-202.

END OF THE LAST INTERGLACIAL:
A PREDICTIVE MODEL OF THE FUTURE?

GEORGE KUKLA

Lamont-Doherty Geological Observatory of Columbia University, Palisades, New York

INTRODUCTION

The present interglacial is in its terminal stage (Kukla & Matthews 1972). In this context, the recent gradual cooling of the middle latitudes in the Northern Hemisphere is of special significance (Kukla et al. 1977). The shift during the last several years is accompanied by numerous record-breaking local weather anomalies such as the first snow in Miami (1977), the lowest rainfall on record for India (1972), and the first snow in living memory in the Sahara (1979). The cooling is fast enough to bring full glacial average hemispheric air surface temperatures in 300-400 years.

Many researchers believe that the cooling is of unknown natural origin, but that it will be reversed at the end of the century by anthropogenic increase of atmospheric CO_2. This is, however, far from certain since CO_2 has been increasing for many decades without any obvious warming effect. Furthermore, the number of subtle feedbacks operating in the climate system have not been considered in the assessment of CO_2 impact. These can dampen or completely reverse the expected CO_2 warming over at least some parts of the world. Given the uncertainty, it is of utmost importance to study available records of the terminal stages of the last interglacial, the cyclic equivalent of the present warm interval, and try to reconstruct the changes of the environment and climate which led to the last glacial stage.

RECORDS OF THE LAST INTERGLACIAL

The last interglacial was the last episode in the earth's history during which the global climate was as warm or warmer than within the past approximately 10 000 years. Both intervals are marked by similar distribution of natural vegetation on land, and by similar regimes of atmospheric and oceanic circulation. The type region of the last interglacial is in Northwest Europe and the type units are pollen bearing strata documenting the presence of mixed hardwood forests in the area (Turner & West 1968, Fairbridge 1972, Kukla et al. 1972).

395

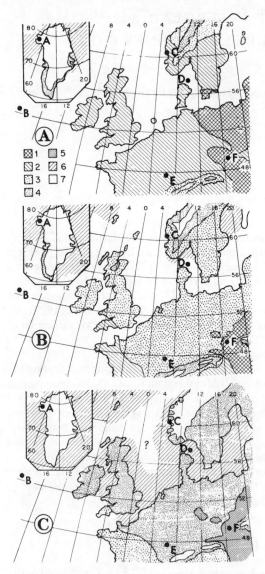

Figure 1. Tentative reconstruction of the environment in Europe and Greenland at phases A (about 118K), B (about 114K), and C (about 110K).
Sites: A. Camp Century; B. Deep sea core V23-82; C. Fjøsanger; D. Skaerumhede; E. Grande Pile; F. Modrice.
Vegetation: 1. hardwood forest; 2. mixed deciduous/coniferous forest; 3. tajga; 4. tundra, with bare spots; 5. badlands and loess steppe. 6. sea ice at the end of April; 7. land based ice and open ocean.

396

Radiometric dating pinpoints the start of the last interglacial to about 127
± 6K years ago (Broecker & von Donk 1969, Mesolella et al. 1969). Counts of
the yearly laminated pollen-deposits in Germany showed that the interglacial
lasted 11 ± 1K years (Muller 1965, Turner 1975). Consequently, it ended about
115 or 116K years ago.

Deposits of the last interglacial are known from innumerable sites on land
as well as in the ocean floor, but few preserve an undisturbed record of the
final stages of the warm interval in continuous depositional sequence with sufficiently high stratigraphic resolution. We will review the principal sites with
sufficiently detailed records.

GRANDE PILE, FRANCE

This site is located at latitude $47°44'N$, longitude $6°30'E$, and is at an elevation
of 330 m. Several cores were raised from a peat bog underlain by lake gyttja,
resting on the tills of the next-to-last glacial age (Seret & Woillard 1976, Woillard 1978). Deposition was continuous over the last approximately 130K years.
The rich pollen content of the strata enables reconstruction of the local vegetation which fluctuated between deciduous forest and treeless tundra. Four
major intervals of closed forest were identified. The earliest one correlates
with the last interglacial, the following two with the temperate interstadials
Brörup and Odderade. Deposits of the interglacial and early glacial are finely
laminated. It is unknown whether or not the laminae are annual, but they
represent a proof of continuous deposition and lack of post-depositional disturbance.

A preliminary estimate by Seret (written communication) based on micro-
scopy of randomly selected segments, indicates about 54 000 ± 2 000 laminae
between the beginning of the last interglacial and the end of the last temperate
interstadial. This figure matches the duration of O^{18} Stage 5, believed to be
the oceanic equivalent of the land based section.

The last interglacial environment, represented by layer A (Figure 2), developed into a diversified closed mixed forest, very similar in composition to
modern natural vegetation on the site. Oak, hornbeam, elder, fir and spruce
were the principal pollen producers. Layer B follows, indicating a closed coniferous forest composed of pine, spruce and birch. Deciduous elements of
temperate forests are present at insignificant levels, and were probably added
by long distance transport. Next layer C is characterized by a dominance of
grass and herbs pointing to an open shrub tundra.

An interesting feature of the Grande Pile diagram is the relatively fast
transition from one type of pollen assemblage to another at the A/B and B/C
boundaries, with a relatively stable environment indicated inbetween.

The shift of vegetational zones from A to B and to C could be compared to
a geographic shift from Northeast France (latitude $48°$) to Northcentral Sweden
(latitude $63°$) and to the Kola Peninsula (latitude $68°$).

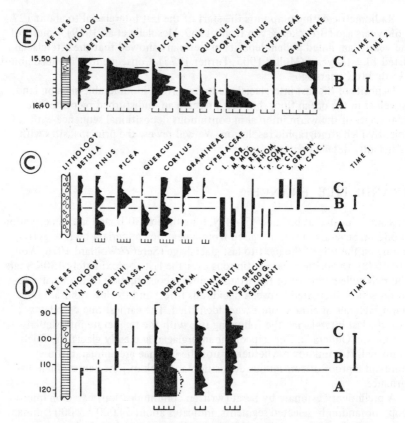

Figure 2. The end of the last interglacial as recorded in Grande Pile (site E), Fjøsanger
(site C), and Skaerumhede (site D). Lithology explained in Figure 3. Location in Figure
1. Pollen as percent of total count, scale in 10 per cent. Relative abundance of molluscs
and foraminifera indicated by thickness of the bar.
L.BOR.: *Lucinoma borealis;* M.MOD.: *Modiola modiolus;* B.RET.: *Bittium reticulatum;*
V.RHOM.: *Benerupis rhomboides;* T.COM.: *Turitella communis;* P.MAX.: *Pecten maxi-
mus;* C.CIL.: *Clinocardium ciliatum;* S.GROEN.: *Serripes groenlandicum;* M.CALC.:
Macoma calcarea; N.DEPR.: *Nonion depressula;* E.MAC.GR.: *Elphidium macellum
granulosum;* C.CRASSA.: *Cassidulina crassa;* I.NORC.: *Islandiella norcrossi.*
Scale for percentage of boreal foraminifera and faunal diversity in tens of percent. Scale
for number of specimens in 2 000 per 100 gram. Phases A, B and C discussed in text.
Bar on the right showing 2 000 year interval according to different models (cf. text). Data
from Woillard (1978), Mangerud et al. (1979) and Bahnson et al. (1973).

The character of the sediment and the distribution of laminae in the lower part of the sequence suggest a relatively homogeneous deposition rate. Indications are, however, that the rate is higher in organic-rich sediments of the warm intervals.

Several models can be constructed to estimate the sedimentation rate in the profile. The first model assumes that the level marking the establishment of a closed mixed hardwood forest in borehole GPX (Woillard 1978) is 126K years old (a thousand years younger than the age of Termination II of Broecker & Van Donk 1970). The length of unit A in Grande Pile, as well as the length of the last interglacial forests in laminated deposits of Bispingen, Northern Germany (Müller 1974), is 11 ± 2K years. The A/B boundary is therefore 115 ± 2K years old. The sedimentation rate within the 2.1 m thick interval representing unit A, in B, and across the B/C boundary is constant at approximately 0.19 mm yr^{-1}. The second model takes the base of the interglacial as 126K years old and the top of the youngest temperate interstadial, Ognon II, as being equivalent to the top of the O^{18} stage 5, approximately 73K years old (cf. Ruddiman & McIntyre 1976). The sedimentation rate within the bracketed interval (5.7 m thick) is constant at approximately 0.108 mm yr^{-1}. The third model differs from the second by taking the sedimentation rate as linearly proportional to the percentage of organic matter (cf. Seret & Woillard 1976, Kukla 1977). The sedimentation rate across the A/B boundary in such model results in approximately 0.20 mm yr^{-1}.

The time represented by unit B, according to different models, varies between 2 100 years (model 1) and 3 700 years (model 2). The transition from A to B as well as from B to C, sampled at 2.5 cm intervals, took less than 230 years (model 2) but probably less than 130 years (model 1 and 3).

Recently, Woillard (1979 and in written communication) analysed the A/B transition in detail with samples taken only 1 mm apart. She found a gradual drop in temperate elements through about 150 years, then a sharp rise of birch and pine and elimination of fir. This final shift toward tajga took only about 20 years.

The almost simultaneous disappearance of oak, alder, elm and hornbeam, and a significant drop of fir rules out pests as well as fires as a possible cause of the change. While several possibilities exist, the most likely explanation postulates some change in the seasonal weather pattern of the growing season preventing the pollination of affected species.

There are several hundred sites scattered throughout Northwest and North Central Europe and England ascribed to the last interglacial. While some may belong to older periods (Kukla 1977, Bowen 1979), most of them probably represent the last interglacial interval. When not truncated, the records are compatible with Grande Pile, showing the closed tajga of phase B replacing the mixed forest. Unfortunately, neither the sedimentation rates, the sampling intervals, nor the lithologic continuity of the record are as favourable as in Grande Pile. Sterile sand is frequently at the layer correlative with unit C. The sites were taken into account in construction of Figure 1.

FJØSANGER, NORWAY

A sequence of shallow marine deposits with variable grain size ranging from clays to clayey gravels is exposed next to Bergen, latitude 60°23'N, longitude 5°27'E. It contains pollen of terrestrial plants as well as marine molluscs and foraminifera. The sedimentation rate is judged to be highly variable; the composite section was published without information on the thickness of the layers (Mangerud et al. 1979).

The lower and middle part of the sequence displays a typical pollen assemblage of mixed hardwood forest, and molluscs and benthos of a bay, as warm and warmer than at present (unit A in Figure 2).

This assemblage is succeeded by unit B, containing pollen of a predominantly coniferous forest composed principally of spruce, pine and birch. Temperate molluscs disappeared or are significantly less abundant. The authors did not interpret the assemblage as marking great change in summer water temperature. They did, however, observe signs of shallowing of the bay. An abrupt change of environment of a treeless tundra and simultaneous drastic cooling of the sea is marked at the B/C boundary. The authors assign this change to the weakening of the flow of Atlantic current into the Norwegian Sea.

Only approximate estimates can be made of the time represented by unit B. Assuming that the thickness of the layer as plotted in Figure 2 of Mangerud et al. (1979) is in the same proportion to time as interglacial unit A, and that unit A was deposited in 8-10K years, then deposition of unit B took approximately 1 500 years and transitions A/B and B/C took at most a few centuries each.

SKAERUMHEDE, DENMARK

Skaerumhede boreholes are located at latitude 57°35'N, longitude 10°25'E, and elevation 24 m.a.s.l. They penetrate marine clays resting on a till and overlain by Weichselian fluvioglacial deposits and till. Warm water molluscs and foraminifera characteristic of Eemian interglacial seas, such as *Turritella terebra* or *Abra nitida*, were found in the original borehole and described by Jessen et al. (1910). Duplicate boring was completed in 1972 and its lithology and foraminifera described in detail by Bahnson et al. (1973). The lowermost section of the new boring, below 103 m, encountered clays of the upper part of the last interglacial (unit A) which contain abundant foraminifera characteristic of Eemian seas, such as *Nonion depressula* or *Elphidium gerthi*. Most of these species are still present in the overlying unit B, which, however, contains an abundant cold water foram *Cassidulina crassa*, a common inhabitant of arctic waters. Diversity is high in layer B. Water depth might have been slightly greater than in A. Unit C follows. It contains rock detritus (probably ice rafted) and arctic foraminifera such as *C.crassa, Elphidium excavatum forma clavata,*

400

Islandiella teretis, I.norcrossi or *Virgulina loeblichi*. Boreal species have almost disappeared and those present are probably reworked. A large drop in the foraminiferal abundance and in faunal diversity occurs in the upper half of unit C (boundary of zones VI and V in original reference). An expressed cold arctic environment is indicated; water was shallower and possibly less saline. Presence of sea ice through much of the year is probable.

Boreal foraminifera reappear twice higher up in the profile at possible equivalents of Brörup and Odderade interstadials.

An estimate of the deposition rate is difficult. Assuming that the A/B boundary at 103 m corresponds to the end of the last interglacial s.s and is approximately 115K years old, and that the bottom of the *Turritela terebra* zone at about 180 m depth is 126K years old, then the 9 m thick unit B would have been deposited in approximately 1 300 years.

Given the complexity of the early glacial environmental changes and the possibility of gaps in the record, the presented model may be, however, in serious error. Oceanographic and climatic changes which took place at the A/B and especially at the B/C boundary, must have been relatively sudden, taking less than 100 years.

MODRICE, CZECHOSLOVAKIA

In the excavation front of an active brickyard near Brno, latitude 49°10'N, longitude 16°35'E, elevation 200 m.a.s.l., a sequence of soils and loesses of the last interglacial and glacial age was studied (Kukla & Koci 1972).

An interglacial deciduous forest left behind a characteristic reddish brown leached soil (parabraunerde). This is overlain by a black biogenic, partly calcareous soil high in humus with frequent worm burrows and scattered charcoal (Figure 3). A band of fine loess (marker) caps the two soils and is followed by hillwash sediments composed of sand-size pellets of redeposited soil aggregates called pellet sands. Loess interbedded with feeble steppe soils and layers of hillwash comes next, followed by two chernozems of early glacial age. As shown by Kukla (1977), the described sequence correlates with the O^{18} oceanic stage 5. The Modrice outcrop is representative of numerous similar sequences known from the vicinity of Prague, Brno and Nitra in Czechoslovakia and Wien in Austria. These sequences contain a rich snail fauna, which, in combination with the soil and sediment types, enable a fairly accurate reconstruction of the history of local environment (Lozek 1964).

The parabraunerde (layer A) and the accompanying fauna testify to the presence of a closed predominantly deciduous forest which must have occupied the sites for at least several millenia. Climate was warmer and more humid than it is at present. The black biogenic soil (layer B) developed under grassland in a mild, or even temperate, but highly continental climate marked by prolonged dry seasons. It may have been similar to that of southern

401

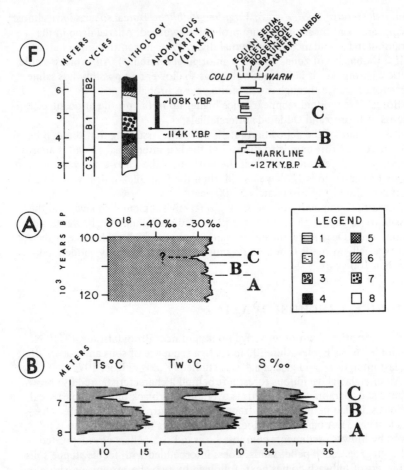

Figure 3. End of the last interglacial in Modrice (F), Camp Century ice core (A), and deep sea core V23-82 (B). Location in Figure 1.

Lithologic symbols: 1: clay and laminated gyttja; 2: silty gyttja; 3: clay with ice rafted stones; 4: biogenic steppe soil; 5: parabraunerde; 6: braunerde; 7: pellet sand; 8: loess. Section of anomalous polarity in Modrice marked with black bar (after Kukla & Koci 1972). Delta O^{18} record from Camp Century with tentative time scale by Dansgaard et al. (1972).

Ts: summer sea surface temperature; Tw: winter temperature; S: salinity reconstructed from faunal assemblages in core V23-82 by Sancetta et al. (1972). Phases A, B and C explained in text.

Ukraine today. Influx of fresh windblown calcareous dust accompanied the shift from forest (A), to steppe (B). This came from distant sources, as the soils in the vicinity were decalcified. At the interface of layers A and B, disseminated charcoal was observed at most of the studied localities. The charcoal originated from pine and larch. Because paleolithic artifacts were occasionally found in the same layer (Kukla 1977, Kukla 1961), anthropogenic origin of the charcoal was suspected. An alternate explanation for the source of the charcoal could be forest fires which decimated the larch and pine woodlands developed in place of the former deciduous forests.

Tentative dating of the Modrice section is based on the interval of anomalous magnetic polarity correlated with the Blake event and found throughout unit C directly overlying layer B. The Blake event was dated to approximately 108-114K. If the correlation is correct, the top of layer B has to be about 114K years old. Steppe soil needed at least several centuries to develop, so the mixed forest must have disappeared at about 115KY. B.P. or earlier.

Overlying sediments of unit C indicate that the vegetational cover was reduced to discontinuous patches of grass. Pellet sands are today formed during torrential rains hitting bare dessicated surfaces of crumbled soils. Loess (including the marker) is windblown dust. While the marker came from a distant source, the impure loess in the upper half of unit C probably originated from local outwash. Remains of wooly rhinoceros were found in unit C in the vicinity of Prague.

CAMP CENTURY, GREENLAND

This 1 300 m deep core drilled to bedrock through the ice at latitude 77°N, longitude 61°W, yielded a continuous record of delta O^{18} variations indicative of average air temperatures at the time of snowfall (Dansgaard et al. 1972).

A time scale is provided by counting the annual layers displaying seasonal variation in the upper part of the core. Age estimates for the lower part are based on ice flow models. Relatively warm snowfall is indicated for the last 10K years and for the layer correlated with the O^{18} oceanic stage 5 (older than about 73K years). Two temporary drops in temperature are recorded within the latter section. One is dated to about 89K years ago and its amplitude is about two-thirds of the full glacial. The second is about 110K years old. Its amplitude is not known because the corresponding core segment was lost. The core probably penetrated into the top of the last interglacial ice (cf. correlation with Grande Pile; in Kukla 1977). While the dating of the bottom part of the core is unsafe, one feature is certain: the episodes of cold snowfall are short and exceptional below the 73K year ago level. If Dansgaard's et al. (1972) and Kukla's (1977) time scales are approximately correct, then the snow temperatures during the deposition of unit B in Europe were approximately at the interglacial level. A significant drop in snowfall temperatures was indicated only in unit C.

403

THE DEEP SEA RECORD

Due to burrowing, low sedimentation rates and mostly insufficient sampling density, the published data on deep sea cores do not have high enough resolution for meaningful correlations with land records.

There are indications in core V23-82 taken at longitude 52°35'N and latitude 21°56'W (Sancetta et al. 1972), that the summer surface temperatures in this part of the ocean dropped at about the time correlative with the A/B boundary in Grande Pile while the winter temperatures remained relatively high and the salinity actually increased. This would suggest a continuing supply of warm waters from the Gulf and increased evaporation in windier weather in summer. Neighbouring core K708-7 has a similar record (Ruddiman & McIntyre 1979).

A sharp but short-lasting drop in summer as well as in winter temperatures and the decrease of salinity, indicating an intrusion of subpolar waters, occurred at about 110K years ago.

However, it must be emphasized that the record is noisy and the presented interpretation should be understood as one of many possible models. A southward penetration of polar and subpolar waters along the north-south transect in the North Atlantic was also reported by McIntyre & Ruddiman (1976) and dated to about 110K.

DISCUSSION

All of the discussed records have one feature in common. They show a long interval of stable conditions with an environment comparable to the present (phase A), followed by a rapid shift into a short approximately 2K year long episode of relatively stable impoverished environment (phase B) indicative of cooler climate. Another rapid shift brings the onset of a typical glacial environment (phase C).

Dating of individual stratigraphic records leaves a large margin of uncertainty. A reliable correlation among the sites is not yet possible. It can be concluded that phase A terminated at about 115-116K years ago, and also that phase B lasted for about 1-3K years. Phase C culminated around 110K years ago. The most-conspicuous feature is the apparently rapid environmental change at the A/B and B/C boundaries. The former took approximately 20 years or less in Grande Pile (Woillard 1979), where the most detailed sampling across the boundary was performed. The B/C took less than 100 years in Skaerumhede. It is as yet impossible to find out whether the A/B and B/C boundaries were synchronous over the described area. There are sufficiently strong reasons to believe they might have been, as the area involved is relatively small and is today similarly affected by major weather anomalies.

Therefore, there are valid reasons for a model reconstruction of the environment in phases A, B and C based on the assumption that these were broadly synchronous. The reconstruction is presented in Figure 1 and should be taken as only one of many possible models. It has the following implications:

1. The terminal stages of phase A were marked by a gradual decline of temperate deciduous trees on land, but by no detectable changes in the marine environment or in sea level. It is doubtful whether a similar slow, gradual impoverishment of natural vegetation, if started at present, would be detectable in the artificially managed forests of Northwestern Europe (Woillard 1979).

2. The transition of phase A into phase B seems to indicate a major and relatively sudden shift in the pattern of atmospheric circulation accompanied by a minor drop of sea level. Sea surface temperatures in the North Sea remained relatively high. The shift, at most, took a few decades.

3. During phase B the Atlantic and Baltic coastal lands were still wet but cooler. Climate in central Europe became continental. Also, the shallow seas off Denmark, Norway and the Northcentral Atlantic became cooler. Snowfall temperatures and probably the principal flow pattern over Western Greenland remained similar to present.

4. The transition of phase B into C was as sudden as the A/B transition. Changes in both continental and marine environment were profound. Woodlands were replaced by shrub tundra over much of South Central and Western Europe. The surface was destabilized and eroded by wind and water action in Northern and Central Europe. The sea along Denmark and Norway dropped by possibly 10-20 m, and cooled by several degrees. Ice rafted debris supplied by Scandinavian mountain glaciers reached Northern Denmark.

5. During phase C, the environment in Northern Europe had a distinct glacial character. A cold climax was reached in the second half of phase C when expansion of sea ice is indicated in the Skategatt area, and around Greenland (by a drop of snowfall temperatures). The sea surface in Central North Atlantic became markedly cooler and fresher.

There is a discrepancy between the estimates of the sea level drop at Barbados and in Scandinavia at the height of phase C. Whereas about 70 m is reported for Barbados (Steinen et al. 1973), less than a 20 m depression is estimated at the two Scandinavian sites. Either the Barbados figure is in error, or the Scandinavian coast was isostatically depressed by the build-up of mountain ice.

The major problem consists of identification of a mechanism responsible for the rapid shifts from phase A to B and later from B to C. There is little doubt that the process was weather related. There are no historical equivalents from European lowlands of a large scale decimation of temperate trees in the area. There are, however, numerous analogs of such rapid events in yearly laminated deposits of earlier interglacials (Müller 1965). For reasons discussed previously, it is not probable that a wide-spread drought and/or fires caused the change in the composition of the forests. The cooling of the North Sea in

405

Skaerumhede and Fjøsanger which accompanied the event, at least according to our model, lends further support to the temperature related cause.

The season in which deciduous trees are most sensitive to temperature is the spring. Damages caused by frosts to fruit trees everywhere, or more recently to coffee plantations in Brazil, are well-known. Today, they rarely cause lasting damage. However, at least on one occasion an exceptional frost wave in South Carolina, USA, was described which decimated citrus and olive trees for decades (Ludlum 1963). According to a description by Governor James Glen, on 18 February 1748 (corrected date), after a spell of mild winter weather the orange trees were ready to blossom. The $-12°C$ frost on that day caused the trees to 'burst all their vessels, for not only the bark of all of them but even the bodies of many of them were split, and all on the side next to the sun'. A similar cold wave on 8 February 1835 wiped out the citrus industry in Florida. The South Carolina frosts of 1748 and 1835 were caused by an unusually deep southward penetration of arctic air masses. Can a similar event happen at the beginning of the growing season in Western Europe? This is not a likely possibility at present, due to the modification of arctic air flow in its passage over the relatively warm North Atlantic. However, expansion of sea ice along the East Greenland coast and around Iceland, and a decrease of sea surface temperatures in the North Sea in late spring could eventually provide a needed bridge for a delayed, exceptionally cold, arctic air outbreak into Western and Central Europe. If repeated a few times in a decade, such event would have a devastating impact on the vegetation.

If some environmental changes regularly occur at the end of each interglacial as the evidence seems to indicate, and if correctly interpreted in our model, the delayed southward spring penetrations of arctic air into Europe can offer a feasible explanation of the large scale decimation of mixed interglacial forests during phase B.

Expansion of sea ice and constriction of the North Atlantic Current are indicated in the final phase C culminating at about 110K years ago.

While at this stage large uncertainty remains concerning the validity of the model, further research is needed. It is advisable to reach a better understanding of the mechanism of formation and of the present frequency of late spring arctic air outbreaks in Europe and the coupling of these phenomena with sea ice and sea surface temperature anomalies.

ACKNOWLEDGEMENTS

We acknolwedge travel funding and support by the Deutsche Forschungsgemeinschaft. G. Woillard, N. Shackleton, W. Ruddiman and W.S. Broecker read the manuscript and suggested numerous improvements. We thank J. Brown for drafting of the figures. This research was partially supported by the National Science Foundation's Office of Climate Dynamics and International Decade of Ocean Exploration Grant OCE77-22893. This is Lamont-Doherty Geological Observatory of Columbia University Contribution No. 2921.

REFERENCES

Bahnson, H., K.S.Petersen, P.B.Konradi & K.L.Knudsen 1973. Stratigraphy of Quaternary deposits in the Skaerumhede II boring: Lithology, molluscs and foraminifera. *Geological Survey of Denmark. Yearbook 1973:*27-62.

Bowen, D.Q. 1979. Quaternary correlations. *Nature 277:*171-172.

Broecker, W.S. & J.van Donk 1970. Insolation changes, ice volumes, and the ^{18}O record in deep-sea cores. *Rev. Geophys. Space Phys. 8:*169-198.

Dansgaard, W., S.J.Johnsen, H.B.Clausen & C.C.Langway 1972. Speculation about the next glaciation. *Quatern. Res. 2*(3):396-398.

Fairbridge, R.W. 1972. Climatology of a glacial cycle. *Quatern. Res. 2*(3):283-302.

Jessen, A., V.Milthers, V.Nordmann, N.Hartz & A.Hesselbo 1910. En boring gennem de kvartaere Lag ved Skaerumhede. *Danm. geol. Unders. II Raekke 25*.

Kukla, G.J. 1961. Stratigraficka posice ceskeho stareho paleolitu. *Pamatky archeologicke 52:*18-30.

Kukla, G.J. 1977. Pleistocene land-sea correlations. I, Europe. *Earth-Science Rev. 13:*307-374.

Kukla, G.J., J.K.Angell, J.Korshover, H.Dronia, M.Hoshiai, J.Namias, M.Rodewald, R.Yamamoto & T.Iwashima 1977. New data on climatic trends. *Nature 270:*573-580.

Kukla, G.J. & A.Koci 1972. End of the last interglacial in the loess record. *Quatern. Res. 2*(3): 374-383.

Kukla, G.J. & R.K.Matthews 1972. When will the present interglacial end? *Science 178:*190-191.

Kukla, G.J., R.K.Matthews & J.M.Mitchell, jr. 1972. The end of the present interglacial. *Quatern. Res. 2*(3):261-269.

Ložek, V. 1964. Quartärmollusken der Tschechoslowakei. *Rozpr. Ustred. Ustavu Geol.:31.*

Ludlam, D.M. 1963. Extremes of cold in the United States. *Weatherwise 16:*275-291.

Mangerud, J., E.Sønstegaard & H.-P.Sejrup 1979. Correlation of the Eemian (interglacial) stage and the deep-sea oxygen-isotope stratigraphy. *Nature 277:*189-192.

Mesolella, K.J., R.K.Matthews, W.S.Broecker & D.L.Thurber 1969. The astronomical theory of climatic change, Barbados data. *J. Geol. 77:*250-274.

Müller, H. 1965. Eine pollenanalytische Neubearbeitung des Interglazialprofils von Bilshausen (Unter-Eichsfeld). *Geol. Jahrb. 83:*327-352.

Müller, H. 1974. Pollenanalytische Untersuchungen und Jahresschichtenzahlungen an der eem-zeitlichen Kieselgur von Bispingen/Luhe. *Geol. Jahrb. 21:*149-169.

Ruddiman, W.F. & A.McIntyre 1976. Northeast Atlantic paleoclimatic changes over the past 600 000 years. In: R.M.Cline & J.D.Hays (eds.), *Investigation of Late Quaternary paleoceanography and paleoclimatology.* Geol. Soc. Amer. Mem. 145, pp.111-146.

Ruddiman, W.F. & A.McIntyre 1979. Warmth of the subpolar North Atlantic Ocean during Northern Hemisphere ice-sheet growth. *Science 204:*173-175.

Sancetta, C., J.Imbrie, N.G.Kipp, A.McIntyre & W.F.Ruddiman 1972. Climatic record in North Atlantic deep-sea core V23-82: comparison of the last and present interglacials based on quantitative time series. *Quatern. Res. 2:*363-367.

Seret, G. & G. Woillard 1976. The glaciations in the 'Vosges Lorraines', Führer zur Exkursionstagung des IGCP Projektes 73/1/24 'Quaternary Glaciations in the Northern Hemisphere', 5-13 September 1976, in den Südvogesen, im Nördlichen Alpenvorland und in Tirol: 1-13.

Steinen, R.P., R.S.Harrison & R.K.Matthews 1973. Eustatic low stand of sea-level between 125 000 and 105 000 B.P.: Evidence from the sub-surface of Barbados, West Indies. *Geol. Soc. Amer. Bull. 84:*63-70.

Turner, C. 1975. The correlation and duration of Middle Pleistocene interglacial periods in Northwest Europe. In: K.W.Butzer & G.Ll.Isaac (eds.), *After the Australopithecines*. Mouton, The Hague. pp.259-308.

Turner, C. & R.G.West 1968. The subdivision and zonation of interglacial periods. *Eiszeitalter und Gegenwart 19:*93-101.

Woillard, G.M. 1978. Grande Pile Peat Bog: A continuous pollen record for the last 140 000 years. *Quatern. Res. 9:*1-21.

Woillard, G.M. 1979. Abrupt end of last interglacial s.s. in NE France. *Nature* 281:558-562.